Advances in Olericulture

Series Editor
Silvana Nicola, Università di Torino, Grugliasco, Italy

The book series Advances in Olericulture provides a state-of-the-art account of research in olericulture, the applied life science of production and utilization of vegetable crops. The series focuses on various aspects of vegetable science and technology covering primarily but not exclusively species where the vegetative organ is the economically important component. The series of books spans current topics from sustainable fertilization to organic production; from open field cultivation to advanced soilless growing techniques; from vegetable seed and seedling physiology to vegetable quality and safety; from environmental stresses to phyllosphere communities interaction with vegetables; from postharvest biology and technology to minimally processing of vegetables. The series is designed to present the most advanced scientific information available linking basic and applied research for serving olericulturists, research workers, teachers and advanced students.

More information about this series at http://www.springer.com/series/11965

Shashank Shekhar Solankey
Meenakshi Kumari • Manoj Kumar
Editors

Advances in Research on Vegetable Production Under a Changing Climate Vol. 1

Editors
Shashank Shekhar Solankey
Department of Horticulture (Vegetable and Floriculture)
Bihar Agricultural University
Sabour, Bhagalpur, Bihar, India

Meenakshi Kumari
Department of Vegetable Science
Chandra Shekhar Azad University of Agriculture & Technology
Kanpur, Uttar Pradesh, India

Manoj Kumar
Division of Vegetable Crops
ICAR-Indian Institute of Horticultural Research
Bengaluru, Karnataka, India

ISSN 2367-4083 ISSN 2367-4091 (electronic)
Advances in Olericulture
ISBN 978-3-030-63499-5 ISBN 978-3-030-63497-1 (eBook)
https://doi.org/10.1007/978-3-030-63497-1

This Springer imprint is published by the registered company Springer Nature Switzerland AG
The registered company address is: Gewerbestrasse 11, 6330 Cham, Switzerland

Preface

Climate change and global warming are one of the prominent global environment issues of the present day which are likely to challenge food security in the future. Climate change influences all four dimensions of food security, that is food availability, food access, food use and utilization, and food stability; hence, there is an urgent need to mitigate its impact on agriculture. Extreme weather events due to climate change cause sudden reduction in agricultural productivity. The concentration of several surface gases such as CO_2, CH_4, N_2O, SO_2 and CFC is continuously increasing, which directly deteriorates crop production as well as quality; moreover, these gases also played an important role in altering the environmental temperature, thus affecting the weather events and causing drought and floods, and are indirectly responsible for soil salinity and alkalinity, which are major threats for agriculture production.

Vegetables are known as protective food among all horticultural crops as they are full of essential vitamins, minerals, fibres and carbohydrates, thus hold a significant part of a healthy diet pattern. Vegetables are the best resource for defeating micronutrient deficiencies and play a vital role for overall health and maintenance of body systems. They also provide much higher income to smallholder farmers and more employment per unit area than staple crops. Vegetables are grown across the world, and their global production has increased twofold in the past quarter century. Yields of vegetables are highest in Asia in the eastern region where climate is temperate or sub-temperate. India has become the second largest producer of vegetables, next to China, since the past two decades; however, the consequences of global warming and drastic changes in weather pattern hit vegetable production worldwide. One of the main reasons for climate change is increase in the amount of greenhouse gases (CO_2, CH_4, N_2O, SO_2, CFC), which increases the global average temperature near Earth's surface and causes global warming. Other than these, burning of fossil fuels and plants, deforestation, and cultivation of agriculture crops, mainly rice, play major role in changing climate.

Probable impacts of climate change is sometimes visualized by negative effects on agricultural production by reducing the growth, quality, potential crop yields. These reduced crop production are due to alterations in crops package of practices *e.g.*, application of irrigation water, fertilizers, pesticides and particularly drainage practices, which possibly leads to loss of fertilizers by leaching and soil erosion. In tropical and subtropical regions, climate change leads to decrease in yield potential but increase at mid and high-mid latitudes. Among other crops, vegetables are highly sensitive to climatic extremes like heat, drought, flood, salinity and alkalinity stresses. Several physiological and biochemical processes such as photosynthetic activity, enzymatic activity, pollination and fruit set are temperature dependent; they are going to be highly influenced by climate vulnerability. The consequences of climate change also influence the incidence of insect pests and diseases as the life cycle of short life cycle groups of insects completes earlier, and fecundity increases with rise in temperature. Hence, they produce more generations in a particular time period than their usual multiplication rate. Fulfilling the demands of a growing population is a major challenge because of this unstable weather pattern. Therefore, there is a need to develop sound adaptation strategies to cope with the adverse impact of climate change on agriculture production. Various agronomical practices such as mulching with crop residues, zero tillage and organic farming help in mitigating the impact of climate change.

Climate change is a topic that has been increasingly grabbing global attention. Now it is an alarming need to change the agriculture production system against changing climate by applying mitigating strategies. There is an urgent need to encourage and focus on providing weather-based advisory, programmes related to NRM and low carbon storage structures as well as reduction in the amount of GHGs. Moreover, the modern vegetable improvement approach, including molecular breeding and genetic engineering, can also be helpful in developing climate-resilient crop varieties for the future.

This book should find a remarkable place on the shelves of new vegetable scientists, plant breeders, biotechnologist, plant pathologist, virologist and entomologist working on climate change, especially with vegetable crops, and in the libraries of all research establishments and companies where this newly exciting subject is researched, studied or taught. We sincerely acknowledge the authors who contributed to this endeavour and helped in the development of this book, entitled *Advances in Research on Vegetable Production Under a Changing Climate Vol. 1*, by providing novel information and in-depth scripts, without which the book would very hard to complete. I crave to recognize all the theme specialists, who were involved in the book and cooperate with their auxiliary, expensive assistance to create this book a success. We feel immense pleasure to express our heartfelt gratitude to Dr. Kirti Singh, former chairman, ASRB, former vice chancellor, A.N.D.U.A. & T., Ayodhya, C.S.K.H.P.A.U., Palampur and I.G.K.V., Raipur, for his inspiring guidance, encouragement and blessings.

Recognitions are due to our academic team members, especially to Dr. Silvana Nicola, series editor (Advances in Olericulture), Springer Nature, and Ms. Melanie van Overbeek, assistant editor, Life Sciences (Agronomy), Springer Nature, who generously supported for completion of this assignment. We are thankful to our family members whose sustained support and encouragement led us to complete this book. We are also thankful to our teachers, seniors, colleagues and friends for their direct and indirect support and valuable suggestions during compilation of this book. We would be glad if the readers recommend any valuable suggestion to make this book more useful. The editors sincerely acknowledge the help and support taken from various resource persons, books and journals, authors, and publishers whose publications have been used while preparing this manuscript. We thank Springer Nature for bringing out this publication in a very systematic manner. Finally, we are grateful to 'Almighty God' for providing strength, inspiration, positive thoughts, insights, and ways to complete this book.

Sabour, India Shashank Shekhar Solankey

Kanpur, India Meenakshi Kumari

Bengaluru, India Manoj Kumar
September, 2020

Contents

About the Editors

Shashank Shekhar Solankey is presently working as assistant professor–cum–junior scientist (horticulture: vegetable science) at Dr. Kalam Agricultural College, Kishanganj, under Bihar Agricultural University, Sabour (Bhagalpur), India. He acquired his master's degree in vegetable science from Acharya Narendra Deva University of Agriculture and Technology, Kumarganj, Ayodhya (India), in 2006 and doctoral degree in horticulture from Banaras Hindu University, Varanasi (India), in 2010. Dr. Solankey has qualified ICAR-ASRB NET examination in 2008. He has worked as research associate at ICAR-IIVR, Varanasi, India. He has more than 7 years of experience in teaching and research. His prime targeted area of research is biotic and abiotic stress resistance as well as quality improvement in vegetables, particularly, solanaceous crops and okra. Dr. Solankey has supervised four M.Sc. students and handled several research projects. He received Best Teacher Award (2016) and Best Researcher Award (2016) for BAU, Sabour; including these, he has 14 international/, national and institutional awards and has published 50 research papers, 07 review papers, 01 souvenir paper, 08 books, 01 abstract book, 36 book chapters, and 15 popular articles. He is also lifetime member of several societies and editorial boards as well as reviewer of many reputed journals.

Meenakshi Kumari is presently working as senior research fellow in the Department of Vegetable Science at Chandra Shekhar Azad University of Agriculture & Technology, Kanpur, Uttar Pradesh, India. She has specialization in vegetable science. Dr. Kumari acquired her master's degree in horticulture (vegetable and floriculture) from Bihar Agricultural University, Sabour (Bhagalpur), Bihar, in 2016 and doctoral degree in vegetable science from Chandra Shekhar Azad University of Agriculture & Technology, Kanpur, Uttar Pradesh (India), in 2019. She has qualified ICAR-ASRB NET examination in the Discipline of Vegetable Science in 2017 and qualified ICAR-SRF exam in 2016. Dr. Kumari has received Best Poster, Oral Presentation Awards (2018), Best Article Award (2018) and Best Thesis Award (2018) for her M.Sc. research work from the Society for Agriculture Innovation & Development, Ranchi (Jharkhand), and was awarded by Bihar Animal Sciences University, Patna (Bihar), at the National Conference on Livelihood and Food Security (LFS-2018). She was selected for DST – Inspire fellowship for her doctoral studies by the Department of Science & Technology, Government of India. Dr. Kumari also got first rank in both master's and doctoral exams. Other than these, she has received international, national and institutional awards for different presentations/thesis as well as published 22 research papers (national and international), 05 review papers, 23 book chapters, 02 books, 1 manual, 05 popular articles and more than 30 abstracts/extended summaries. She is also a lifetime member of several societies and reviewer of many reputed journals.

Manoj Kumar is presently working as doctoral research scholar in the Department of Vegetable Science at ICAR – Indian Institute of Horticultural Research, Bengaluru, (Karnataka), under ICAR – Indian Agricultural Research Institute, New Delhi, India. He has specialization in vegetable science. Dr. Kumar completed his graduation from the University of Horticultural Sciences, Bagalkot, Karnataka, India, in 2014. He acquired his master's degree in horticulture (vegetable science) from the University of Agricultural & Horticultural Sciences,

Shivamogga, Karnataka, India, in 2016. He has qualified ICAR-ASRB NET examination in Discipline of Vegetable Science in 2017. Dr. Kumar received Best Poster Award (2016) for his M.Sc. research work from the University of Agricultural & Horticultural Sciences, Shivamogga, Karnataka, in the 'National Conference on Post Graduate Research in State Agricultural Universities'. He has also received ICAR-National Talent Scholarship (NTS) for under-graduation and post-graduation. Dr. Kumar was selected for ICAR – Senior Research Fellowship (SRF), ICAR-IARI Fellowship and UGC- National Fellowship for Higher Education for his doctoral studies by the Government of India. Other than these, he has published 11 research papers (national and international), 7 book chapters, 03 popular articles and more than 15 abstracts/ extended summaries.

Contributors

Shirin Akhtar Department of Horticulture (Vegetable and Floriculture), Dr. Kalam Agricultural College, Kishanganj, Bihar Agricultural University, Sabour, Bhagalpur, Bihar, India

Satyaprakash Barik Department of Vegetable Science, College of Agriculture, Odisha University of Agriculture and Technology, Bhubaneswar, Orissa, India

D. R. Bhardwaj Division of Crop Improvement, ICAR-Indian Institute of Vegetable Research, Varanasi, Uttar Pradesh, India

Sunil Kumar Das Department of Vegetable Science, College of Agriculture, Odisha University of Agriculture and Technology, Bhubaneswar, Orissa, India

Sarvesh Pratap Kashyap Division of Crop Improvement, ICAR-Indian Institute of Vegetable Research, Varanasi, Uttar Pradesh, India

Anjani Kumar ICAR-Agricultural Technology Application Research Institute (ATARI), Patna, Bihar, India

Manoj Kumar Division of Vegetable Crops, ICAR-Indian Institute of Horticultural Research, Bengaluru, Karnataka, India

Randhir Kumar Department of Horticulture (Vegetable & Floriculture), Bihar Agricultural University, Sabour, Bhagalpur, Bihar, India

Meenakshi Kumari Department of Vegetable Science, Chandra Shekhar Azad University of Agriculture & Technology, Kanpur, Uttar Pradesh, India

Durga Prasad Moharana Department of Horticulture, Institute of Agricultural Sciences, Banaras Hindu University, Varanasi, Uttar Pradesh, India

Division of Crop Improvement, ICAR-Indian Institute of Vegetable Research, Varanasi, Uttar Pradesh, India

Abhishek Naik Marketing Department, Ajeet Seeds Pvt. Ltd. (Aurangabad, Maharastra), Kolkata, West Bengal, India

Pallavi Neha Division of Post Harvest Technology & Agri. Engineering, ICAR-Indian Institute of Horticultural Research, Bengaluru, Karnataka, India

Silvana Nicola Department of Agricultural, Forest and Food Sciences, DISAFA, Vegetable Crops and Medicinal & Aromatic Plants, VEGMAP, University of Turin, Grugliasco, TO, Italy

Menka Pathak Department of Vegetable Science, College of Agriculture, Odisha University of Agriculture and Technology, Bhubaneswar, Orissa, India

H. G. Prakash Directorate of Research, Chandra Shekhar Azad University of Agriculture and Technology, Kanpur, Uttar Pradesh, India

Nagendra Rai Division of Crop Improvement, ICAR-Indian Institute of Vegetable Research, Varanasi, Uttar Pradesh, India

Pankaj Kumar Ray Krishi Vigyan Kendra, Saharsa, Bihar, India

Bihar Agricultural University, Sabour, Bhagalpur, Bihar, India

K. Madhusudhan Reddy Department of Horticulture (Vegetable & Floriculture), Bihar Agricultural University, Sabour, Bhagalpur, Bihar, India

Anand Kumar Singh Department of Horticulture, Institute of Agricultural Sciences, Banaras Hindu University, Varanasi, Uttar Pradesh, India

D. P. Singh Directorate of Research, Chandra Shekhar Azad University of Agriculture and Technology, Kanpur, Uttar Pradesh, India

Hemant Kumar Singh Krishi Vigyan Kendra, Kishanganj, Bihar, India

Bihar Agricultural University, Sabour, Bhagalpur, Bihar, India

R. N. Singh Directorate of Extension Education, Bihar Agricultural University, Sabour, Bhagalpur, Bihar, India

Ramesh Kumar Singh Division of Crop Improvement, ICAR-Indian Institute of Vegetable Research, Varanasi, Uttar Pradesh, India

Shashank Shekhar Solankey Department of Horticulture (Vegetable and Floriculture), Bihar Agricultural University, Sabour, Bhagalpur, Bihar, India

Saurabh Tomar Department of Vegetable Science, Chandra Shekhar Azad University of Agriculture & Technology, Kanpur, Uttar Pradesh, India

D. K. Verma Department of Soil Science and Agricultural Chemistry, Dr. Kalam Agricultural College, Kishanganj, Bihar Agricultural University, Sabour, Bhagalpur, Bihar, India

Abbreviations

%	per cent
@	at the rate of
°C	degree Celsius
°F	degree Fahrenheit
ABA	abscisic acid
AmA1	amaranth seed albumin
APX	ascorbate peroxidase
CA	controlled
CAT	catalase
CFC	chlorofluorocarbon
CH_4	methane
CO	carbon monoxide
CO_2	carbon dioxide
COR	coronatine
DNA	deoxyribonucleic acid
EC	electrical conductivity
eCO_2	elevated atmospheric CO_2
FACE	free-air carbon dioxide enrichment
GEn	genetically engineered
GHG	greenhouse gas
GM	genetically modified
GR	glutathione reductase
GS	genomic selection
H_2O	water vapour
HITF	heat inducible transcription factor
HSE	heat shock element
HSF	heat shock factors
HSGs	heat shock genes
HSPG	heat shock protein genes
HSR	heat shock response
HSTFP	heat shock transcription factor proteins

HT	high temperature
IPCC	Intergovernmental Panel on Climate Change
LEA	late embryogenesis abundant
MA	modified
MABB	marker-assisted backcross breeding
MAGIC	multi-parent advanced generation inter-cross
MARS	marker-assisted recurrent selection
MAS	marker-assisted selection
MDA	malondialdehyde
MeJA	methyl jasmonate
N_2O	nitrous oxide
NAR	net assimilation rate
NICRA	National Innovations on Climate Resilient Agriculture
O_3	ozone
PEG	polyethylene glycol
POX	peroxidase
ppm	parts per million
PS	photo system
QTL	quantitative trait loci
RGR	relative growth rate
RLK	receptor-like kinase
RMPs	recommended management practices
ROS	reactive oxygen species
RuBisCO	ribulose bisphosphate carboxylase
SCS	soil carbon sequestration
SIC	soil inorganic carbon
SNP	single nucleotide polymorphisms
SO_2	sulphur dioxide
SOD	superoxide dismutase
SOM	soil organic matter
TF	transcription factor
TGT	temperature gradient tunnel
UNFCCC	United Nations Framework Convention on Climate Change
UV	ultraviolet
WAKs	wall-associated kinases
WUE	water-use efficiency

The Role of Research for Vegetable Production Under a Changing Climate Future Trends and Goals

Shashank Shekhar Solankey, Meenakshi Kumari, Manoj Kumar, and Silvana Nicola

1 Introduction

During the period of 1960–2018, the agricultural production is more than tripled due to invention of new Green Revolution technologies which enhance the productivity of agricultural crops and also the use of land, water and other natural resources got expanded for agricultural purposes. Despite this, in most of global countries malnutrition and hunger remain a major challenge and seems to persistent due to wider change in global climate and population scenarios. The current rate of progress of agricultural produce will not enough to feed the growing population by 2030, and not even by 2050 due to slow adoption of mitigation techniques for climate resilience.

Since, 1970s due to climate change natural disasters has increased fivefold. The natural environment got deteriorate day by day due to expanding of food production and economic growth. The level of groundwater is depleting at faster rate and deep erosion in biodiversity is another challenge (FAO 2017). The increased demand of

S. S. Solankey
Department of Horticulture (Vegetable and Floriculture), Bihar Agricultural University, Sabour, Bhagalpur, Bihar, India

M. Kumari
Department of Vegetable Science, Chandra Shekhar Azad University of Agriculture & Technology, Kanpur, Uttar Pradesh, India

M. Kumar
Division of Vegetable Crops, ICAR-Indian Institute of Horticultural Research, Bengaluru, Karnataka, India

S. Nicola (✉)
Department of Agricultural, Forest and Food Sciences, DISAFA, Vegetable Crops and Medicinal & Aromatic Plants, VEGMAP, University of Turin, Grugliasco, TO, Italy
e-mail: silvana.nicola@unito.it

S. S. Solankey et al. (eds.), *Advances in Research on Vegetable Production Under a Changing Climate Vol. 1*, Advances in Olericulture,
https://doi.org/10.1007/978-3-030-63497-1_1

agricultural product due to higher population growth put up the current agriculture under pressure (Ziervogel and Ericksen 2010; Godfray et al. 2010). At present, two main challenges to global food system are climate change and malnutrition and many studies have been attempted to achieve the global demand of food. Several studies have been conducted to find out the yield change due to climate change in major food crops and specific agronomic measures to counteract these impacts (Lobell et al. 2011; Lobell 2014).

Due to continuous climate change farmers will move away from low yielding crops and substitute them with better adapted crops to the new conditions due to continue progress in climate change (Seo and Mendelsohn 2008; Burke et al. 2009). Thc sensitive nature of agriculture produce and continuous changing climate are the major challenges.

Agricultural productivity, food security and other sectors affected by continuous change in climate. In tropical region, high temperature, declining rainfall patterns and increasing frequency of drought and floods are the expected future climate change (IPCC 2007; Mitchell and Tanner 2006). US National Centre for Atmospheric Research reported that in 2050 the rainfall trend is continuously declined and in compare to previous 50 years the region is expected to be 10–20% drier (Mitchell and Tanner 2006). The economic impacts of climate change on agriculture have been measured in various studies. Vegetables are usually succulent and sensitive plants therefore, severely affected by minor changes in the climate. The main focus of this chapter is to measure the effect of climate change in vegetable and role of research for vegetable production under changing climate. In developing countries vegetables are the main source of livelihood for most of communities because vegetables are loaded with several vitamins, carbohydrate, salts and proteins. Now a day's vegetables become an integral part of average household's daily meals because of increasing awareness towards their health.

2 Innovative Research Techniques for Vegetable Production Under Changing Climate

1. Organic farming
2. Irrigation management
3. Grafting techniques
4. Protected cultivation
5. Conservation tillage
6. Cropping system
7. Mulching
8. Post harvest technology
9. Genetic improvement
10. Biotechnology

3 Challenges of Vegetables Research During Climate Change

Climate changes pose several challenges and negative impacts upon both quality and production of vegetables. Several climate change especially temperature, rainfall, salinity, drought will reduce the productivity of vegetables. Vegetables are sensitive to these climatic changes, and sudden change in temperature and other climatic factors affects its growth, pollination, flowering, fruit development and thus reducing both average yields and quality of most major vegetables (Afroza et al. 2010). In potato, water stress during tuber formation stage, leads to higher susceptibility of tubers to postharvest black spot disorder (Hamouz et al. 2011). In carrot, preharvest water stress results in greater weight loss during storage (Shibairo et al. 1998). Smaller fruits with high soluble solids in tomato are due to high salinity condition of growing soil. Shelf life of leafy vegetables are affected by low light during growing period like in lettuce, shelf life of fresh cut lettuce grown in low light is much shorter than lettuce produced under optimal conditions (Witkowska and Woltering 2010). Vegetables arc highly sensitive to drought condition which is the primary cause of crop loss and reducing average more than 50% for most of the crop (Sivakumar et al. 2016). Vegetative stage of chilli is not much affected by heat fluctuation but reproductive stage is most affected. Production and quality of vegetables are much affected by high temperature as high temperature affect flower and fruit set, length, width and weight of fruit, number of fruits per plant and ultimately overall fruit yield (Tables 1, 2 and 3).

4 Role of Research Techniques

4.1 Protected Cultivation Technology: A Boon for Bio-technological Works

Under changing climate, cultivation of vegetables under protected condition is one of the best ways to protect our vegetables mainly from adverse environmental conditions such as temperature, hail, heavy rains, sun scorch, snow etc. It is an advanced agro-technology, which allow regulation of macro and micro environment, facilitating earliness, plant performance, duration of crop with higher and better quality yields (Gruda and tanny 2015). Nursery raising of vegetables under protected structures gives many folds benifits and also protects our crop from biotic and abiotic stresses (Sanwal et al. 2004). Better microclimate under polyhouse gives higher yield of different vegetables (Cheema et al. 2004). The yield and income of farmers increased as compare to open field conditions in tomato, capsicum (Kumar et al. 2016a, b). In comparison to open field condition, yield was increased by 80% under shade net and water saving of about 40% in covered cultivation (Rao et al. 2013). It provides an excellent opportunity to produce high value cash crops, vegetables and flowers and managed under controlled conditions with higher per unit productivity

Table 1 Impact of climate change on vegetable crops

Sl. No.	Climate change	Crops	Impact	References
1.	Heat	Tomato	Failure pollination, poor pollen production, poor viability of pollen, dehiscence, ovule abortion, abnormal reproduction, abnormal flower drop and at last failure to set fruit.	Hazra et al. (2007)
		Chillies	Affect fruit set *i.e*, fertilization, affect color development in ripe fruit, ovule abortion, flower drop, poor fruit set and fruit drop.	Arora et al. (2010)
		Cucumber, melons,	Germination affected, delay ripening of fruits and also reduce sweetness.	Kurtar (2010)
		Ash gourd, bottle gourd and pumpkin	Affect production of female flower	Ayyogari et al. (2014)
		Cauliflower	Bolting (not desirable for vegetable purpose)	Thamburaj and Singh (2011)
2.	Drought			
		Tomato	Flower abscission	Bhatt et al. (2009)
		Onion and okra	Affects germination	Arora et al. (2010)
		Potato	Sprouting of tubers	Arora et al. (2010)
3.	Salinity			
		Cabbage	Reduced germination, root and shoot length and fresh root and shoot weight	Jamil and Rha (2014)
		Chilli	Reduce dry matter production, leaf area, relative growth rate and net assimilation rate	Lopez et al. (2011)
		Cucurbits	Reduction in fresh and dry weight	
		Beans	Suppressed growth and photosynthesis activity and changed stomata conductivity thus reduces transpiration	Kaymakanova et al. (2008)
4.	Flooding			
		Tomato	Damage plants due to accumulation of endogenous ethylene	Drew (2009)
		Flooding + high temperature	Rapid wilting which results in death of plants	Kuo et al. (2014)
		Onion	Yield loss 30–40%	Kumar SN (2017)

and profitability (Choudhary 2016). In case of tomato indeterminate tomato hybrids (ID-32, ID-37, Rakshita, Himsona, Himsikhar, Snehlata, Naveen etc.) gives on an average 2–3 times higher quality yield and income as compared to traditional open farming systems (Table 4).

Table 2 Varieties of vegetable with various abiotic stress tolerance

Crops	Variety/lines	Tolerance to abiotic stress
Tomato	Pusa Sheetal	Set fruit at low night temperature *i.e,* 8 °C
	Pusa Hybrid 1	Set fruit at high night temperature *i.e,* 28 °C
	Pusa Sadabahar	Set fruit at both low and high night temperature
	Sabour Suphala	Tolerant to salt at seed germination stage
	Arka Vikas	Moisture stress tolerance
Chilli	DLS-10-02, DLS-20-11, DLS—160-1 and DLS-152-1	Heat tolerant
Eggplant	SM-1, SM-19 and SM-30	Drought tolerance
	Pragati and Pusa Bindu	Salt tolerance
Okra	Pusa Sawani	Salinity tolerant
Musk melon	Jobner 96-2	Tolerant to high soil pH
Spinach beet	Jobner green	Tolerant to high soil pH (upto 10.5)
Cucumber	Pusa Barkha	Tolerant to high temperature
	Pusa Uday	Grow throughout year
Bottle gourd	Pusa Santusthi	Hot and cold set variety
Onion	Hisar-2	Salinity tolerant
Carrot	Pusa Kesar	Tolerant to high temperature
Radish	Pusa Himani	Grow throughout year
Sweet potato	Sree Nandini	Drought tolerance
Potato	Kufri Surya	Heat tolerant upto 25 °C night temperature
	Kufri Sheetman, Kufri Dewa	Frost tolerance
Cassava	H-97, Sree Sahya	Drought tolerance
French bean	Arka Garima	Tolerant to heat and low and low moisture stress
Dolichos bean	Arka Jay	Tolerant to low moisture stress
Garden pea	Arka Tapas, Arka Uttam, Arka Chaitra	Tolerant to high temperature (upto 35 °C)
Carrot	Arka Suraj	Flowers and seeds sets under tropical condition.

Source: Koundinya et al. (2018)

4.2 *Molecular Breeding Approaches for Resistance Breeding*

In plant and animal breeding, use of DNA markers has opened a new realm in agriculture called molecular breeding (Rafalski and Tingey 1993). The use of DNA marker in this technology could speed up the selection process in comparison to traditional breeding method. Selection of primary trait link age between marker and trait is known as Marker assisted selection. It is an important tool to increase the

Table 3 Varieties of vegetables with various biotic stress tolerances

Sl. no.	Crops	Variety/F_1 hybrid	Abiotic stress
1.	Tomato	Arka Vishesh (H-391)	Triple disease resistance to tomato leaf curl, bacterial wilt and early blight
		Arka Apeksha (H-385)	Triple disease resistance to leaf curl, bacterial wilt and early blight
		Arka Abhed	Multiple disease resistance to tomato leaf curl, early blight, bacterial wilt and late blight
		Arka Rakshak	Triple disease resistance to ToLCV, BW and early blight
		Arka Samrat	Triple disease resistance to ToLCV, BW and early blight
		Arka Ananya	Combined resistance to ToLCV and bacterial wilt.
		Arka Alok and Arka Abha	Resistance to bacterial wilt
2.	Brinjal	Arka Unnathi, Arka Harshitha, Arka Avinash, Arka Neelkanth, Arka Nidhi, Arka Anand,	Resistant to bacterial wilt
3.	Chilli	Arka Khyati, Arka Harita, Arka Suphal	Tolerant to powdery mildew & viruses
		Arka Meghana	Field tolerant to viruses and sucking pest
4.	Capsicum	Arka Athulya	Tolerance to powdery mildew
5.	Watermelon	Arka Manik	Triple resistance to powdery mildew, downy mildew and anthracnose
6.	French bean	Arka Anoop	Resistance to both rust and bacterial blight
7.	Dolichos bean	Arka Sukomal	Rust resistant
		Arka Prasidhi	Resistance to rust
8.	Garden pea	Arka Ajit, Arka Karthik, Arka Sampoorna, Arka Priya	Resistance to powdery mildew and rust
9.	Carrot	Arka Suraj	Tolerant to powdery mildew and nematodes
10.	Onion	Arka Pitamber	Tolerant to purple blotch, basal rot diseases and thrips
		Arka Kirtiman, Arka Lalima	Field tolerance to diseases and pests

productivity of vegetables under changing climate condition. It increases the efficiency of selection by including various approaches like marker assisted backcross breeding (MABB), forward marker assisted selection, marker assisted recurrent selection (MARS) and genomic selection (GS).

Table 4 Vegetables varieties suitable for protected cultivation

Sl. No.	Crops	Variety/hybrids
1.	Tomato	ID-32, ID-37, Rakshita, Himsona, Himsikhar, Snehlata, Naveen, GS-600
2.	Coloured capsicum	Natasha, Swarna, Indra, Bombi, Orobelle, Bachata, Inspiration
3.	Parthenocarpic cucumber	Isatis, Kian, Hilton, Sun Star, Multistar, Fadia, Mini Angel
4.	Summer squash	Pusa Alankar, Pusa Pasand, Australian Green, Seoul Green, Kora, Yellow Zucchini, Himanshu
5.	Bitter gourd	Pusa Rasdar
6.	Musk melon	Pusa Sarda

4.3 SNP (Single Nucleotide Polymorphism) in Vegetable Crops

An individual nucleotide base difference between any two homologous DNA sequences representing the same locus in a genome is known as single nucleotide polymorphism. SNPs are the ultimate and most abundant molecular markers. It can be broadly classified into two ways namely hybridization based techniques and sequencing based techniques. In vegetable crops, SNPs can be widely used in tomato, carrot, potato, cucumber, brinjal etc.

4.4 Role of Biotechnology

Biotechnology plays a major role in the improvement of vegetables to make them suitable for altering climatic situation. Many biotechnology tools like tissue culture (micro propogation, meristem, endosperm culture, embryo, protoplast culture, haploid and callus & cell suspension), genetic engineering, genome editing and molecular markers of vegetables are useful tools that can cope with stress factors. Some of the important challenges which can be addressed by biotechnological tools are enlisted below:

1. In both biotic and abiotic stress condition its increases productivity of crop
2. Manage tolerance of herbicide
3. Manage diseases resistance
4. Improvement of genetic engineering technologies to enhance public perception

4.4.1 Biotechnology Based Approaches for Next Generation Agriculture

1. Tissue culture industry
2. Genomics
3. Molecular breeding

4. Genetic Engineering
5. Crops with novel traits

4.4.2 Need of Biotechnology for Vegetable Improvement

1. Eliminate unreliable phenotypic evaluation
2. There is no linkage drag.
3. Produced true to types
4. Overcome distant hybridization barriors (no species/genus gene transfer barrior)
5. Eliminates long term field trails
6. Shorten breeding cycles
7. 100% achievement of gene transfer

5 Improvement of Vegetable Through Genetic Engineering

Alteration of genome of an organism by introducing one or few specific foreign genes is known as genetic engineering. The crop which is modified by this tool is known as transgenic crops or genetically modified crops and the gene introduced is referred as transgene. Genetic modified crops are resistance to various biotic stresses (disease and insect resistance) and abiotic stresses like drought resistance, salt resistance, heavy metal resistance, cold tolerance, frost tolerance etc. Nutritional content of potato *i.e.,* protein and essential amino acids is increased by a seed-specific protein, AmA1 (amaranth seed albumin) of Grain Amaranthus (*A. hypochondriacus*) (Chakraborty et al. 2000). Beta carotene precursor of vitamin A increased more than three times than normal control.

6 Grafting

Under this climate change situation, various environmental stress became more crucial for vegetable production. Grafting of commercial cultivars onto selected rootstock offers an adoptive mechanism to overcome several biotic and abiotic stresses (Koundinya and Kumar 2014). This technique is widely exploited in comparison to relatively slower breeding methods to enhance environmental- stress tolerance of fruit vegetables (Flores et al. 2010). Now a days, grafting techniques has increased in crops like Tomato, brinjal, pepper, melon, cucumber, watermelon and pumpkin (Lee et al. 2010). Heat stress tolerance in temperature sensitive tomato was achieved by grafting onto more resistant rootstock (Abdelmageed et al. 2014). High yield of brinjal was achieved by grafting onto *Solanum torvum* because it enhances fruit size (Moncada et al. 2013). Grafting of eggplant (S. *melongena* cv. Yuanqie) onto a heat-tolerant rootstock (cv. Nianmaoquie) prolonged its growth stage and also give upto

10% increase in yield (Ahmedi et al. 2007). Grafting increased not only fruit production and marketable fruits but also gave higher phenolic antioxidant content (Sabatino et al. 2016).

In Tomato, for adjusting under suboptimal root-temperature cold tolerant rootstock gives higher capacity to their root/shoot (Venema et al. 2008). Bacterial wilt and flooding tolerance in tomato was achieved by grafting onto *Solanum melongena* (Palada and Wu 2007).

7 Tissue Culture

For improvement of vegetable crops, tissue culture industry is a fast growing sector. Micropropagation of superior genotypes is being practiced in India for the last three decades across a variety of vegetable crops such as potato, carrot, broccoli etc. For multiplication of plants by embryogenesis, organogenesis and by non-adventitious shoot proliferation mainly in vitro techniques is widely used (Table 5).

8 Embryo Rescue Technique

In horticultural crops to overcome the post-zygotic barriers such as endosperm abortion and embryo degeneration, embryo rescue technique is widely used and several hybrids have been developed in several vegetable crops like capsicum, tomato, muskmelon etc. (Kumari et al. 2018). In this technique, immature seed is harvested and induced to germinate on culture medium, with or without the addition of plant growth regulators, to negate the waiting time for seed to mature. In lettuce, haploid plants were developed through embryo culture techniques (Zenkteler and Zenkteler 2016). Hybrids in between *Lycopersicum esculentum* X *Lycopersicum peruvianum* were developed by using embryo rescue.

Table 5 Achievements of tissue culture in vegetable crops

Sl. No.	Crops	Purpose/ Achievement	Refs.
1.	Onion/ Shallot	Elimination of onion yellow dwarf virus	Walkey et al. (1987)
2.	Pea	Elimination of pea seed borne mosaic virus	Kartha and Gamborg (1978)
3.	Brinjal	Elimination of mosaic virus	Raj et al. (1991)
4.	Chilli	Haihua 3 variety	Li and Jiang (1990)

9 Conclusion

The main significant cause of yield loss in plants is abiotic stress which reduces yields by as much as 50%. In last five decades, India has achieved a lot in terms of agriculture but the major challenge is to feed its growing population. Due to continuous change in climate, abiotic stress becomes a major area of concern for plant scientists, affecting both production and quality of crop worldwide. Heat stress directly changes the physical properties of bio-molecules. The change in climate is one of the biggest worries because as most of the India's farming is still depend on the monsoon. Various breeding programme in different crops helps to cope up with these challenges. The first most important goal of any researcher is to develop resistant varieties which give high quality production under any conditions. By adopting advanced technologies farmers get higher yield and better quality that ensure more income for improving livelihood and nutritional security. But to cope with these challenges some technological inventions are seriously needed for the drastic improvement in the crop production scenario.

References

Abdelmageed AH, Gruda N, Geyer B (2014) Effects of temperature and grafting on the growth and development of tomato plants under controlled conditions. Rural poverty reduction through research for development and transformation

Afroza B, Wani KP, Khan SH, Jabeen N, Hussain K (2010) Various technological interventions to meet vegetable production challenges in view of climate change. Asian J Hortic 5:523–529

Ahmedi W, Nawaz MA, Iqbal MA, Khan MM (2007) Effect of different rootstocks on plant nutrient status and yield in Kinnow mandarin (*Citrus reticulata* Blanco). Pak J Bot 39:1779–1786

Arora S k, Partap PS, Pandita ML, Jalal I (2010) Production problems and their possible remedies in vegetable crops. Indian Hortic 32:2–8

Ayyogari K, Sidhya P, Pandit MK (2014) Impact of climate change on vegetable cultivation- a review. Int J Agric Environ Biotechnol 7:145

Bhatt RM, Rao NKS, Upreti KK, Lakshmi MJ (2009) Hormonal activity in tomato flowers in relation to their abscission under water stress. Indian J Hortic 66:492–495

Burke MB, Lobell DB, Guarino L (2009) Shifts in African crop climates by 2050, and the implications for crop improvement and genetic resources conservation. Glob Environ Chang 19:317–325

Chakraborty S, Chakraborty N, Datta A (2000) Increased nutritive value of transgenic potato by expressing a nonallergenic seed albumin gene from *Amaranthus hypochondriacus*. Proc Natl Acad Sci USA 28:97(7)

Cheema DS, Kaur P, Kaur S (2004) Off season cultivation of tomato under net house conditions. ISHS Acta Hortic 659:177–181

Choudhary AK (2016) Scaling of protected cultivation in Himachal Pradesh, India. Curr Sci 111(2):272–277

Drew MC (2009) Plant responses to anaerobic conditions in soil and solution culture. Curr Adv Plant Sci 36:1–14

FAO (2017) The future of food and agriculture – trends and challenges. Rome

Flores FB, Sanchez-Bel P, Estan MT, Martinez-Rodriguez MM, Moyano E, Morales B, Campos JF, Garcia-Abellán JO, Egea MI, Fernández-Garcia N, Romojaro F, Bolarín MC (2010) The effectiveness of grafting to improve tomato fruit quality. Sci Hortic 125:211–217

Godfray C, Beddington JR, Crute IR, Haddad L, Lawrence D, Muir JF, Preety JN, Robinson S, Thomas SM, Toulmin (2010) Food security: the challenge of feeding 9 billion people. Science 327:812. https://doi.org/10.1126/science.1185383

Gruda N, Tanny J (2015) Protected crops-recent advances, innovative technologies and future challenges. Acta Hortic (ISHS) 1107:271–278

Hamouz K, Becka D, Morava J (2011) Effect of environmental conditions on the susceptibility to mechanical damage of Potatoes. In: Agris On-line papers in economics and informatics. Available from http://www.agris.cz/vyhledavac/detail.php?id=116921&iSub=518&sHighlight=agria&PHPSESSID=997764de5509230f23dc06268b7b8344

Hazra P, Samsul HA, Sikder D, Peter KV (2007) Breeding tomato (*Lycopersicon esculentum* Mill) resistant to high temperature stress. Int J Plant Breed 1:31–40

IPCC (2007) Climate change–2007: the physical science basis. In: Solomon S, Qin D, Manning M, Chen Z, Marquis M, Averyt KB, et al., (eds), Contribution of Working Group I to the Fourth Assessment Report of the Intergovernmental Panel on Climate Change. Cambridge University Press, Cambridge

Jamil M, Rha ES (2014) The effect of salinity (NaCL) on the germination and seedling of sugar beet (*Beta vulgaris* L.) and cabbage (*Brassica oleraceae capitata* L.). *Korean*. J Plant Res 7:226–232

Kartha KK, Gamborg OL (1978) In diseases of tropical food crops. (H. Maraite and J.S. Meyer, eds.) Proceedings of the international symposium U. C. L. Louvain-la-Neuve, Belgium, p 267

Kaymakanova M, Stoeva N, Mincheva T (2008) Salinity and its effects on the physiological response of bean (*Phaseolus vulgaris* L.). J Cent Eur Agric 9:749–756

Koundinya AVV, Kumar VS (2014) Vegetable grafting: a step towards production of quality seedlings. In: Munsi PS, Ghosh SK, Bhowmick N, Deb P (eds) Innovative horticulture: concepts for sustainable development, recent trends. New Delhi Publishers, New Delhi, pp 217–222

Koundinya AVV, Kumar PP, Ashadevi RK, Hegde V, Kumar PA (2018) Adaptation and mitigation of climate change in vegetable cultivation: a review. J Water Climate Change. https://doi.org/10.2166/wcc.2017.045

Kumar SN (2017) Climate change and its impacts on food and nutritional security in India. In: Belavadi VV, Karaba NN, Gangadharappa NR (eds) Agriculture under climate change: threats, strategies and policies. Allied Publishers Pvt. Ltd., New Delhi, p 48

Kumar P, Chauhan RS, Grover RK (2016a) Economic analysis of capsicum cultivation under poly house and open field conditions in Haryana. Int J Farm Sci 6(1):96–100

Kumar P, Chauhan RS, Grover RK (2016b) Economic analysis of tomato cultivation under poly house and open field conditions in Haryana, India. J Appl Nat Sci 8(2):846–848

Kumari, P., Thaneshwari. And Rahul. (2018). Embryo rescue in horticultural crops Int J Curr Microbiol Appl Sci, 7(6): 3350–3358

Kuo DG, Tsay JS, Chen BW, Lin PY (2014) Screening for flooding tolerance in the genus *Lycopersicon*. Hortic Sci 17:76–78

Kurtar ES (2010) Modelling the effect of temperature on seed germination in some cucurbits. Afr J Biotechnol 9(9)

Lee JM, Kubota C, Tsao SJ, Bie Z, HoyosEchevarria P, Morra L, Oda M (2010) Current status of vegetable grafting: diffusion, grafting techniques, automation. Sci Hortic 127:93–105

Li C, Jiang Z (1990) The breed successful of 'Hai-hua-no 3' sweet pepper new variety by anther culture. Chin Acta Hortic Sinica 17(1):39–45

Lobell DB (2014) Climate change adaptation in crop production: beware of illusions. Glob Food Sec 3:72–76. https://doi.org/10.1016/j.gfs.2014.05.002

Lobell DB, Schlenker W, Costa-Roberts J (2011) Climate trends and global crop production since 1980. Science 333:616–620

Lopez MAH, Ulery AL, Samani Z, Picchioni G, Flynn RP (2011) Response of chile pepper (*capsicum annuum* L.) to salt stress and organic and inorganic nitrogen sources: *i.e*, growth and yield. Trop Subtrop Agroecosyst 14:137–147

Mitchell T, Tanner TM (2006) Adapting to climate change: challenges and opportunities for the development community. Tearfund, Middlesex

Moncada A, Miceli A, Vetrano F, Mineo V, Planeta D, Anna FD (2013) Effect of grafting on yield and quality of eggplant (*Solanum melongena* L.). Scientia Hortic 149:108–114

Palada MC, Wu DL (2007) Increasing off-season tomato production using grafting technology for peri-urban agriculture in Southeast Asia. Acta Hortic (742):125–131

Porter JR, Xie L, Challinor AJ, Cochrane K, Howden SM, Iqbal MM, Lobell DB, Travasso MI (2014) Chapter 7: Food security and food production systems. In: Climate change 2014: impacts, adaptation, and vulnerability. Part a: global and sectoral aspects. Contribution of Working Group II to the fifth assessment report of the Intergovernmental Panel on Climate Change. Cambridge University Press, Cambridge, pp 485–533

Rafalski JA, Tingey SV (1993) Genetic diagnostics in plant breeding: RAPDs, microsatellities and machines. Trends Genet 9:275–279

Raj SK, Aminuddin, Aslam M, Singh BP (1991) Elimination of eggplant mottled crinkle virus using virazole in explant cultures of *Solanum melongena* L. Indian J Exp Biol 29(6):594–595

Rao KVR, Agrawal V, Chourasia L, Keshri R, Patel GP (2013) Performance evaluation of capsicum crop in open field and under covered cultivation. Int J Agric Sci 9(2):602–604

Sabatino L, Lapichino G, Maggio A, Anna ED, Bruno M, Anna FD (2016) Grafting affects yield and phenolic profile of *Solanum melongena* L. landraces. J Integr Agric 15(5):1017–1024

Sanwal SK, Patel KK, Yadav DS (2004) Vegetable production under protected conditions in NEH region. ENVIS Bull Himalayan Ecol 12(2):1–7

Seo SN, Mendelsohn R (2008) An analysis of crop choice: adapting to climate change in South American farms. Ecol Econ 67:109–116

Shibairo SI, Upadhaya MK, Toivonen PMA (1998) Influence of preharvest stress on postharvest moisture loss of carrots (*Daucus carota* L.). J Hortic Sci Biotechnol 73(3):347–352

Sivakumar R, Nandhitha GK, Boominathan P (2016) Impact of drought on growth characters and yield of contrasting tomato genotypes. Madras Agric J 103

Thamburaj S, Singh N (2011) Textbook of vegetables, tubercrops and spices. Indian Council Agric Res, New Delhi

Venema JH, Dijk BE, Bax JM, Hasselt PRV, Elzenga TM (2008) Grafting tomato (*Solanum lycopersicum*) onto the rootstock of a high-altitude accession of *Solanum habrochaites* improves suboptimal temperature tolerance. Environ Exp Bot 63(1–3):359–367

Walkey DGA, Webb MJW, Bolland CJ, Miller A (1987) Production of virus-free garlic (*Allium sativum* L.) and shallot (*A. ascalonicum* L.) by meristem tip culture. J Hortic Sci 62:211–220

Witkowska I, Woltering EJ (2010) Preharvest light intensity affects shelf life of fresh cut Lettuce. Acta Hortic (877):223–227

Zenkteler E, Zenkteler M (2016) Development of haploid embryos and plants of *Lactuca sativa* induced by distant pollination with *Helianthus annuus* and *H. Tuberosus*. Euphytica 208(3):439–451

Ziervogel G, Ericksen PJ (2010) Adapting to climate change to sustain food security. Wiley Interdiscip Rev Clim Chang 1:525–540

Challenges and Opportunities in Vegetable Production in Changing Climate: Mitigation and Adaptation Strategies

Shashank Shekhar Solankey, Meenakshi Kumari, Shirin Akhtar, Hemant Kumar Singh, and Pankaj Kumar Ray

1 Introduction

At present global warming and climate change distress the whole world and it become the most focusing global environmental issue. Various abiotic stresses directly affect the production of crop and about 70% reduction in crop yield is estimated due to these changes (Acquaah 2009). Over the past 20 years South Asia has the robust economic growth, yet it is home of the above 25% of the worlds hungry and 40% of the world's malnourished women and children. Constant climatic variability, which outcomes in frequent drought and flood, is among the major reasons for this phenomenon (Shirsath et al. 2016).

In different countries high temperature cause melting of glaciers and raising the level of sea and ultimately cause change in the precipitation pattern which leads to

S. S. Solankey (✉)
Department of Horticulture (Vegetable and Floriculture), Bihar Agricultural University, Sabour, Bhagalpur, Bihar, India

M. Kumari
Department of Vegetable Science, Chandra Shekhar Azad University of Agriculture & Technology, Kanpur, Uttar Pradesh, India

S. Akhtar
Department of Horticulture (Vegetable and Floriculture), Dr. Kalam Agricultural College, Kishanganj, Bihar Agricultural University, Sabour, Bhagalpur, Bihar, India

H. K. Singh
Krishi Vigyan Kendra, Kishanganj, Bihar, India

Bihar Agricultural University, Sabour, Bihar, India

P. K. Ray
Krishi Vigyan Kendra, Saharsa, Bihar, India

Bihar Agricultural University, Sabour, Bihar, India

S. S. Solankey et al. (eds.), *Advances in Research on Vegetable Production Under a Changing Climate Vol. 1*, Advances in Olericulture,
https://doi.org/10.1007/978-3-030-63497-1_2

droughts, floods and extreme storms. Various studies reported that until the end of this century climate change will deleterious effects on the productivity of crop (Medellın-Azuara et al. 2011; Deschenes and Greenstone 2012).

The productivity of agricultural crops are highly affected due to change in the climate and it continuously increases the pressure on agriculture. Record breaking events of weather offered a glimpse of the challenges in the whole world. CO_2 concentration of the earth's atmosphere is continuously increasing and from the beginning of industrial revolution era about 25% of CO_2 increased. Globally the concentration of CO_2 in 2010 has increased 113 ppm from the last year which is 280 ppm and in twenty-first century reported to be doubled (IPCC 2007).

During late 1970s there has been increase in uncertain temperature and variation in rainfall patterns in tropical and subtropical climate and also increase in the percentage of extreme drought and extreme moisture surplus. The main threats for vegetable production, particularly in the tropical zone are continuous change in the climatic factors like increase in atmospheric temperature, high UV radiation, irregular precipitation, floods and drought (Tirado et al. 2010; Spaldon et al. 2015). Changing pattern of rainfall and rising temperature are some of the climatic factors which results in drought and flood hence, they become worldwide issue (Bates et al. 2008). In Asia and Africa, one of the serious concerns is dependence of low-income families. There is urgent need to adapt climate vulnerability i.e. higher temperature and shifting precipitation patterns by Indian farmers having small land holdings. The increasing level of CO_2 enhances the plant growth by increasing the photosynthesis process and depressing the plant respiration but it ultimately affecting the various metabolic processes in plant system. Several physiological processes of plant are also affected by various factors like temperature change, radiation comes from UV rays, availability of nutrients and water which are often due to change in climatic factors. According to the analysis of several workers, in the last 100 years the average global temperature has increased by 0.8 °C and by the year 2020 it has expected to increased 4.0 °C. While in region of India (South Asia), the temperature has predicted to rise 0.5–1.2 °C by the year 2020, 0.88–3.16 °C by 2050 (CO_2 concentration in atmosphere are likely to be >550 ppm) and 1.56–5.44 °C by 2080, according to the Intergovernmental Panel on climate change but it depends on the future development (Rao et al. 2009; IMD 2010). Rapid rising of greenhouse gas is one of the reasons of climate change which make great challenge to the food security. The interactions between atmospheric and oceanic process and their complex system are responsible for the climate change. Increasing concentration CO_2, CH_4 and N_2O, SO_2, CFC and losses of stratospheric ozone is relevant process of atmosphere. Other than these, O_3; SO_2; and CO_2 directly affects crop yield, while CO_2, CH_4 and N_2O play major role in increasing the harmful effects of weather and in altering air temperature CO_2, CH_4 and N_2O play major role and affect the production of agricultural crops mainly in the tropical zones of the world (Tirado et al. 2010). Therefore, in Asian region the reduction of crop yields by 2.5–10% from 2020 onwards and by 5–30% after 2050, with most horrible reduction assumed in South and Central part of Asia (Cruz et al. 2007). For increasing production of vegetable crops under these situation of climate change several steps like selection of genotypes which are resistance to these climate change, manipulation of genetic constitution of crops to tolerant against several abiotic stress of climate, several measures which improve the efficiency of crop towards available nutrient, water, nitrogen fixation and advantages of CO_2 for crop growth and yield.

Vulnerability Impact of Climate Change on Indo-Gangetic Plains

In this context, Chaudhary *et al.* (2012) ranked 161 districts as per their vulnerability index (V.I.) value of the Indo-Gangetic Plains (IGP). On the basis of these index, 161 districts were distributed into 4 different classes *i.e.* extreme level of vulnerable (−7.35 to −2.00), highly vulnerable (−1.99 to $0.00^{1/2}$), moderately vulnerable (0.01–2.00), and low vulnerable (2.01–4.69). On the basis of this IGP map based analysis it is very clearly indicates that in case of Bihar most of the districts are comes under extreme level of vulnerable class whereas, in case of Punjab mostly districts are comes under less vulnerable class. One of the district of Bihar known as Kishanganj was comes under most vulnerable with V.I. value of 4.687 whereas, district of Punjab i.e., Fatehgarh Sahib was comes under least vulnerable with V.I. value of −7.346 among all IGP districts.

Generally, vulnerability index normalized on 0–1 scale for diverse districts of individual state in research area and after that the index map were chalk out and drawing normalized vulnerability for each state. Due to moderately higher productivity and higher HDI, districts like Kaimur, Aurangabad, Rohtas and Khagaria had less than 0.25 normalized vulnerability index whereas, districts like Purbi Champaran, Madhubani, Sitamarhi and Sheohar had low average land holding size and low HDI (low capacity of adaption) due to which comes under extreme normalized vulnerability ranging between 0.8 and 1 (Fig. 1). Rest of all districts were comes under moderate to high normalized vulnerability with value between 0.26 and 0.75 (Chaudhary et al. 2012).

Significance of Change in Climate on Vegetable Production

In comparison to staple crops, vegetables are more knowledge and capital intensive. At present time vegetable growing sector is more dynamic than any other sector in

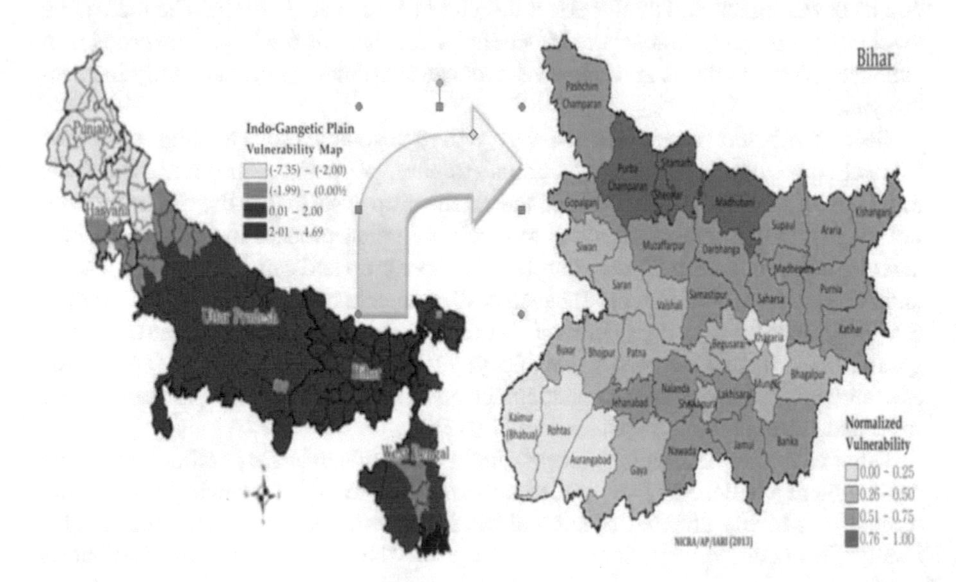

Fig. 1 Vulnerability map of Indo-Gangetic Plains of Bihar. (Adopted from Chaudhary et al. 2012)

the whole country. To cope with increasing demands of foods of growing population, annual production of 162.89 million tonnes of vegetables (Anonymous 2015) have to be increased and also expand domestic and external markets. Vegetables are full of essential nutrients and vitamins and also provide higher income than staple food crops. In different parts of country different vegetables are grown among which tomato, potato, onion, chilli is the main important vegetables in India. In India, West Bengal is highest in production followed by Uttar Pradesh, Bihar, Andhra Pradesh and Orissa (Anonymous 2015). Vegetables mostly require mild temperature for their growth and development hence in hot and humid regions of India productivity are generally low. The main cause of low yields and quality of vegetables are high temperature and limited soil moisture and even slight elevations in average day and/or night temperature affect the yields. High temperature mainly affects the crops like tomato, potato, broccoli and leafy vegetables (Anonymous 2000).

In favour of vegetable crops the present demand is to change the production system for better adaption of crops to heat stress and rainfall patterns. Moderate elevations in day/night temperature in tomato have negative effects on reproductive capacity hence results in lower fruit quantity and quality (Sato et al. 2006). Pollination of crop is more affected by climate change than any other constituents of agriculture but at present time whole pollination of agricultural crops depends upon the single pollinator species of honey bees (Kremen 2013).

For ensuring food and nutritional security, vegetables play a major role but they are highly affected by droughts and floods due to their perishable nature. The consequence of climate change is more on small and marginal farmers, particularly vegetable growers (FAO 2009). In compare to summer vegetables, winter vegetables are more sensitive to harsh weather. Climate vagaries like heat, drought and water logging directly affect the vegetable production at any stage of crop growth cycle i.e., from initial stage of crop growth, pollination, flowering, setting of fruit and its development and finally affect the yield (Afroza et al. 2010). The increasing level of CO_2 in earth atmosphere is beneficial for most of the vegetable crops as it enhances photosynthesis and improves water-use efficiency, consequently increasing yield.

Seed germination and reproductive growth of vegetable crops are highly affected by heat stress like in soalnaceous crops (tomato, pepper), cabbage and potatoes, reduction of crop yield will continue unless the crop is shifted to the cooler production regions. High temperature will reduce pollination process in many vegetables and also affect initiation process of tubers in tuber crop and cause tip burn, sun scald and blossom end rot in tomato. It cause bolting in cole crops and affect the production of anthocyanin in capsicum and brinjal. Daily ozone range when reached to greater than 50 ppb reduced vegetable yield by 5–15% (Narayan 2009). Ozone exposure impaired to stomatal conductance and reduces root growth, in carrots, beet roots and sweet potatoes (Spaldon et al. 2015).

Bihar and other developing states of India that tends to share parallel sustainable development challenges like limited resources, continuous growing population, remoteness, highly affected by natural disasters, vulnerability to external shocks, highly dependent on international trade, etc. The rainfall pattern of Bihar is regularly changing which leads to both shortage of water and excess of water (flooding). Continuous negative effect of such weather extremes will also cause negative impact on human body, lives and associated environment (Singh et al. 2014). This content mainly focused on the vegetable crops and the effect of change in global climate on different growth stage, production, quality and productivity of vegetables crops and also about the different strategies and approaches (conventional and non-conventional) to cope with these harmful challenges. The effect of weather extremes on different vegetable crops that may occur due to global warming, green house gases etc. are also reviewed.

2 Major Constraints Responsible for Limited Productivity of Vegetables

In whole world, main reason of annual crop loss is some of the abiotic stresses. Various stages of plant growth and development are affected by abiotic stress due to change in various cycle of plant growth (Rai et al. 2013). Some of the climatic factors such as drought, heat and salinity expressed osmotic stress thus cause disturbance in cellular homeostatsis as well as with ionic distribution (Kumar and Verma 2018).

Both the yield and quality of crop is even affected by mild stress of climatic factors however its totally depends on life cycle of crop. The overall impact of these climatic factors on crop growth and cycle are described below.

2.1 Heat Stress

At present continuous rising in temperature, heat stress becomes a serious agricultural problem worldwide. Both morpho-anatomical and physic-chemical process of plants is affected by constantly high temperatures and these changes not only affect the growth and development of plant as well as drastic reduction in both yield and quality of the produce (Wahid et al. 2007). As per the IPCC (2007) it has been reconfirmed that the global climate is continuously changing and it has been reported that by the end of the twenty-first century across the whole world, the average temperature of atmosphere would rise by 1.8–4.0 °C and this situation is depends upon the adoption of developed pathways by the countries. In past century it is reported that 0.74 °C rose in global atmospheric temperature. Vegetables cultivation under optimum climatic condition gives three to four folds higher productivity. Major climatic extremes which are responsible for low vegetable productivity and quality are some of the abiotic stresses *i.e.,* flooding, unavailability of irrigation water, increase in atmospheric temperature, high salinity and alkalinity. To stand against few abiotic factors plant may respond by their morphological as well as biochemical mechanisms (Capiati et al. 2006). Stress response of plants significantly influences the degree of the impact due to climate change by environmental interactions. The yield of vegetables crops decreases with increase in heat that reduces crop vegetative phase and prior reproductive phase, increase respiration and decreased time for sun rays interception (Rawson et al. 1995). Thermo variation can directly influence plant photosynthesis, and an increased worldwide temperature is assume to put deteriorating impact on fruit quality after harvest by changing several quality parameters like synthesis of sugar, carboxylic acids, major antioxidants, and firmness.

Prolonged disclosure of potato in high CO_2 condition induce sugar content and malformation in tuber of potato and affect nutritional *i.e*, protein and minerals content and sensory quality in case of pea and beans. Post-harvest quality of vegetables is highly affected by day-by-day increased levels of ozone. To evaluate the effect of elevated temperature on growth, pollination, quality and yield of crops various technologies have been developed. To evaluate the reaction of crop to certain range of temperature in the field temperature gradient tunnel (TGT) have been developed which helps the scientist in their research (Okada et al. 1992; Nakagawa et al. 1993; Rawson et al. 1995). The TGT system offers wide combination of temperature range in the field which are important for various farming practices i.e. a big advantage of it (Chakrabarti et al. 2012).

The metabolism of plants are adjusted according to the stress situation like in high temperature stress plants release some solutes that organize the structure of protein and cellular and also maintain cell osmotic adjustment (turgor pressure), and re-establish the balance of homeostasis and cellular redox by doing alteration in antioxidant system (Janska et al. 2012; Munns and Tester. 2008; Valliyodan and Nguyen 2006). The expression of various genes that are involved in the protection of plant from heat stress got altered at high temperature at molecular level (Chinnusamy et al. 2007; Shinozaki and Yamaguchi-Shinozaki 2007). Whereas, at genetic level the expression of gene got changed which helps in the finding of heat tolerance that results modification of physiological and biochemical processes of plants for adaption to high temperature (Hasanuzzaman et al. 2010; Moreno and Orellana 2011).

In some cases, several crops may got benefit by the increasing level of $CO_{2,}$ in addition for forecast elevated temperatures effects are more complex in comparision to provided temperature (Peet and Wolf 2000). Besides these, heat stress can affect several functions of the plant *viz*., rate of photosynthesis and respiration, relation of water and stability of membrane along with imbalance in amount of various plant harmones, primary and secondary metabolites during plant growth and development stages (Magan et al. 2011). Depends on the crop species and extend of heat stress it also affect the germination of seed *i.e*, decreased or inhibited the rate of germination (Bewley 1997; Lobell et al. 2011; Ramos and Miller 2013). In *Arabidopsis thaliana* under high temperature stress in a simulated heat treatment under field conditions early flowering and larger flower size has been noticed (Springate and Kover 2013). The rate of mortality of several insects which helps in pollination process mortality rate of pollinators and populations of pest, diseases and natural enemies got increased due to heat stress (Chakraborty and Newton 2011).

Moreover, the warmer spring temperatures have negative impact on pollen tube growth, flowering, and ovule size, which results in less fruit yield (Beppu and Kataoka 2011; Pope 2012). The response of duration and intensity of heat stress depends upon various factors and growth stages of the crop, the type of plant tissue (Porter and Semenov 2005).

Źróbek-Sokolnik (2012) classified the plants on the basis of optimum temperature for growth into three different groups:

(a) Psychrophiles: Those plants which can be grow normally at temperature between 0 and 10 °C.
(b) Mesophyles: Those plants which can be grow at moderate temperature i.e., between 10 and 30 °C.
(c) Thermophyles: The plants which need slightly high temperature i.e., between 30 and 65 °C or above. Under this group of heat stress there are three types of plants *viz*., sensitive, relatively resistant and tolerant/resistant.

The response and tolerance level of different plant species is different against heat stress. Adaptive mechanisms of different plant help in survival of plants against warm and arid environment (Fitter and Fitter 2002).

The major vegetable crops like tomatoes and chillies are extremely sensitive to heat stress during their growth and reproductive stages. In which, tomato plants are especially more sensitive to heat stress, that affects its function particularly at flower initiation and pollination stages (Peet et al. 1998; Li et al. 2012; Mattos et al. 2014). Cool season crops like, *Pisum sativum* L. is affected by day temperature even if air temperature exceed about 35 °C for sufficient duration plant death occurs (Hemantaranjan et al. 2014). Muller et al. (2016) reported the effect of mild heat conditions on tomato flowers and he observed that the number and viability of pollen prains get reduced and also the deformed anther resemble a partial conversion to pistil identity. The heat stress affects the development of reproductive parts of many crop spices and as a result they don't produce any flower or even in case if flower produce they unable to set seeds or fruit. Normally, a temperature range between 0 and 40 °C is optimum for physiological processes of plant system. Vegetables require lower cardinal temperature for their growth and development but it depend on the species and ecological origin. For temperate spices such as carrot and lettuce these temperature can be shifted towards 0 °C.However, for tropical regions crop such as cucurbitaceous vegetables these temperatures reach 40 °C (Went 1953). For higher yield, the ratio of photosynthesis/respiration should be higher. The ratio of photosynthesis/respiration rate is usually higher than 10 at 15 °C temperature hence in temperate regions plants survive more than tropical ones. At higher temperature the chain of photosynthetic process *i.e.*, enzyme activity and electron transport got affected (Sage and Kubien 2007) and indirectly increases leaf temperature. One of the chief factors that influenced conductance of stomata and photosynthetic activity is the extent of the leaf-to-air vapor pressure difference (D) (Lloyd and Farquhar 2008).

Above a certain temperature threshold, several changes occur in tissue of plant tolerance to heat stresses due to change in major function of plant tissue (Bieto et al. 1996). The higher the temperature during the early vegetative phase of crop, will sooner the crop reproductive phase i.e., flowering, fruiting and maturity. Cauliflower, lettuce and celery matured earlier when grown under higher temperature as compare to lower temperature (Wurr et al. 1996 and Hall et al. 1996). Temperature regimes due to changes in climate also have some positive effects such as gives opportunities to develop some crops in different areas, whereas crop quality *i.e.*, size of fruit, yields and appearance decreases by undesirable events (Ackerman and Stanton 2013).

If the tomato crops grown in direct sunlight just for 3 days at above 36 °C temperature results in colour (lycopene) development i.e., ripening, evolution of ethylene and climactric respiratory. Moreover, high temperature above 30 °C suppress the parameters related to fruit ripening in tomato i.e., development of lycopene, softening, rate of respiration and production of ethylene (Buescher 1979; Hicks et al. 1983). Exposure of fruits to higher temperature *i.e*, 40 °C affect metabolic disorders of plant and also cause bacterial and fungal invasion. At the end of storage period various visual appearances like sun scald, injury of heat and disease incidence commonly occurred. In tomato heat injury or sun scald cause yellowish-white side patches on fruits because of damage of lycopene pigments (Mohammed et al. 1996; Ghini et al. 2011; Pangga et al. 2011).

In contrast, high temperature stress (32.5 °C) tolerance of tender fruit of cucumber increases with no change in vitro aminocyclopropane carboxylate oxidase activity (Chan and Linse 1989). Exposure of vegetables to high temperature and exposures to direct sun light results in several internal and external symptoms and physiological disorders.

Mohammed et al. (1996) studied the effect of exposure of "Dorado" tomatoes in direct sunlight (34.62 °C) for 5 h and he observed that 73% higher electrolyte leakage in those fruits as compare to fruits grown in shaded (29.62 °C) conditions. Though, during 18 days of storage at 20 °C there were no changes recorded for electrolyte leakage and also for direct sunlight exposure. Tomato fruits exposure to high temperature affects fruit visual quality and biochemical characteristic. The direct sun light exposed tomatoes have high titratable acidity (20%) and lower total soluble solids content (10%) (Mohammed et al. 1996).

Pani (2008) stated that in some parts of India the impact of climate change have now been experienced. Delayed onset of winter in Sambalpur, India due to increased temperatures results in tremendous declined of cauliflower yields. Now, weight of each cauliflower curd was reduced from 1 kg to 0.25–0.30 kg. The yield reduction because of weather extremes directly affect the production expenses of several vegetables like tomato, chilli, brinjal, radish etc. and high temperature also affect the quality parameters i.e., development of colour, tenderness of fresh vegetable, total soluble solids, visual appearance and also nutritive value.

Leafy greens (lettuce, spinach and beet leaf) and most Cole crops (cabbage, cauliflower, knoll khol, broccoli, peas) and root crops (radish, carrot, turnip and beet root) are considered as low temperature requirement crop hence, high temperature affects the development of these crops. In lettuce, spinach and beet leaf higher temperature cause bolting and in case of cole crops and root crops lower temperature induce bolting. Over winter cool period is required for the sedd production of cole crops and root crops.

2.1.1 Temperature Gradient Tunnel (TGT)

In several parts of the globe due to increase in temperature of atmosphere heat stress become a foremost agricultural trouble. High temperature results in alterations of morpho-anatomical and morphological features like seed germination, early growth, pollen viability, flower drop, gametic fertilization, fruit setting, fruit shape, fruit colour, fruit quality and other traits. These disorders can be reduced by adopting improved cultural practices and new breeding approaches. Several kinds of morph-physiological characters that is useful in tolerance to heat in the traditional approaches of breeding by using TGTs. In reality, TGTs is a best way which provides a novel way to study a array of temperatures of huge amount of plants. TGT has a semicircular bows which are covered with UV stabilized polythene sheet or polycarbonate to create similar atmosphere to the greenhouse (Fig. 2).

The polycarbonate or UV stabilized polythene sheet should have about 90% transparency level. The TGTs based heat stress screening is very much effective for study of heat tolerance in vegetable crops like tomato, chillies, cauliflower and other fruit vegetables.

2.1.2 Morphological Traits for Heat Tolerance

- Long root length: Both nutrient and water absorbed easily by plants from the soil
- Short life-span: The effect of temperature on plant should be minimized.
- Hairy/pubescence: Repels sun rays by providing partial shade to cell wall as well as cell membrane.
- Small leaf size: Reduction of stomata hence resists evaporation.

Fig. 2 External and internal view of Temperature gradient tunnel. (Source: Indian Institute of Vegetable Research, Varanasi, U.P., India)

- Orientation of leaf and leaf angle: Photosynthetic activity enhanced and avoids direct exposer to sun light.
- Glossy and Waxy leaf: Sun rays repellent.

In tomato several factors associated with yield is highly affected by higher temperature i.e. reduced pollination, fruit set, fruit size and poor fruit quality (Stevens and Rudich 1978). Heat stress results flower drop, unusual flower development, anthredial cone splitting, poor pollen yield, dehiscence, viability and germinability, abortion of embry prohibited availability of carbohydrate, and other reproductive abnormalities of fruit formation. As well as, considerable photosynthesis inhibition at high temperature, resulting in significant yield loss. Moreover, the vegetative as well as reproductive functions are sturdily modified by heat alone or in combination with other climatic factors (Abdalla and Verderk 1968).

In comparison to vegetative phase, reproductive phase is more sensitive to heat stress in tomato. For proper growth and development of tomato plants during the photoperiod, temperature between 25 and 30 °C is optimum wheras, 20 °C during the dark period as well as temperature above 35 °C affect the major stages of crop growth. Hence, slight increase in optimum temperature *i.e.*, 2–4 °C results in reduced crop yield and fruit quality due to poor gamete development and reduced ability of pollinated flowers to produce seeded fruits (Peet et al. 1997; Firon et al. 2006). The development of floral bud also affected by heat stress results in abortion of flower and ultimately flower drop (Kalia and Yadav 2014). The temperature regulation in different vegetable crops is depicted in Table 1.

Moreover, high temperature also affects the morphological, physiological and biochemical process of plant as well as different phases of plant growth and development (Table 2). High temperature affect the maturity of different vegetable crops like under high temperature crops like lettuce, celery, cauliflower, etc. results prematurity as compare to the same crops grown under lower temperature (Kalia and Yadav 2014).

By using conventional breeding approach two improved tomato cultivars having genes which are tolerant to heat and gave high yield were selected and bred for obtaining desired plants from segregating populations. For introducing heat tolerance traits in tomato molecular breeding strategy like genetic engineering can be used by which heat tolerant gene can be transfered to plant cell (Akhtar et al. 2014). For tolerance to heat stress marker-assisted breeding method along with cloning and characterization of underlying genetical components possibly highly valuable (Maheswari et al. 2015).

In plant cell, thermo tolerance has been directly inserted by over expressing the heat shock protein genes (HSPG) or indirectly induced by the levels of heat shock transcription factor proteins (HSTFP). Tolerance to high temperature also increased by increasing the levels of enzymes of cell detoxification or by elevating the levels of osmolytes and also by modifying membrane fluidity (Kalia and Yadav 2014). To develop a plant from transformed cell plant tissue culture technique is also used. Moreover, to identify the gene in plant cell or in F_2 population molecular breeding technique may be also used to identify the tolerance ability against heat stress. In the

Table 1 Regulation of temperature in different vegetable crops

Sl. No.	Crops	Effect of temperature on production	Temperature range (°C)
1	Tomato	Optimum night temperature for fruit set	15–20 °C
		Fail to set fruits	13° or below
		Optimum temperature for lycopene formation (highest)	21–24 °C
		Production of lycopene pigments drops	Above 27 °C
		Disappearance of red colour of fruit which results in yellowish red Red colour.	Above 30 °C
		Affect fruit set adversely	>32 °C day and 22 °C night
		Lycopene completely destroyed	>40 °C
2	Chilli	High temperature leads to poor fruit set	40 °C
		Optimum temperature for chilli cultivation	15–35 °C
3	Capsicum	Colour developments of fruits drops	Above 27 °C
4	Cabbage	Average temperature for tight head formation	25 °C
		Optimum temperature for growth and heading	15–20 °C
		Tropical heat tolerant varieties able to set head	30–35 °C
5	Cauliflower	Bolting for snow ball (late) types	Low temp (−1 to −2 °C)
		Optimum temperature for curd formation	17 °C
		Snow ball cauliflower comes flower under	10 °C
		Optimum temperature for curd initiation and development	
		Early- 1 group	20–27 °C
		Early- 2 group	20–25 °C
		Mid early group	16–20 °C
		Mid late group	12–16 °C
		Late group	10–16 °C
6	Carrot	Best colour development	15.5–21.1 °C
		Temperature range for seed germination	7.2–23.9 °C
		Optimum temperature for root formation	18–22 °C
		Tropical carrot for flower initiation	15–25 °C, 1–2 months
		Temperate carrot for flower initiation	5–8 °C for 40–60 days
7	Beetroot	Flower initiation (bolting)	4.5–10 °C for 1 month
8	Radish	Optimum temperature for root flavour, texture and size bolting, high pungency	10–15 °C high temperature >26 °C
9	Onion	Optimum temperature for seed germination	20–25 °C
		Temperature for bulb development	15.6–25.1 °C
		Temperature for before bulbing	13–21 °C
		Optimum temp for flower initiation	10–12 °C
		Best storage temperature for mother bulb for seed production	12 °C

(continued)

Table 1 (continued)

Sl. No.	Crops	Effect of temperature on production	Temperature range (°C)
10	Okra	Optimum temperature for seed germination	25–35 °C
		Fastest seed germination	35 °C
		Seed germination failed at	Below 17 °C
		Temperature for flower drop	Above 42 °C
11	Cucurbits	Most of the cucurbits need day temperature for seed germination	Above 25 °C
		Optimum soil temperature for seed germination	18–25 °C
12	Musk melon	Temperature for fruit developmental stage	35–40 °C
13	Cucumber	Female flower production reduced at	Above 30 °C
14	Water melon	Average temperature for normal vegetative growth	25–30 °C

Source: Modified from Selvakumar (2014) and Solankey et al. (2015)

current decade, biotechnology/molecular breeding is significantly contributing for depth understanding of high temperature tolerant gene and their mechanism in tomato plants.

2.1.3 Heat Stress Resistant Sources of Vegetable Crops

A remarkable contribution for heat tolerant tomato and Chineses cabbage lines (*Brassica rapa* subsp. *pekinensis* and *chinenesis*) was done by vegetables scientists. The crosses between tropical lines having tolerance to heat and temperate or winter varieties having disease resistance were used for development of heat tolerant varieties (Opena and Lo 1981).

Landraces of different countries and heat tolerant breeding lines were used for development of heat-tolerant tomato lines. However, the heat-tolerant lines have lower yields potential that is a serious concern for breeders. The heat and cold tolerant line of *Solanum habrochaites* (EC 520061) can withstand under high (40 ± 2 °C) and low (10 ± 2 °C) temperatures. IARI, New Delhi also developed some of the heat and cold tolerant tomato varieties for commercial cultivation *i.e.*, Pusa Sadabahar (heat and cold tolerant) and Pusa Sheetal (cold tolerant) and two hybrid Pusa Hybrid-1 and Pusa Hybrid-8 (heat tolerant) (Akhtar et al. 2014; Kalia and Yadav 2014; Solankey et al. 2015). However, various Institutes/Universities have been released several varieties and genotypes which are tolerance to heat stress (Table 3). In case of radish, Pusa Chetaki has been developed for grown in north Indian plains from April–August under elevated temperature regime whereas, carrot variety Pusa Vrishti able to tolerate high temperature as well as high humidity (March–August). Early cauliflower variety Pusa Meghna can able to form curd at high temperature (Akhtar et al. 2014).

Table 2 High temperature stress that causes physiological disorders of different vegetables

Sl. No.	Vegetables	Disorder	Climatic factor
1	Asparagus	Amount of fiber in stalks and spheres increased	High temperature
2	Asparagus	Feathering and lateral branch growth	Temperature >32 °C, especially if picking frequency then it is not increased
3	Bean	In pods fibre increased	High temperature
4	Snap bean	Brown and reddish water-soaked spot on the pod	High temperature
5	Carrot	Low carotene content	Temperature <10 °C or >20 °C
6	Cauliflower	Blindness, buttoning, riceyness	Fluctuations in temperature
7	Cauliflower, broccoli	Stems become hollow, head turned leafy, no heads, bracting	High temperature
8	Cabbage	Bleached and papery outer leaves	High temperature
9	Lettuce	Tip burn, soft rot	Drought, combined with high temperature, high respiration
10	Lettuce	Bolting	Temperature >30 °C temperature
11	Tomato, pepper, watermelon	Blossom end root	Excessive temperature, particularly if associated with drought and high transpiration,
12	Tomato	Sun burn, sun scaled	Heat injury and exposure of fruits in direct sun light
13	Potato	Black heart	Excessively hot weather in saturated soil

Source: Modified from Swarup (2006), Moretti et al. (2010) and Spaldon et al. (2015)

These varieties are revolutionary varieties and can be used in breeding programme for developing heat tolerant varieties. Under high temperature stress the severity of tomato leaf curl virus (ToLCV), early and late blight were also observed therefore the future breeding programme depends upon the transfer of resistance genes of these diseases into a heat tolerant lines with wide adaptability through gene pyramiding by using wild relatives (*S. pimpinellifolium* and *S. habrochiates*) (Solankey et al. 2015).

2.2 Cold Stress

Vegetables like cauliflower, cabbage, sprouting broccoli, knol khol, peas, potato, onion, garlic, carrot, turnip, radish, spinach, lettuce etc. are able to tolerant low temperature and even frost is known as Cool season vegetable crops. These crops

Table 3 Vegetables species and genotypes tolerant to heat

Sl. No.	Vegetable crops	Genotypes tolerant to heat	References
1	Tomato	Pusa Hybrid-1, PusaSadabahar, Pusa Hybrid-8, Arka Meghali, Arka Vikas, Punjab Barkha Bahar-1, Punjab Barkha Bahar-2 *Solanum cheesmani*, EC-41824, EC-164855, EC-7764, EC-169308, EC-122063, EC-164653, EC-37311, EC-126955, EC-31515, EC-130042, EC-154660, EC-35220, EC-163709, EC-164677, EC-164322, Sparton Red-8, OK-7-2, OK-7-3, Saladette, TAMU Saladette, P-126934, BL-6807, S-6916, C1A-5161, VF-36, Punjab Tropic, Marjano, P-4, Analanche, Chiko-III, Ment, C-28, Red Rock, Red Cherry, KS-1, KS-2, Walter, EC-162935, AC-325, AC-326, PS-1, EC-1127, EC-130042, EC-168064	Spaldon et al. (2015)
		VC11-3-1-8, VC 11-2-5, Divisoria-2, Tamu Chico III, PI289309, *Solanum habrochaites* (EC-520061).	Opena et al. (1992)
		S. pimpinellifolium, S. pennellii, S. habrochaites, S. chmielewskii and *S. cheesmaniae,* Pusa Hybrid-1, Pusa Sadabahar, Sun leaper, Solar Fires, Heat Master and Equinox.	Alam et al. (2010), Gulam et al. (2012), and Nahar and Ullah (2011)
		S. pennellii, IIHR-14-1, IIHR-146-2, IIHR-383, IIHR-553, IIHR-555, IIHR-2274	Ram (2005) and Chavan (2007)
		Red Bounty, **STM2255**	Anonymous (2020)
2	Potato	Kufri Surya	Selvakumar (2014)
3	Brinjal	Kashi Sandesh, Kashi Taru	Spaldon et al. (2015)
4	Capsicum	IIHR Sel.-3	
5	Okra	Kashi Pragati, Kashi Kranti	
6	Cauliflower	Sabour Agrim, Pusa Meghna, IIHR 316-1, IIHR-371-1	
7	**Broccoli**	**Abrams**	Anonymous (2020)
8	Bottle gourd	Thar Samridhi, Pusa Santushti	Spaldon et al. (2015)
9	Cucumber	Pusa Barkha	Selvakumar (2014)
10	Garden pea	Matar Ageta-6, Azad Pea G 10, IIHR-1, IIHR-8	Spaldon et al. (2015)
11	French bean	IIHR-19-1	
12	Cluster bean	RGC-197, RGC-936	
13	Cowpea	Kashi Kancha, Kashi Nidhi	
14	Onion	NP53, Raseedpura local	
15	Radish	Pusa Chetki	
16	Carrot	Pusa Kesar, Pusa Vrishti	Selvakumar (2014)
17	**Snap beans**	**PV 857**, **Annihilator**, **Dominator**, **Usambara**	Anonymous (2020)
18	**Lettuce**	**Forlina**, **Salanova, Green Butter**, **Salanova, Red Butter**, **Skyphos**, **Starfighter**; **Dov**, **Arroyo**.	Anonymous (2020)

Source: Modified from Hazra and Som (1999), Maheswari et al. (2015) and Solankey et al. (2015)

Table 4 Classification of vegetable crops according to their adaptation to field temperature

Cool-season vegetable crops		Warm-season vegetable crops	
Hardy (can withstand moderate frosts)	Half-hardy (can withstand light frosts)	Tender (sensitive to frost and low temperatures)	Very Tender (very sensitive to low temperatures)
Asparagus, garlic, radish, broad bean, horseradish, rhubarb, broccoli, kohlrabi, spinach, Brussels sprouts, turnip, cabbage, parsley, chive and garden pea (flowers & pods are more sensitive to frost)	Beetroot, Chinese cabbage, potato, carrot, globe artichoke, Swiss chard, cauliflower, lettuce, celery and parsnip.	New Zealand spinach, sweet corn, tomato and Green bean.	Chilli, okra, sweet pepper, cucumber, pumpkin, sweet potato, eggplant, squash, vegetable marrow, Lima bean, sweet melon and watermelon.

are sensitive to heat as compare to heat tolerant crop likes tomato, brinjal, chilli, pumpkin, etc. as well as some of the crops like sweet potato, water melon, musk melon are able to tolerant temperature upto 40 °C (Table 4).

In north India, particularly during Rabi season frost and chilling injury occurs due to various change in climate which arises problem in growing of several crops like tomato, chilli, brinjal etc. The growth and development of tomato (*Solanum lycopersicum*) got affected at below 12 °C temperatures (Hu et al. 2006a). Moreover, plants are damaged by the chilling injury at temperatures range of 0–12 °C. Damage severity of any plant is directly relative to the length of time spent in range of temperature.

The commercial cultivated tomatoes are generally cold sensitive therefore the crop were planted late in the field to escape chilling injury along with minimize the risk. However the varieties which are tolerant to cold are planted earlier and got high returns due to early harvesting. However, the several wild species of tomato (*S. habrochaites*, *S. chilense* and *S. peruvianum*) recover rapidly after exposure to high temperatures i.e. below 10 °C. The resistance genes of these wild species can be transferred to the cultivated tomato through molecular markers assisted back-crosses and molecular markers assisted selection, as well as also used for genetic maps construction and identification of agronomic importance genes (Goodstal et al. 2005). Wild species of tomato i.e., *S. Habrochaites has high photosynthesis rate with dense hair stem, leaves and fruit and seedlings have ability to survive at even* 0 °C temperature. The cold tolerant line of *Solanum habrochaites* (EC-520061) can withstand at low (10 ± 2 °C) temperatures. Moreover, cold tolerant varieties of tomatoes are Pusa Sadabahar and Pusa Sheetal is recommended for commercial cultivation (Akhtar et al. 2014; Kalia and Yadav 2014; Solankey et al. 2015). One low temperature tolerant species of capsicum (*Capsicum cardenasii*) has been identified and can be used as tolerant source for developing of chilling/low temperature tolerant variety/genotypes in capsicum (Swarup 2006). Moreover, the okra species *A. Angulosus* and bottle gourd variety Pusa Santushti are tolerant to low temperature and frost (Selvakumar 2014).

2.3 Drought Stress

Drought and heat are generally accompanied each other, that promotes evapo-transpiration and influences kinetics of photosynthetic, consequently increases moisture stress and ultimately affect the crop productivity (Mir et al. 2012). Linsley et al. (1959) suggested that a period during which significant rainfall is not occurred over a long time is known as drought. As per the opinion of Quizenberry (1982) such precipitation represent drought in such condition when growth and development of crop plants is restrict due to water scarcity otherwise doesn't represent drought.

In crop plants the climate change affect the availability of moisture and the moisture availability will directly correlated with productivity and quality of vegetable crops. In association with sub-optimal temperatures, reduced precipitation could cause reduction in availability of irrigation water and increase in evapo-transpiration, leading to severe crop moisture-stress conditions (IPCC 2001).

In compare to many other crops, vegetables are more sensitive to drought and consists more than 90% water (AVRDC 1990). In India, an uneven and erratic distribution of rainfall occurs and two thirds of the geographic area receives low precipitation (<1000 mm). Many Indian states comes under drought prone area like Rajasthan, some parts of Gujarat, Haryana, Andhra Pradesh and Bihar drought stress is the major abiotic stress (Mitra 2001). In Bihar, flood-prone area is mainly in north Bihar whereas, drought prone area is in south Bihar therefore the major challenge for vegetable growers is to overcome these climatic hazards (Singh et al. 2014). Thus, moisture stress drastically reduces the yield and quality of vegetable crops. Moisture stress increases solute concentration in the plant system and promotes water out of plant cells by osmotic flow that's eventually reducing the water potential and disrupting membranes as well as cell processes during photosynthesis.

2.3.1 Mechanism of Drought Resistance

The adaptive mechanisms by which the ability of genotypes to survive in drought stress condition is known as drought resistance (Jones et al. 1980) that can be grouped into three categories (Fig. 3), *i.e.*, drought escape, drought avoidance and drought tolerance (Leonardis et al. 2012; Kumar et al. 2012). Moreover, in general plants adapt several mechanisms at a time to tolerate moisture stress (Gaff 1980).

1. *Drought escape:* The ability of a plant to complete its life cycle before development of severe moisture deficits *viz.* early flowering and maturity, duration of growth period variations that depends upon the degree of water scarcity (Kumar et al. 2012).
2. *Drought avoidance*: It may define as the ability of a plant to endure periods without significant precipitation even as maintaining a high plant status at high plant water potential, *i.e.*, drought avoidance or dehydration postponement (Krammer

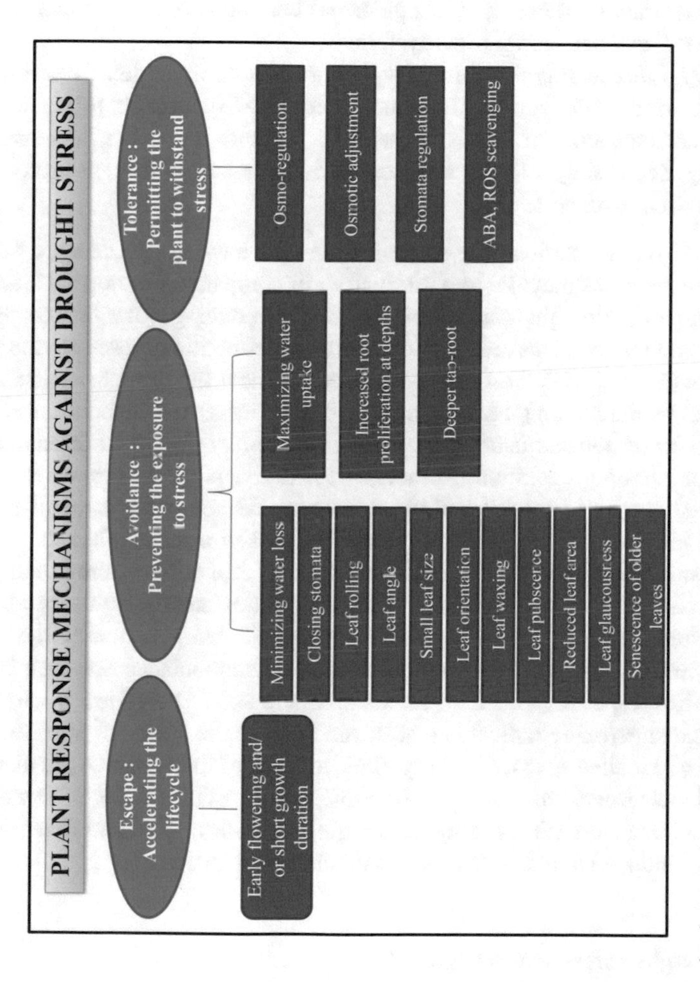

Fig. 3 Plant response mechanisms against drought stress. (Modified from Leonardis et al. 2012 and Kumar et al. 2012)

1980). The avoidance mechanisms of plants for drought is linked with whole physiological plant mechanisms viz., canopy tolerance and reduction of leaf area that decrease radiation transpiration and adsorption, stomatal closure and cuticular wax formation, and adjustments of sink-source relationships through varying root depth and spreading, development of root hair moreover root hydraulic conductance (Rivero et al. 2007; Kumar et al. 2012). The physiological traits for drought evasion are high relative moisture content, low membrane injury index, high transpiration efficiency, low epidermal conductance, early vigour, fibrous lateral roots and deeper tap roots system.

3. *Drought tolerance*: It is an ability of a plant to tolerate water deficits at low tissue water potential or dehydration tolerance (Levitt 1972). Drought tolerance mechanisms are balancing of turgor by osmotic pressure alteration, increasing cell elasticity, decreasing cell size and desiccation resistance through protoplasmic tolerance (Ugherughe 1986).

In general the adaptations for drought tolerance have no advantages for short duration genotypes as they are less productive in comparison to normal situations. The mechanism of drought resistance conferred by reducing water loss (by closing of stomata and reduced leaf area) which results in reduction of assimilation of carbon dioxide (Mitra 2001). Through osmotic adjustment the drought resistance can be improved by maintaining turgor pressure, but the increased concentration of solute is liable for osmotic adjustment may have damaging effect in addition to energy requisite for osmotic regulation (Turner 1979). Consequently, crop adaptations to moisture deficit may be established via. a balance between escape, avoidance and tolerance while maintaining optimum yield. The moisture stress tolerance traits are high harvest index, more water use efficiency, waxy leaf surface, more leaf pubescence, stomata partial closing and leaf osmotic adjustment (Kumar et al. 2012; Chatterjee and Solankey 2014). Drought stress significantly decreased the physiological process i.e. leaf water potential and stomatal conductance however, in terms of biochemical responses, considerable accumulation of carbonyl groups, hydrogen peroxide and superoxide radical has been recorded in the leaves of stressed plants representing oxidative stress. Similarly, the activities of total superoxide dismutase (SOD) and their isoenzymes viz., Cu/Zn-SOD, Cu/Zn-SOD and Fe-SOD were also observed to have increasing radical scavenging activities, ROS and defending the cells against induced oxidative stress (Chakhchar et al. 2016).

2.3.2 Drought Tolerance in Vegetables

Vegetables are sensitive to drought conditions but some of the vegetables like brinjal, cow pea, amaranth and cassava are able to tolerant drought to a certain extent. Generally vegetable crops required about 6 inches (radishes) to 24 inches (tomatoes and watermelons) of water per season (Kalia and Yadav 2014). By calculating the value of water use and effective precipitation values we can predict precise irrigation requirements of the crop. Vegetables are in general highly sensitive to moisture stress during three critical periods: flowering, fruit setting and fruit enlargement.

The solanaceous fruit vegetables like tomato, brinjal and chillies require frequent and shallow irrigation. However, the brinjal crop is little bit drought hardy but tomato and capsicum is highly sensitive to water stress. Moisture stress not only reduces the fruit yield but also highly affected the fruit external and internal qualities. Irregular and in adequate water supply also cause of fruit cracking, shrinking and blossom end rot in tomatoes and capsicum. Flower and fruit drop is common in these vegetables under limited water supply that can cause severe yield loss. The wild species of tomato, *viz.*, *S. cheesmanii*, *S. lycopersicum*, *S. chilense*, *S. lycopersicum* var. *cerasiforme*, *S. peruvianum*, *S. pimpinellifolium* and *S. pennellii* possess moderate to high level of genetic potential to drought stress tolerance (Solankey et al. 2015). The renowned tomato geneticist C.M. Rick (1973) reported that *Solanum chilense* has able to well adapt in desert areas and sometime also in the areas with no vegetation growth (Maldonado et al. 2003) because of its high leaf margin and deep root system (Sánchez Peña 1999). As well as in comparision to cultivated tomato, *Solanum chilense* has enlarged primary root aa well as more extensive and more fibrous secondary roots (O'Connell et al. 2007). *S. chilense* is found to have five-times more drought tolerance ability than cultivated tomato. Moreover, under drought conditions one of the wild species of tomato i.e., *S. pennellii* has thick and round waxy leaves which makes it capable to increase its water-use efficiency (O'Connell et al. 2007) and these leaves are able to produce acyl-sugars in its trichomes, as well as its leaves are able to take up dew unlike the cultivated tomato (Rick 1973).

Crops having shallow and fibrous root system like lettuce, beet leaf and spinach got benifits by giving frequent irrigation throughout the growing season and are generally sown/transplanted at or near field capacity. During the head formation to harvest stage, cabbage and lettuce are more sensitive to drought stress besides excess irrigation results in burst heads. At all stage of crop growth cauliflower and broccoli are sensitive to drought stress, and reduced growth, loos curds and early bolting occur due to drought. Moreover, the yield of root, tuber and bulb crop depends on the production and translocation of carbohydrate from the leaf to the root/tuber/bulb. Root elongation or bulb formation and enlargement stage are the most sensitive stage of these vegetables against moisture stress. Proper amount of irrigation from sowing to harvesting is required for root vegetables like radish and carrots and moisture stress results in small, woody, pithy, forked and bad flavoured roots. In carrot and radish uneven irrigation results in misshapen or split roots while secondary growth in potato and sweet potato and in onion and garlic cause early bulbing (Kalia and Yadav 2014).

Several vegetable crops like in cucurbitaceous crops cucumber, melons, pumpkin, squashes and in leguminous, crop like lima beans, snap beans, lablab bean, garden pea are more sensitive to drought stress mostly at flowering, fruiting and pod development stage. Under limited water condition pollination, fruit setting and fruit development are seriously hampered. The rate of plant survival and vigour during high rate of evapo-transpiration and marginal soil moisture can be increased by giving irrigation especially after seeding and transplanting. To soften the field crusting of okra and leguminous vegetables always used less amount of irrigation water for easy emergence of seedlings.

2.3.3 Critical Stages of Drought Tolerance in Vegetables

Due to succulent nature of vegetables, moisture stress especially at crucial growth time reduces the fruit yield and quality (Table 5). Water stress during early growth phase directly declines the number of fruits, flower drop and low fruit setting, reduced fruit size and poor fruit quality, as well as reduces seed number, viability and vigour in seed crops. Plants exhibit various defence mechanisms against moisture stress at the physiological, biochemical, molecular, cellular and whole plant levels. Singh and Sarkar (1991) stated that association of different physic-chemical characters of direct relevance, beside a single trait, should be taken as selection criteria for moisture stress.

For early detection of moisture stress genotypes the seedlings are treated with polyethelene gycol (PEG) similarly Aazami et al. (2010) drawn an experiment with culture medium supplemented with PEG led to highest proline accumulation in tomato cv. Roma. IIHR, Benaluru has identified a drought tolerant tomato line (IIHR-2274) on the basis of number of fruits under different moisture stress regime (Chavan 2007).

2.3.4 Screening Procedure for Drought Resistance

Genetic diversity present in the plant species may exploit as prime source for screening against moisture stress. Genetic diversity in terms of morphological, physiological and biochemical characters are main significant and thus, used as a selection criteria to find out the appropriate plant ideotype (Kumar et al. 2012). For development of effective screening procedure, many researchers have drawn a variety of procedures for screening of drought (Table 6). By adopting selection for suitability of performance over diverse environmental conditions *via* extensive morphological screening, evaluation and biometrical approaches the plant breeders can get the solution for environmental stress (Blum 1988a).

2.3.5 Genetic Mechanism for Drought Resistance

The combined effect of a variety of morphological characters *viz.*, early growth, reduced leaf size, curling and wax coating of leaves, deep root system, more number of branches, stability in yield and physiological traits *viz.*, stomatal activity, transpiration rate, water-use efficiency, osmotic adjustment and biochemical traits *viz.*, accumulation of polyamine, proline, trehalose, etc. as well as increasing the

Table 5 Critical stages of moisture stress and its impact on important vegetable crops

Vegetable crops	Water requirement (mm)	Critical period for watering	Impact of water stress
Tomato	336	Early flowering, fruit set, enlargement and colour development	Shedding of flower, poor fertilization, reduced fruit size, fruit weight, fruit cracking, puffiness and development of calcium deficient disorder *i.e.* blossom end rot and poor seed viability.
Brinjal	486	Flowering, fruit set and fruit and colour development	Reduces fruit size with poor colour development in fruits and poor seed viability.
Chillies	500–640	Flowering and fruit set	Shedding of flowers and young fruits, decline in dry matter production and nutrient uptake, poor seed viability
Potato	–	Tuberization and tuber enlargement	Poor tuber growth and yield, splitting, internal brown spot
Okra	345	Flowering and pod development	Considerable yield loss, development of fibres, poor seed viability
Cole crops	541	Head/ curd formation and enlargement	Tip burning and splitting of head in cabbage; browning and buttoning in cauliflower and broccoli
Root crops	441	Root enlargement	Distorted, rough and poor growth of roots, strong and pungent odour in carrot, accumulation of harmful nitrates in roots, early bolting in beet root.
Onion and garlic	500	Bulb formation and enlargement	Splitting and doubling of bulb, low dry matter content, poor storage life
Cucumber	310–436	Flowering as well as throughout fruit development	Deformed and non-viable pollens, bitterness and deformity of fruits, poor seed viability
Melons	469–506	Flowering and evenly throughout fruit development	Poor fruit quality, low TSS, low sugar and ascorbic acid content, increase nitrate content in fruits and poor seed viability.
Summer squash	–	Bud development and flowering	Deformed and non-viable pollen grains, flower shedding misshapen fruits
Leafy vegetables	–	Throughout growth and development of plant	Fibrous leaves, poor vegetative growth, accumulation of nitrates
Asparagus	–	Spear production and fern (foliage) development	Reduce spear quality through declined spear size and high fibre content, leading to tougher and poor quality spears.
Lettuce	–	Consistently throughout development	Toughness of leaves, poor plant growth, tip burning
Garden pea	240	Flowering and pod filling	Poor root nodulation, vegetative growth, pod filling and seed viability

(continued)

Table 5 (continued)

Vegetable crops	Water requirement (mm)	Critical period for watering	Impact of water stress
Lima bean	–	Pollination and pod development	Leaf colour takes on a slight grayish cast, blossom drop, poor seed viability
Snap bean	–	Flowering and pod enlargement	Blossoms drop with inadequate moisture levels and pods fail to fill, poor seed viability
Sweet corn	–	Silking, tasseling and ear development	Poor pollination, missing rows of kernels, low yields, or even eliminate ear production, poor seed viability
Sweet potato	–	Root enlargement	Reduced root enlargement, cracking and low yield.
Cassava	–	Early growth phase and root enlargement	Reduction of leaf area and poor yield

Source: Modified from Bahadur et al. (2011), Kumar et al. (2012) and Selvakumar (2014)

Table 6 Screening procedure for drought resistance

S. No.	Instruments/techniques used	Purpose of screening	References
1	Infrared thermometry	Efficient water uptake	Blum et al. (1982)
2	Banding herbicide Metribuzin at a definite depth of soil, and use of Iodine-131 and soil-less culture under stress of 15 bar	Root growth	Robertson et al. (1985) and Ugherughe (1986)
3	Adaptation of psychometric procedure	Evaluation of osmotic	Morgan (1980, 1983)
4	Diffusion porometry technique	Leaf water conductance	Gay (1986)
5	Mini-rhizotron technique	Root penetration, distribution and density in the field	Bohm (1974)
6	Infrared aerial photography	Dehydration postponement	Blum et al. (1978)
7	Carbon isotope discrimination	Increased water-use efficiency	Farquhar and Richards (1984)
8	Drought index measurement	Number of fruits, fruit weight and fruit yield	Clarke et al. (1984) and Ndunguru et al. (1995)
9	Visual scoring or measurement	Maturity, antheredial cone splitting, leaf molding, leaf length, angle, orientation, root morphology, and other morphological traits	Mitra (2001), Ram (2005), and Kumar et al. (2012)

Source: Adopted from Kumar et al. (2012) and Chatterjee and Solankey (2014)

intensity of nitrate reductase activity and carbohydrates are responsible for expression of drought tolerance in any crop species (Kumar et al. 2012). The morphological and physiological traits of some crops are due to inheritance pattern and gene action.

Likewise, characters related to root are polygenically inherited (Ekanayake et al. 1985) however, long and more numbers of roots were governed by dominant alleles whereas, recessive allels governed thick root tip (Gaff 1980). Moreover, osmotic adjustment (O'Toole and Moya 1978) and leaf molding (Turner 1979) have revealed monogenic inheritance pattern. Heritability and genetic advance in leguminous vegetables and okra for number of pods per plant under drought condition showed good narrow sense, yet leaf water potential have higher heritability (Ben-Ahmad et al. 2006; Naveed et al. 2009). Plants tend to maintain water content during moisture stress condition by accumulating various compatible solutes (fructan, trehalose, polyols, glycine betaine, proline and polyamines) i.e., safe and do not interfere with plant processes (Mitra 2001). The genes which are liable for a variety of enzymes biosynthesis have been identified and cloned from different organism viz., bacteria, yeast, plants and animals as well as isolated and characterized by different workers in a variety of crop species (Table 7).

2.3.6 Drought Stress Resistant Sources of Vegetable Crops

Pandey et al. (2016) conducted an experiment on muskmelon genotypes to evaluate the genotypes which are moisture stress tolerance and also for standardization of screening technique and they reported that few drought tolerant genotypes give less yield. However, these genotypes were not found economical for commercial cultivation even they are considered as drought tolerant genotype. Positive significant correlation was observed between drought tolerant efficiency and relative water content in moisture deficit condition, *Fv*/*Fm*, chlorophyll content index and root length. Although, several workers have identified many drought resistance/tolerance species of different vegetable crops (Table 8). Likewise in tomato, variety Red Rock and a few lines of *S. pimpinellifolium* (Stoner 1972; Rana and Kalloo 1989) have physiological basis of moisture stress tolerance abilityat vegetative growth and later stages as well as wild accessions of *Solanum chilense* (deep vigorous root system) and *S. pennellii* (Rai et al. 2011). However, the moisture stress tolerant lines, LA716 (*S. pennellii*) has a shallow as well as fibrous root system which have more use efficiency (WUE) under drought stress and ability to conserve moisture in succulent leaves than *S. esculentum* (Martin and Thorstenson 1988). The elevated WUE in this genotype is because of reduced leaf conductance i.e. less and tiny stomata, enlarged trichomes, lower chlorophyll content and Rubisco enzyme activity per unit of leaf area, and big mesophyll cell surface which are exposed to intercellular air space (Martin et al. 1999).

A potential source of moisture stress-tolerant characters in *Phaseolus vulgaris* was done by Lazcano-Ferrat and Louatt (1999) through interspecific hybridization with *P. acutifolius* (morpho-physiological characteristics for drought tolerance). In

Table 7 Genes conferring drought tolerance and their salient features

Genes	Function	Mechanism of action	References
DREBs, Osmotin, ZAT12 and *BADH2*.	Stress induced transcription factors	Signalling, control of transcription, proteins and membrane protection, compatible solute (betaines, sugars, polyols, and amino acids) synthesis, and free-radical and toxic-compound scavenging activity.	Krishna et al. (2019)
DREBs/CBFs; ABF3	Stress induced transcription factors	Enhanced expression of downstream stress related genes confers stress (drought, cold, salt) tolerance. Efeectively over expression can lead to stunting plant growth.	Oh et al. (2005) and Ito et al. (2006)
SNAC1	Stress induced transcription factor	SNAC1 expression reduces water loss increasing stomatal sensitivity to ABA	Hu et al. (2006b)
OsCDPK7	Stress induced Ca-dependent protein kinase	Enhanced expression of stress responsive genes	Saijo et al. (2000)
Farnesyl-transferase *(ERA1)*	Negative-regulator of ABA sensing	Down-regulation of farnesyl transferase increase the plant's response to absicic acid and moisture stress tolerance reducing stomatal conductance	Wang et al. (2005)
Mn-SOD	Mn-superoxide dismutase	Overexpression improves stress tolerance also in field conditions	McKersie et al. (1996)
AVP1	Vacuolar H ± pyrophosphatase	Overexpression facilitate auxin fluxes leading to increased root growth	Gaxiola et al. (2001) and Park et al. (2005)
HVA1; OsLEA3	Stress induced LEA proteins	Over-accumulation of LEA increases drought tolerance also in field conditions	Bahieldin et al. (2005) and Xiao et al. (2007)
ERECTA	A putative leucine-rich repeat receptor-like kinase is a major contributor to a locus for D on *Arabidopsis* chromosome 2	ERECTA acts as a regulator of transpiration efficiency with effects on stomatal quantity, epidermal cell expansion, mesophyll cell proliferation.	Masle et al. (2005)

(continued)

Table 7 (continued)

Genes	Function	Mechanism of action	References
otsA and *otsB*	Escherichia coli trehalose biosynthetic genes	Enhance trehalose accumulation associated with higher soluble CHO levels, elevated photosynthetic capacity and increased tolerance to photo-oxidative damage	Garg et al. (2002)
P5CS	d-Pyrroline-5-carboxylate synthetase	Enhanced accumulation of proline leads to increased osmotolerance	Kavi Kishor et al. (1995) and Zhu et al. (1998)
mtlD	Mannitol-1-phosphate dehydrogenase	Mannitol accumulation promotes increased osmotolerance	Abebe et al. (2003)
GF14l	14-3-3 protein	Lines over expressing GF141 have a 'stay Green' phenotype, increase water stress tolerance and higher photosynthesis under moisture deficit conditions.	Yan et al. (2004)
NADP-Me	NADP-malic enzyme	The over expression decreased stomatal conductance and improves WUE	Laporte et al. (2002)
AREB	bZIP transcription factor in tomato	Over expression increasing dehydrin expression.	Hsieh et al. (2010)
Cupida	Leaf necrosis in tomato	Over wilting or stomatal defect	Anonymous (2006)
Dehydrin	Increased distances from dehydrins activated by abscisic acid in tomato	Possibly the dehydrins protect membranes during stresses. Dehydrins are upregulated by abscisic acid.	Weiss and Egea-Cortines (2010)
Chloroplast drought-induced stress protein	thiol-disulfide exchange intermediate activity in potato	Preservation of the thiol-disulfide redox potential of chloroplastic proteins during moisture deficit.	Anonymous (2006)
CDSP 32	Thiol-disulfide exchange intermediate activity in potato	Preserve chloroplastic structures against oxidative injury upon moisture stress.	Anonymous (2006)
CDSP 34	Increases in CDSP 34 transcript and protein abundances were also observed in potato plants subjected to high illumination.	The CDSP 34 protein is proposed to play a structural role in stabilizing stromal lamellae thylakoids upon osmotic or oxidative stress.	Beyly et al. (1998)

(continued)

Table 7 (continued)

Genes	Function	Mechanism of action	References
Wilty	Dominant TGRC gene in tomato	Leaves over wilt when drought stressed. Wilting under field or greenhouse conditions; marginal leaf narcrosis.	Anonymous (2006)
Wilty dwarf	Recessive TGRC gene in tomato	Grayish-green, droopy leaves; stunted plants; leaves droop when drought stressed.	Anonymous (2006)
Water stress-induced *ER5* protein	Stress induced CaLEA6 (for *Capsicum annuum* LEA) is 709 bp long with an open reading frame encoding 164 amino acids	Predicted to produce a highly hydrophobic, but cytoplasmic, protein.	Kim et al. (2005)
Abscicic acid stress ripening 2	Putative DNA binding and chaperon like activity	A part of the *Asr* gene family that basically induced by abiotic stress like, water and is expressed in the leaf phloem companion cells.	Giombini et al. (2009)
ZAT12	Stress-induced transcription factors	Increased expression of downstream stress-related genes confers moisture stress.	Rai et al. (2013)
SlNAC4	Stress-responsive transcription factor	Modulation of ABA-independent signaling system.	Zhu et al. (2014)

Source: Modified from Cattivelli et al. (2002), Kumar et al. (2012), Krishna et al. (2019)

leguminous vegetables/nodulated plants that higher amount of accumulation of trehalose directly improved drought tolerance (Farlas-Rodriguez et al. 1998). Potato wild species *S. demissum* and *S. acaule* under in vitro and glasshouse pot trial showed drought tolerance under (Arvin and Donnelly 2008).

2.3.7 Physiogenetic Approach for Drought Tolerance

Drought stress extent depends upon appearance of symptoms and its effects on physiology of plant metabolisms and yield of fruit. The plant water content and tissues water potential are two physiologically pertinent symptoms of moisture stress (Jones 2007). Under drought stress condition various morphological symptoms are clearly seen *viz*., rolling of leaves (Kadioglu et al. 2012), molding of leaves, yellowing (chlorosis) and browning of leaves and wilting of leaves and plant. Resistance to drought and yield are different traits and different genes and gene system governed it (Turner 1986). Besides it is also reported that both yield and drought are independently handled and the degree of independence for each resistance mechanism must

Table 8 Moisture stress tolerant species and genotypes of different vegetable crops

Sl. No.	Vegetable crops	Genotypes/varieties	References
1	Tomato	*Solanum habrochaites* (EC-520061), *S. pimpinellifoloium* (PI-205009, EC- 65992, PanAmerican), *S. pennelli* (IIHR 14-1, IIHR 146-2, IIHR 383, IIHR 553, IIHR 555, K-14, EC-130042, EC-104395, Sel-28), *S. cerasiforme, S. cheesmanii, S. chilense, S. sitiens.*	Rai et al. (2011)
		Arka Vikas, Arka Meghali, RF- 4A	Singh (2010)
		L. pennellii (LA0716), *L. chilense* (LA1958, LA1959, LA1972), *S. sitiens* (LA1974, LA2876, LA2877, LA2878, LA2885), *S. pimpinellifolium* (LA1579)	Razdan Maharaj and Mattoo (2007), Symonds et al. (2010), and Rai and Rai (2006)
		S. pennelli, IIHR-14-1, IIHR-146-2, IIHR-383, IIHR-553, IIHR-555, EC-130042, IC-35992, Sel-28	
		S. pimpinellifoloium, S. chilense, S. pennellii, Arka Vikas	Swarup (2006) and Selvakumar (2014)
		Paiyur-1	
2	Brinjal	*S. microcarpon, S. macrosperma, S. gilo S. integrifolium,* Bundelkhand Deshi	Rai et al. (2011)
		S. sodomaeum (syn. S. linneanum)	Toppino et al. (2009)
		SM- 1, SM- 19, SM- 30, Violette Round, Supreme, PKM-1, Kashi Sandesh, Kashi Taru	Kumar and Singh (2006) and Selvakumar (2014)
3	Chilli	*C. chinense, C. Eximium, C. baccatum* var. *pendulum*	
		Arka Lohit, IIHR – Sel.-132	Singh (2010)
		Samrudhi, Kashi Anmol	
4	Potato	*S. acaule, S. demissum* and *S. stenotonum,* Alpha, Bintje	Arvin and Donnelly (2008)
		S.curtilobum, S.xjuzepczukii, S.ajanhuiri,	Ross (1986)
		Kufri Sheetman	
		Solanum chacoense, Kufri Sindhuri	
5	Okra	*A. caillei, A. rugosus, A. tuberosus*	Charrler (1984)
6	Onion	*Allium fistulosum, A. munzii,* Arka Kalyan, MST 42, MST 46	Singh (2010)
		Agrifound Dark Red, Arka Kalyan, Raseedpura Local	
7	Cowpea	IT-38956-1, TVX-944-02E, RC-19, RC-101	Hazra and Som (1999) and Rai and Yadav (2005)
8	French bean	*P. acutifolius*	Kavar et al. (2011)
9	Clusterbean	RGC-936, RGC-1003, RGC-1017, RGC-1066, Thar Bhadavi	Hazra and Som (1999) and Rai and Yadav (2005)

(continued)

Table 8 (continued)

Sl. No.	Vegetable crops	Genotypes/varieties	References
10	Water melon	*Citrullus colocynthis (L.) Schrad.*	Dane and Liu (2007)
11	Cucumber	INGR-98018 (AHC-13)	Rai et al. (2008)
12	Bottle gourd	Thar Samridhi, Kashi Ganga	Rai and Yadav (2005)
13	Winter Squash	*Cucurbita maxima*	Chigumira and Grubben (2004)
14	Ash gourd	Kashi Dhawal	Rai and Yadav (2005)
15	*Cucumis* Spp.	*Cucumis pubescens*, INGR-98013 (AHK-119), *Cucumis melo var. momordica* (VRSM- 58), *Cucumis melo* var. *callosus* INGR-98015 (AHS-10), CU 159, CU 196, INGR-98016 (AHS-82), AHK- 200, SKY/DR/RS-101, *Cucumis melo*, SC- 15, *Cucumis melo var. chat*, Arya	Rai et al. (2008), Kusvuran (2012), and Pandey et al. (2011)
16	Carrot	Ooty-1	Rai and Yadav (2005)
17	Cassava	CE-54, CE-534, CI-129, CI-260, CI-308, CI-848, TP White, Narukku-3, CI-4, CI-60, CI-17, CI-80	Singh (2010) and Selvakumar (2014)
		Shri Sahya	
		Co-3, Co-4	
18	Sweet potato	Sree Bhadra, VLS6, IGSP 10, IGSP 14	Singh (2010) and Selvakumar (2014)
		Sree Nandhini	
19	Colocasia	White Goriya, Haloo Kesoo	Hazra and Som (1999) and Rai and Yadav (2005)
20	Elephant Foot Yam	NDA-5, Gajendra	Hazra and Som (1999)

Source: Modified from Kumar et al. (2012), Maheswari et al. (2015) and Solankey et al. (2015)

be evaluated (Blum et al. 1983). For exploiting the impact of traits specific breeding, detailed informations of the climatic requirement of the crop, G × E interactions, adaptation in drought environments by osmotic regulation, accumulation of carbohydrates in stem, improved photosynthesis, heat as well as drought tolerant enzymes synthesis, etc. are essential physiological parameters (Mir et al. 2012). In brinjal, moisture stress considerably increases membrane permeability but reduces the chlorophyll concentrations of leaf, plant growth, fruit yield (Kirnak et al. 2001).

In tomato crops, under water stress polyphenols play a major function, thus more synthesis of phenolics compound and presence of flavonoids would play a major role under water stress damage (Sánchez-Rodríguez et al. 2011). This characteristic confers drought adaptation, and there are few evidences where a minor gene influenced the appearance of osmotic adjustment (Basnayake et al. 1995). Different stages of crop growth show different symptoms to drought stress. Physiological traits are especially more significant for response to water deficits (Table 9).

Table 9 Response of physiological traits to drought conditions in vegetables

Plant traits	Effects relevant for yield	Modulation under stress	References
Stomatal conductance/ leaf temperature	More/less frequent water consumption. Leaf temperature represents the evaporation and consequently is a function of stomatal conductance	Stomatal tolerance increases under stress	Jones (1999) and Lawlor and Cornic (2002)
Photosynthetic capacity	Modulation of concentration of Calvin cycle enzymes and elements of the light reactions	Reduction under stress	Lawlor and Cornic (2002)
Timing of phenological phases	Early/late flowering, maturity and growth duration, dichogamy and anthesis, reduced grain number.	Wheat and barley advanced flowering, rice delayed, maize asynchrony	Slafer et al. (2005) and Richards (2006)
Anthesis-silking interval (ASI) in maize	ASI is negatively correlated with yield under moisture stress condition.	Drought stress at flowering causes a delay in silk emergence relative to anthesis	Bolanos and Edmeades (1993) and Edmeades et al. (2000)
Starch availability during ovary/embryo development	Reduced starch availability ultimately causes abortion and reduced in grain size and number.	Inhibition of photosynthetic activity reduces starch availability	Boyer and Westgate (2004)
Partitioning and stem reserve utilization	Up and down remobilization of reserves from stems for grain-filling, reducing kernel weight.	Compensation of reduced current leaf photosynthesis by increased remobilization	Blum (1988b) and Slafer et al. (2005)
Stay green	Delayed leaf senescence.	–	Rajcan and Tollenaar (1999)
Single plant leaf area	Reduced plant size and decreased productivity	Reduced under stress (wilting, senescence, abscission)	Walter and Shurr (2005)
Rooting depth	Uneven tapping of soil water resources	Reduced total mass but increased root/shoot ratio, growth into wet soil layers, re-growth on stress release	Hoad et al. (2001) and Sharp et al. (2004)
Cuticular tolerance and surface roughness	Irregular water loss, modification of boundary layer and reflectance	–	Kerstiens (1996)

(continued)

Table 9 (continued)

Plant traits	Effects relevant for yield	Modulation under stress	References
Photosynthetic pathway	C_3/C_4/CAM, higher WUE and higher sub-optimal temperature tolerance of C_4/ CAM plants.	–	Cushman (2001)
Osmotic adjustment	Accumulation of solutes like, ions, monosaccharide and disaccharide, amino acids, glycinebetaine, etc.	Slow response to water potential	Serraj and Sinclair (1996)
Membrane composition	Improved membrane stability and changes in aquaporine function.	Regulation in response to water potential changes	Tyerman et al. (2002)
Antioxidative defence	Defence mechanism against active oxygen species.	Acclimation of defence systems	Reddy et al. (2004)
Accumulation of stress-related proteins	Helpful in the defence of cellular structure and protein metabolic activities.	Accumulated under stress	Ramanjulu and Bartels (2002) and Cattivelli et al. (2002)

Source: Adopted from Cattivelli et al. (2002) and Kumar et al. (2012)

2.4 Flood Tolerance

In Indian condition, the vegetable growers are still awaiting of monsoon for sowing/transplanting of many vegetables, but production take place in every seasons. During rainy season, heavy rains cause high moisture which restricted the growth and development of vegetables. Among all vegetables colocasia can tolerate flood at high extant while tomato and chillies are highly sensitive. Most of the vegetables are in general flood sensitive and genetic variation for this trait is limited, especially in cultivated tomato and early group of cauliflowers. Excess water logging prohibits aerobic processes, due to improper/non availability of oxygen in the root zone which cause wilting (Kalia and Yadav 2014). Accumulation of endogenous ethylene in flooded tomato plants was reported that causes injure to the plants tissues and ultimately wilting of plants (Drew 1979). The characteristic feature of water logged tomato plants is rapid development of epinastic growth of leaves (Kawase 1981). In tomato, the wilting of flooded plants will be earlier if temperatures will be rises (Kuo et al. 1982). Many genotypes of brinjal are highly flood tolerant (Midmore et al. 1997), accordingly, can be used as a rootstock to improve the tolerance against excess soil moisture/flood and identified with fine grafting compatibility with tomato. Beside these, protections against excess soil moisture, some brinjal accession are drought tolerant so that such brinjal rootstocks offer tolerance against limited soil moisture stress. Ezin et al. (2012) observed that flood tolerant species are able to synthesize macromolecules such as ADH and sucrose, and capable of protection against post-flooding injury; in the experiment the tomato genotypes *viz.*, CA4 and CLN2498E were tolerant to flooding.

Table 10 Optimum soil pH levels for different vegetables

Vegetable crops	Optimum soil pH
Asparagus, garlic	6.0–8.0
Plea beans, beets, cabbage, spinach, Brussels sprouts, kale, muskmelon, peas, spinach, Palak, summer squash	6.0–7.5
Celery, chive, endive, lettuce, cauliflower, horseradish, radish, onion	6.0–7.0
Sweet corn, pumpkins, tomatoes	5.5–7.5
Snap bean, Lima bean, peppers, parsnip, carrot, cucumber	5.5–7.0
Eggplant, watermelon	5.5–6.5
Potato	5.0–6.5
Soil pH: Acidic <7.0–5.0; neutral 7.0; alkaline >7.0–8.0 or more	

Source: Modified from Swarup (2006)

2.5 *Salinity Tolerance*

Soil reaction, or pH, is a measure of the hydrogen ion concentration as an indication of the soil's degree of acidity or alkalinity. At pH scale value above 7 considered alkaline whereas at 7 is neutral and above 7 are acidic. Most vegetable crops tolerate a soil pH within the range of 5–8 moreover, vegetables in general prefer slightly acidic soil (pH 6.0–6.8) for getting optimum yield along with good quality of produce (Table 10). Soil pH regulates nutrient availability.

Soil salinity refers to the presence of excessive amounts of soluble salts (Na^+, Mg^{2+}, Ca^{2+}, Cl^-, SO_4^{2-}) in the soil. High soil salinity inhibits water extraction by plants. Plants differ in their tolerance of soil salinity. Soil salinity is more of a problem under hot and dry weather conditions. Due to high saline soils worldwide around 20 and 30% of cultivated and irrigated lands are depresses, respectively (Szabolcs 1992; Foolad 2005). As well as, the salinized areas are annually expending at a rate of 10%; and the major contributors are low rainfall, high surface evaporation, native rocks weathering, use of saline water in irrigation, and non-scientific agronomical practices (Kalia and Yadav 2014). Out of these, major constraints for normal plant growth and development are physiological causes like toxicity of ion, water deficit, improper nutrition balance, high salinity in the root zone cause both quality and yield reduction (Ghassemi et al. 1995). Plants are more susceptible to salinity stress at early stage of seedlings and flower initiation than fruiting plants (Lutts et al. 1995). Vegetable crops like cucumber, tomato, brinjal are slightly sensitive to salinity in comparision to onion and carrot (Table 11). Response of plants to salinity may vary among species as it depends on their genetic tolerance mechanism.

Lunin et al. (1963) proposed a very basic rule for studying crop response to salinity: (1) tolerance of a plant against salinity will depend on the stages of growth of plant at which salinization occurred and the final level of salinity observed; (2) Values of salinity tolerance also take concern the economic portion of plant. In beetroot, due to salinity more reduction is reported in roots than in the tops while in case of onion reduction for bulbs were lesser than their tops. Moreover, genes for

Table 11 Soluble salt test values and relative salt tolerance of fruit and vegetable crops

Non-tolerant (0–2 mmhos/cm)	Slightly tolerant (3–4 mmhos/cm)	Moderately tolerant (5–7 mmhos/cm)	Tolerant (8–16 mmhos/cm)
Carrot	Cabbage	Beets	Asparagus
French bean	Lettuce	Cucumber	Swiss chard
Onion	Peppers	Muskmelon	
Radish	Potato	Squash	
	Sweet corn	Okra	
		Tomato	
		Egg plant	
		Spinach	

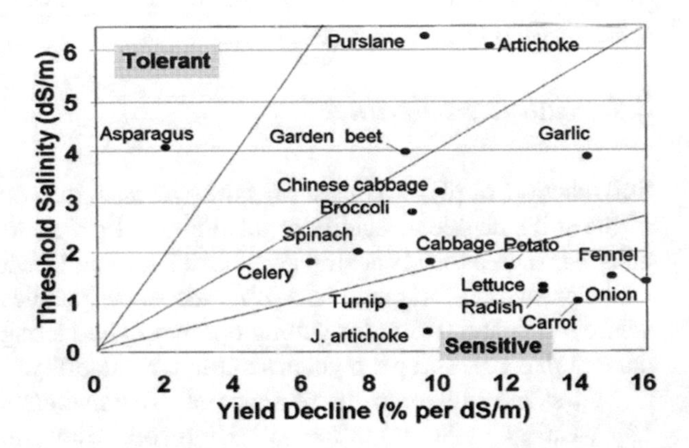

Fig. 4 Vegetable species tolerance to salt, according to salinity threshold and percent yield decline. (Shannon and Grieve 1999)

salinity tolerance and other genes that reflect both quantitative traits in addition to environmental interactions are functions together (Shannon and Grieve 1999). Therefore, it is not identified that tolerance to salt is governed by more genes and is a complex trait which is responsible for quantitative and genetic character (Shannon 1996). Statistically, the value at which yield is decreased by 50% (C50) is most dependable value for crop salinity tolerance response of plant (Shannon and Grieve 1999). In the development of salinity tolerance, traditional breeding programme have a very limited success because of genetic and physiologic complexity of this trait (Flowers and Yeo 1986). Moreover, high genetic diversity, effective methods for screening and transfer of resistance genes to the species of interest is still required.

The effective screening for salt tolerance is required to be performed in hydroponic system under this, salt concentration of known nutrient solution can be adjusted (Cuartero and Fernandez-Munoz 1999). In general, bulb crops (*Allium* spp.) like, onion, garlic, leek and chive are more prone to salinity stress as their yield decline drastically (Fig. 4), but significant information observed for only major bulb crops *viz*., onion and garlic (Shannon and Grieve 1999). However, few vegetables can able to tolerant some extent of salt such as, celery, beet leaf, spinach, etc. The okra variety Pusa Sawani and chilli genotypes *viz*., Demre, Ilica 250, 11-B-14, Yalova Carliston, Bagci Carliston, Mini Aci Sivri and Yaglik 28 have potential to improved germination under salt stress condition thus, can be exploited as resistant genes source for development of varieties/advanced breeding lines (Yildirim and Guvenc 2006, Selvakumar 2014). In tomato the accessions of *S. pimpinellifolium i.e.,* LA1579 and LA1606, and *S. lycopersicum* var. *cerasiforme* aacesion, LA4133 have been identified as salt tolerance from AVRDC, Taiwan. Other than these, salt tolerance potential have been found in several wild species of tomato *viz., S. cheesmani, S. peruvianum, S pennellii,* and *S. habrochaites* (Cuartero et al. 2006, Selvakumar 2014).

3 Vegetables Grafting for Management of Environmental Stress

In twentieth century, vegetables grafting was originated in East Asia for management of soil borne pathogens (diseases) of tomato, eggplant, and cucurbitaceous vegetables. By availability of suitable rootstocks, grafting provides tolerance against several stresses like drought, flood, salinity and chilling and also to low soil temperatures. In vegetables, grafting was started during 1950s in eggplant after that in cucumber in 1960s and in 1970s in tomato (Edelstein 2004). Grafting of melons onto rootstock of squash provide more tolerance against salt than the non-grafted melons (Romero et al. 1997). Rootstocks of *Lagenaria siceraria* is not much tolerant to salt as compare to rootstocks of *Cucurbita* spp. (Matsubara 1989).

Moreover, *Cucurbita ficifoila* is low temperature tolerant rootstock for cucumber (Selvakumar 2014). Eggplants rootstock *Solanum integrifolium* have ability to tolerate 18–21 °C temperature while, in case of tomato rootstocks of *Solanum lycopersicum* x *S. habrochaites* have ability to tolerant low (10–13 °C) soil temperature (Kalia and Yadav 2014).

Grafted of tomato cultivar Big Red onto cv. Heman (*S. lycopersicum* × *S. habrochaites*) and Primavera (*S. lycopersicum*) yielded more number of fruits than the control under protected structure, though, under low temperature situation in the field the positive effect nearly moved out (Khah et al. 2006). In cucumber, figleaf gourd (*Cucurbita ficifolia*) and bur cucumber (*Sicos angulatus*) can be used as compatible rootstocks. Figleaf gourd is also unique among cucurbits with an adjustable root temperature at around 15 °C, i.e., 6 °C lower than cucumber roots (Rivero et al. 2003b). Grafting of cucumber scion onto a squash rootstock (*Cucurbita moschata* Duch) could able to tolerate suboptimal temperatures as compared to self grafted cucumber, when the cut rootstock of squash was exposed to a bottom heat treatment, which implicated soaking the cut end of the scion cutting in hot nutrient solution of 30 °C for one day (Shibuya et al. 2007). However, in tomato rootstocks of the high-altitude *i.e.*, LA 1777 (*S. habrochaites*), 'KNVF' (*S. lycopersicum* × *S. Habrochaites*), and Bloom et al. (2004) reported chilling tolerant lines, LA 1778 (*S. Habrochaites*) × *S. lycopersicum* cv. T5 were able to alleviate cold stress for various scions. In watermelon scion grafted onto Shin-tosa-type (*Cucurbita maxima*

× *C. moschata*) rootstocks i.e. exploited to extend the planting duration in cool weather conditions (Davis et al. 2008).

In tomato, rootstock (*L. esculentum* cv. RX-335) has been identified as tolerant to heat, showed decreased hydrogen peroxide (catalyzed by antioxidant enzymes) concentration which indicates lower oxidative stress (Rivero et al. 2003a). Schwarz et al. (2010) tested several eggplant rootstocks and found at 28 °C vegetative growth enhanced, they had no advantage rather they decreased total fruit dry weight. Moreover, tomato heat-tolerant rootstock (cv. Summer Set) also failed to improve the yield. Though, brinjal (*S. melongena* cv. Yuanqie) jointed onto a heat temperature tolerant rootstock (cv. Nianmaoquie) seemed to be promising and resulted in a prolonged growth stage and yield increase up to 10% (Wang et al. 2007). AVRDC, Taiwan, have recommended high stress tolerant rootstocks with high yields for capsicum (*C. annuum* cv. Toom-1 and 9852-54) (Palada and Wu 2008). Under supra-optimal temperature stress, eggplants rootstock reduced electrolyte leakage of tomato plants thus reduce membrane damage and improve ability to retain solutes and water (Abdelmageed and Gruda 2009). Moreover, increased activity of other antioxidative enzymes from the ascorbate/gluthation cycle but lower the activity of SOD. In a case of tomato, eggplant used as a rootstock of tomato scion exhibited high concentration of ascorbate but low proline level as compared to self-grafted tomato. Grafting improves more than 60% of marketable yield as compared to non-grafted melons when mini watermelons grafted onto a commercial rootstock of hybrids between *Cucurbita maxima* (PS 1313) Duchesne × *C. moschata* Duchesne grown under limited irrigation conditions (Rouphael et al. 2008).

Bitter gourd (*Momordica charanthia* L. *cv*. New Known You # 3) grafted onto rootstock of sponge gourd (*Luffa cylindrica* cv. Cylinder#2) improved tolerance to flood (Liao and Lin 1996). In flooding condition, grafted plants of watermelons formed parenchyma and adventitious root but not by non-grafted plants. For increasing tolerance against several biotic stresses and excessive moisture growing of tomato plants on brinjal rootstock 'EG195' or 'EG203' and capsicum on chilli rootstock 'PP0237-7502', PP0242-62′ and 'Lee B' (AVRDC 2003, 2009) advocate by the World Vegetable Centre, Taiwan (Schwarz et al. 2010).

4 Use of Biotechnological Tools in Environmental Stress Management

According to the concept of Darwin's "survival of the fittest" the lines which develop several changes for adaption *i.e.*, phenotypically and genetically are survived. For adaption of living organisms under this situation of climate change at molecular as well as biochemicals levels developed various adaptive changes to adjust in the new environment (Kumar and Goswami 2012). Several essential processes *i.e.*, physic-chemical traits for plant growth and development is influenced by abiotic stresses and during stress plants develop different defence mechanism at molecular, cellular and also at whole plant levels.

In vegetable crops, understanding of genetic basis of environmental stresses (heat, cold, drought, flood, salinity) resistance is very important for development of superior genotypes. Stress tolerance of various crops can be promoted by preconditioning of plants under diverse climatic stresses or exogenous application of osmoprotectants like, proline and glycinebetaine (Wahid et al. 2007). Moreover, as per Hemantaranjan et al. (2014) the temperature alteration are sensed through cellular responses because of signal transduction into the cell. Moreover, activation of Rubisco decreases by several metabolic reactions and inside conductance to CO_2, and this reduction of Rubisco activase and consequently retard net photosynthesis (Pn). Significance of inside conductance to CO_2, heat inducible transcription factor (HITF) for heat shock response (HSR) engages various HSR genes. In this situation molecular level research is very essential to develop genetically modified (GM) crops which are able to fit against high CO_2 condition across the globe. The variation found in the wild and cultivated germplasm of plants offers valuable platform which helps in the detection of markers for marker assisted selection (MAS) and for cloning as well as also helps in transfer of resistance gene against climatic factors (Obidiegwu et al. 2015). Newly, by using high throughout expression assays, numerous gene correlated with different stress tolerance have been determined. For easy understanding of stress tolerance various techniques like QTL analysis, modelling of genetic networks play important role in the development of markers for MAS and also for genetic transformation it identity candidate genes (de la Peña and Hughes 2007).

Gene transferred from stress tolerant species and varieties of vegetable crops will enhance tolerance level, although wide crosses sometimes produces fertile progenies and sometimes tissue culture techniques *i.e*, embryo rescue is also helpful to produce progeny plants (Kumar and Goswami 2012). Aggregating of recent and innovative information of various genomic researches accelerated the improvement of crop plants. For identification of different tools molecular markers play an important role by reducing the environmental affect thus enhancing the efficiency of crop breeding programmes, helping in pre-selection, and short out successive population sizes for field testing.

The outcome of recent and innovative techniques which involves the use of "omics" techniques includes various approaches *i.e.*, biochemical, physiological, molecular and plant biotechnological which provide useful innovations for maintaining better livelihood, poverty mitigation and also decrease the risk of farmer towards changing climatic conditions (Obidiegwu et al. 2015).

In various countries, for development of stress tolerant varieties better alleles of wild species can be transferred into the cultivated species by the use of DNA markers which helps in pyramiding of different genes by controlling quantitative traits. Stress tolerance QTLs have been identified to a variety of crop species such as in tomato for salt tolerance two wild species namely *S. pennellii and S. pimpinellifolium.* In case of tomato the engineered plants show better use of limited water due to stronger and larger root system. Drought tolerance genes *i.e., CBF/DREB1* in tomato and other crops have been successfully implemented by the vegetable scientists (Hsieh et al. 2002). In Tomato, under stable mild heat the expression of B-class genes, *TOMATO APETALA3*, *TOMATO MADS BOX GENE6* (*TM6*) and *LePISTILLATA* was reduced in anthers. Plants in which *TM6* was moderately silenced expressed hypersensitivity to thermo elevation in reference to the occurrence of pistilloid anthers, pollen viability and pollen amount. These heat stress induced down regulations *B*-class genes of tomato contributes to anther deformations and reduced male fertility (Muller et al. 2016). Findings on the physiology of stress tolerance have represented that tolerance to a particular stress is determined by different traits as well as regulated by subsequent genes. A combination of a genome wide scan of expression, using molecular arrays and QTL analysis could offers significant information in detecting the major genes correlated with abiotic stress tolerance mechanism (de la Peña and Hughes 2007).

5 Conclusion

Many genes of plant species got triggered under adverse climatic stress such as heat, cold, drought, flood and salinity which helps them to stand under several stress condition. It is noticed that the recent climate changes have adverse effect on plant growth, development and yield. Although, the plants also arose various defence mechanisms to produce different biochemical molecules to survive under the adverse climatic condition. Improvement in traits which is related to biochemical involves high activity of biochemical molecules e.g., antioxidants, phenolic compounds, osmolite, NADH oxidase's activity, reducing sugars, unsaturated fatty acids, kinases, etc. In contrast, the activity of various phyto-chemical molecules turns down as ROS such as H_2O_2. The leaves biochemical traits like, malondialdehyde content, and antioxidant enzymes activities perhaps take part as a important role towards the increasing capacity tolerance of the plants. Application of biotechnological tools has revolutionized the process of conventional plant breeding. The priority should be given to the novel combination of temperature manipulation in the field under TGTs and the employment of MAGIC lines, to determine flexibility and assess the function of plant adaptability to survive under adverse climatic changes. Climate changes also alter the genetic makeup of population. Even though, most focus should be given to maintain the germplasm of such varieties that are able to tolerate such adverse climatic conditions. The phonological sensitivity can be a useful indicator of genotypes to be exploiting in re-establishment. The upcoming work should be focused on exploring the plant genetic and physico-chemical defence mechanisms for evolving of climate tolerant vegetable crop varieties for vulnerable climatic conditions.

References

Abdalla AA, Verderk K (1968) Growth, flowering and fruit set of tomato at high temperature. Neth J Agric Sci 16:71–76

Abdelmageed AHA, Gruda N (2009) Influence of grafting on growth, development and some physiological parameters of tomatoes under controlled heat stress conditions. Eur J Hortic Sci 74(1):16–20

Abebe T, Guenzi AC, Martin B, Cushman JC (2003) Tolerance of mannitol-accumulating transgenic wheat to water stress and salinity. Plant Physiol 131:1748–1755

Ackerman F, Stanton E (2013) Climate impacts on agriculture: a challenge to complacency? 13-04. Global Development and Environment Institute Working Paper. Tufts University, Medford

Acquaah G (2009) Principles of plant genetics and breeding. Wiley, New York

Afroza B, Wani KP, Khan SH, Jabeen N, Hussain K, Mufti S, Amin A (2010) Various technological interventions to meet vegetable production challenges in view of climate change. Asian J Hortic 5(2):523–529

Akhtar S, Naik A, Hazra P (2014) Harnessing heat stress in vegetable crops towards mitigating impacts of climate change. Chapter 10: 173–200. In: Chaudhary ML, Patel VB, Siddiqui MW, Mahdi SS (eds) Climate change: the principles and applications in horticultural science, vol I. CRC Press, Boca Raton, pp 1–396. isbn:9781771880312

Alam MS, Sultana N, Ahmad S, Hossain MM, Islam AKMA (2010) Performance of heat tolerant tomato hybrid lines under hot, humid conditions. Bangladesh J Agric Res 35(3):367–373

Anonymous (2000) United States global change program research. Global climate change impacts in the United States: Agriculture: Washington, DC, pp 71–78

Anonymous (2006). http://solgenomics.net

Anonymous (2015) Indian horticulture database – 2014, Ministry of Agriculture, Government of India, Gurgaon

Anonymous (2020) Great American Media Services & Vegetable Growers News- 2020. https://vegetablegrowersnews.com/news/some-heat-tolerant-vegetable-varieties-to-consider/

Arvin MJ, Donnelly DJ (2008) Screening potato cultivars and wild species to abiotic, stresses using an electrolyte leakage bioassay. J Agric Sci Technol 10:33–42

AVRDC (1990) Vegetable production training manual. Asian Vegetable Research and Training Center. Shanhua, Tainan, 447.

AVRDC (2003) Guide: grafting tomatoes for production in the hot-wet season. Asian Vegetable Research and Development Center, Publ. No#03-551, Shanhua, Tainan, Taiwan, 6pp. www.avrdc.org/fileadmin/pdfs/graftingtomatoes.pdf

AVRDC (2009) Guide: grafting sweet peppers for production in the hot-wet season. Asian vegetable Research and Development Center, Publ.-No 09-722-e, Shanhua, Tainan, Taiwan, 8pp. www.libnts.avrdc.org.tw/fulltext pdf/FLYER/f0002.pdf

Bahadur A, Chatterjee A, Kumar R, Singh M, Naik PS (2011) Physiological and biochemical basis of drought tolerance in vegetables. Vegetable Sci 38(1):1–16

Bahieldin A, Hesham HT, Eissa HF, Saleh OM, Ramadan AM, Ahmed IA, Dyer WEEL, Itriby HA, Madkour MA (2005) Field evaluation of transgenic wheat plants stably expressing the *HVA1* gene for drought tolerance. Physiol Plant 123:421–427

Basnayake J, Looper M, Ludlow MM, Henzell RG, Snell PJ (1995) Inheritance of osmotic adjustment to water stress in three sorghum crosses. Theoretical Appl Genetics 90(5):675–682

Bates BC, Kundzewicz ZW, Wu S, Palutikof JP (2008) Climate change and water. IPCC Technical Paper VI, Geneva, 210 pp

Ben-Ahmad C, Ben-Rouina B, Athar HUR, Boukhriss M (2006) Olive tree (*Olea europaea* L. CV. "Chemlali") under salt stress: water relations and ions content. Pak J Bot 38(5):1477–1484

Beppu K, Kataoka I (2011) Studies on pistil doubling and fruit set of sweet cherry in warm climate. J Japn Soc Hortic Sci 80:1–13

Bewley JD (1997) Seed germination and dormancy. Plant Cell 9:1055–1066

Beyly GB, Peltier A, Rey GP (1998) Molecular characterization of *CDSP-34*, a chloroplastic protein induced by water deficit in *Solanum tuberosum* L. plants, and regulation of *CDSP-34* expression by ABA and high illumination. Plant J Cell Molecular Biol 16(2):257–262

Bieto JA, Talon M, Fisiologia Y (1996) Bioquimica vegetal. McGraw-Hill Inter-Americana, Madrid, p 581

Bloom AJ, Zwieniecki MA, Passioura JB, Randall LB, Holbrook NM, St. Clair DA (2004) Water relations under root chilling in a sensitive and tolerant tomato species. Plant Cell Environ 27:971–979

Blum A (1988a) Plant breeding for stress environments. CRC Press, Boca Raton, 223p

Blum A (1988b) Improving wheat grain filling under stress by stem reserve mobilisation. Euphytica 100:77–83

Blum A, Schertz KF, Toler RW, Welch RI, Rosenow DT, Johnson JW, Clark LE (1978) Selection for drought avoidance in sorghum using aerial infrared photography. Agron J 70:472–477

Blum A, Mayer J, Golan G (1982) Infrared thermal sensing of plant canopies as a screening technique for dehydration avoidance in wheat. Field Crop Res 57:137–146

Blum A, Piorkova H, Golan G, Mayer J (1983) Chemical desiccation of wheat plants as simulator of post-anthesis stress I. Effects on translocation and kernel growth. Field Crop Res 6:51–58

Bohm W (1974) Mini-rhizotrons for root observations under field conditions. J Agron Crop Sci-Zeitschrift Für Acker Und Pflanzenbau 140:282–287

Bolanos J, Edmeades GO (1993) Eight cycles of selection for drought tolerance in lowland tropical maize.1. Responses in grain-yield, biomass, and radiation utilization. Field Crop Res 31:233–252

Boyer JS, Westgate ME (2004) Grain yields with limited water. J Exp Bot 55:2385–2394

Buescher RW (1979) Influence of high temperature on physiological and compositional characteristics tomato fruits. Leben-Wissen Technol 12:162–164

Capiati DA, País SM, Téllez-Iñón MT (2006) Wounding increases salt tolerance in tomato plants: evidence on the participation of calmodulin-like activities in cross-tolerance signaling. J Exp Bot 57:2391–2400

Cattivelli L, Baldi P, Crosatti C, Di Fonzo N, Faccioli P, Grossi M, Mastrangelo AM, Pecchioni N, Stanca AM (2002) Chromosome regions and stress-related sequences involved in resistance to abiotic stress in Triticeae. Plant Mol Biol 48:649–665

Chakhchar A, Lamaoui M, Aissam S, Ferradous A, Wahbi S, El Mousadik A, Ibnsouda-Koraichi S, Filali-Maltouf A, El Modafar C (2016) Differential physiological and antioxidative responses to drought stress and recovery among four contrasting Argania spinosa ecotypes. J Plant Interact 11(1):30–40. https://doi.org/10.1080/17429145.2016.1148204

Chakrabarti B, Singh SD, Anand A, Singh MP, Pathak H, Nagarajan S (2012) Impact of high temperature on crop and soil (Chapter 8), 88–95. In: Pathak H, Aggarwal PK, Singh SD (eds) Climate change impact, adaptation and mitigation in agriculture: methodology for assessment and applications. Indian Agricultural Research Institute, New Delhi, p xix + 302. isbn:978-81-88708-82-6

Chakraborty S, Newton AC (2011) Climate change, plant diseases and food security: an overview. Plant Pathol 60:1–14

Chan HT, Linse E (1989) Conditioning cucumbers to increase heat resistance in the EFE system. J Food Sci 54:1375–1376

Charrler A (1984) Genetic resources of *Abelmoschus* (okra). IBPGR Secretariat, Rome, 5pp

Chatterjee A, Solankey SS (2014) Functional physiology in drought tolerance of vegetable crops-an approach to mitigate climate change impact, Chapter 9, 149–172. In: Chaudhary ML, Patel VB, Siddiqui MW, Mahdi SS (eds) Climate change: the principles and applications in horticultural science, vol I. CRC Press, Boca Raton, pp 1–396. isbn:9781771880312

Chaudhary A, Sehgal VK, Das M, Pathak H (2012) Vulnerability of the Indo-Gangetic plains to climate change (Chapter 19), 263–269. In: Pathak H, Aggarwal PK, Singh SD (eds) Climate change impact, adaptation and mitigation in agriculture: methodology for assessment and applications. Indian Agricultural Research Institute, New Delhi, p xix + 302. isbn:978-81-88708-82-6

Chavan ML (2007) Drought tolerance studies in tomato (*Lycopersicon esculentum* Mill.). Department of Crop Physiolgoy College of Agriculture, Dharwad University of Agricultural Sciences, Dharwad, pp 133–135

Chigumira NF, Grubben GJH (2004) *Cucurbita maxima* Duchesne. In: Grubben GJH, Denton OA (eds) PROTA2: vegetables/legumes. PROTA, Wageningen

Chinnusamy V, Zhu J, Zhou T, Zhu JK (2007) Small RNAs big role in abiotic stress tolerance of plants. In: Jenks MA, Hasegawa PM, Jain SM (eds) Advances in molecular breeding toward drought and salt tolerant crops. Springer, Dordrecht

Clarke JM, Townley-Smith TF, McCaig TN, Green DG (1984) Growth analysis of spring wheat cultivars of varying drought resistance. Crop Sci 24:537–541

Cruz RV, Harasawa H, Lal M, Wu S, Anokhin Y, Punsalmaa B, Honda Y, Jafari M, Li C, Huu N (2007) Asia climate change-2007: impacts, adaptation and vulnerability. In: Parry ML, Canziani OF, Palutikof JP, van der Linden PJ, Hanson CE (eds) Contribution of Working Group-II to the fourth assessment report of the Intergovernmental Panel on Climate Change. Cambridge University Press, Cambridge, pp 469–506

Cuartero J, Fernandez-Munoz R (1999) Tomato and salinity. Sciencia Hortic 78:83–125

Cuartero J, Bolarin MC, Asins MJ, Moreno V (2006) Increasing salt tolerance in tomato. J Exp Bot 57:1045–1058

Cushman JC (2001) Crasulacean acid metabolism. A plastic photosynthetic adaptation to arid environments. Plant Physiol 127:1439–1448

Dane F, Liu J (2007) Diversity and origin of cultivated and citron type watermelon (*Citrullus lanatus*). Genet Resour Crop Evol 54:1255–1265

Davis AR, Perkins-Veazie P, Dakata Y, Lopez-Galarza S, Maroto JV, Lee SG, Hyh YC, Sun Z, Miguel A, King SR, Cohen R, Lee JM (2008) Cucurbit grafting. Crit Rev Plant Sci 27:50–74

de la Peña R, Hughes J (2007) Improving vegetable productivity in a variable and changing climate. SAT eJ 4(1):1–22

Deschenes O, Greenstone M (2012) The economic impacts of climate change: evidence from agricultural output and random fluctuations in weather: reply. Am Econ Rev 102:3761–3773

Drew MC (1979) Plant responses to anaerobic conditions in soil and solution culture. Curr Adv Plant Sci 36:1–14

Edelstein M (2004) Grafting vegetable crop plants pros and cons. Acta Hortic 65

Edmeades GO, Bolanos J, Elings A, Ribaut JM, Banziger M, Westgate ME (2000) The role and regulation of the anthesis silking interval in maize. In: Westgate ME, Boote KJ (eds) Physiology and modelling kernel set in maize. CSSA special publication no 29. CSSA, Madison, pp 43–73

Ekanayake IJ, O'Toole JC, Garrity DP, Masajo TM (1985) Inheritance of root characters and their relations to drought resistance in rice. Crop Sci 25:927–933

Ezin V, Vodounon CA, de la Peña R, Ahanchede A, Handa AK (2012) Gene expression and phenotypic characterization of flooding tolerance in tomato. J Evol Biol Res 4(3):59–65

FAO (2009). Global Agriculture towards 2050. Issues brief. High level expert forum. Rome, 12–13 October. www.fao.org/wsfs/forum2050/wsfs-background-documents/hlef-issuesbriefs/en/. Accessed Mar 2010

Farlas-Rodriguez R, Melllor RB, Arias C, Peña CJ (1998) The accumulation of trehalose in nodules of several cultivars of common bean (*Phaseolus vulgaris* L.) with resistance to drought stress. Physiol Plants 102:353–359

Farquhar GD, Richards RA (1984) Isotopic composition of plant carbon correlates with water-use efficiency of wheat genotypes. Aust J Plant Physiol 11:539–552

Firon N, Shaked R, Peet MM, Phari DM, Zamsk E, Rosenfeld K, Althan L, Pressman NE (2006) Pollen grains of heat tolerant tomato cultivars retain higher carbohydrate concentration under heat stress conditions. Sci Hortic 109:212–217

Fitter AH, Fitter RSR (2002) Rapid changes in flowering time in British plants. Science 296(5573):1689–1691

Flowers TJ, Yeo AR (1986) Ion relations of plants under drought and salinity. Aust J Plant Physiol 13:75–91

Foolad MR (2005) Breeding for abiotic stress tolerances in tomato. In: Ashraf M, Harris PJC (eds) Abiotic stresses: plant resistance through breeding and molecular approaches. The Haworth Press Inc., New York, pp 613–684

Gaff DF (1980) Protoplasmic tolerance of extreme water stress. In: Turner NC, Kramer PJ (eds) Adaptation of plants to water and high temperature stress. Wiley, New York, pp 207–230

Garg A, Kim J, Owens T, Ranwala A, Choi Y, Kochian L, Wu R (2002) Trehalose accumulation in rice plants confers high tolerance levels to different abiotic stresses. Proc Natl Acad Sci USA 99:15898–15903

Gaxiola RA, Li J, Undurraga S, Dang LM, Allen GJ, Alper SL, Fink GR (2001) Drought and salt tolerant plants result from over expression of the AVP1 H^{+-} pump. Proc Natl Acad Sci USA 98:11444–11449

Gay AP (1986) Variation in selection for leaf water conductance in relation to growth and stomatal dimensions in *Lolium perenne* L. Ann Bot 57:361–369

Ghassemi F, Jakeman AJ, Nix HA (1995) Salinisation of land and water resources: human causes, extent management and case studies. The Australian National University/CAB International, Canberra/Wallingford, 526p

Ghini R, Bettiol W, Hamada E (2011) Diseases in tropical and plantation crops as affected by climate changes: current knowledge and perspectives. Plant Pathol 60:122–132

Giombini MI, Frankel N, Iusem ND, Hasson E (2009) Nucleotide polymorphism in the drought responsive gene *Asr-2* in wild populations of tomato. Genetica 136(1):13–25

Goodstal F, Kohler G, Randall L, Bloom A, St. Clair D (2005) A major QTL introgressed from wild *Lycopersicon hirsutum* confers chilling tolerance to cultivated tomato (*Lycopersicon esculentum*). Theor Appl Genet 111:898–905

Gulam F, Prodhan ZH, Nezhadahmadi A, Rahman M (2012) Heat tolerance in tomato. Life Sci J 9(4):1936–1950

Hall AJ, McPherson HG, Crawford RA, Seager NG (1996) Using early season measurements to estimate fruit volume at harvest in kiwifruit. N Z J Crop Hortic Sci 24:379–391

Hasanuzzaman M, Hossain MA, Fujita M (2010) Selenium in higher plants physiological role, antioxidant metabolism and abiotic stress tolerance. J Plant Sci 5:354–375

Hazra P, Som MG (1999) Technology for vegetable production and improvement. Naya Prokash, Kolkata, India

Hemantaranjan A, Nishant BA, Singh MN, Yadav DK, Patel PK (2014) Heat stress responses and thermotolerance. Adv Plants Agric Res 1(3):00012. https://doi.org/10.15406/apar.2014.01.00012

Hicks JR, Manzano-Mendez J, Masters JF (1983) Temperature extremes and tomato ripening. Proc Fourth Tomato Quality Workshop 4:38–51

Hoad SP, Russell G, Lucas ME, Bingham IJ (2001) The management of wheat, barley and oat root systems. Adv Agron 74:193–246

Hsieh TH, Lee JT, Charng YY, Chan MT (2002) Tomato plants ectopically expressing *Arabidopsis CBF1* show enhanced resistance to water deficit stress. Plant Physiol 130:618–626

Hsieh TH, Li CW, Su RC, Cheng CP, Sanjaya T, Y.C. and Chan, M.T. (2010) A tomato *bZIP* transcription factor, *SlAREB*, is involved in water deficit and salt stress response. Planta 231(6):1459–1473

Hu WH, Zhou YD, Du YS, Xia XJ, Yu JQ (2006a) Differential response of photosynthesis in greenhouse and field ecotypes of tomato to long-term chilling under low light. J Plant Physiol 163:1238–1246

Hu H, Dai M, Yao J, Xiao B, Li X, Zhang Q, Xiong L (2006b) Over expressing a *NAM*, *ATAF*, and *CUC (NAC)* transcription factor enhances drought resistance and salt tolerance in rice. Proc Natl Acad Sci USA 103:12987–12992

IMD, Annual Climate Summary (2010) India Meteorological Department, Pune. Government of India, Ministry of Earth Sciences, p 27

IPCC (2001) Climate change 2001 impacts, adaptation and vulnerability. Intergovernmental Panel on Climate Change, New York

IPCC (2007) Climate change–2007: the physical science basis. In: Solomon S, Qin D, Manning M, Chen Z, Marquis M, Averyt KB et al (eds) Contribution of Working Group I to the fourth assessment report of the Intergovernmental Panel on Climate Change. Cambridge University Press, Cambridge

Ito Y, Katsura K, Maruyama K, Taji T, Kobayashi M, Seki M, Shinozaki K, Yamaguchi-Shinozaki K (2006) Functional analysis of rice *DREB1/ CBF*-type transcription factors involved in cold-responsive gene expression in transgenic rice. Plant Cell Physiol 47:1–13

Janska A, Marsik P, Zelenkova S, Ovesna J (2012) Cold stress and acclimation: what is important for metabolic adjustment? Plant Biol 12:395–405

Jones HG (1999) Use of thermography for quantitative studies of spatial and temporal variation of stomatal conductance over leaf surfaces. Plant Cell Environ 22:1043–1055

Jones HG (2007) Monitoring plant and soil water status: established and novel methods revisited and their relevance to studies of drought tolerance. J Exp Bot 58:119–130

Jones MM, Osmond CB, Turner NC (1980) Accumulation of solutes in leaves of sorghum and sunflower in response to water deficits. Aust J Plant Physiol 7:181–192

Kadioglu A, Terzi R, Saruhan N, Saglam A (2012) Current advances in the investigation of leaf rolling caused by biotic and abiotic stress factors. Plant Sci 182:42–48

Kalia P, Yadav RK (2014) Climate change and its impact on productivity and bioactive health compounds of vegetable crops. Chapter 8, pp 117–147. In: Chaudhary ML, Patel VB, Siddiqui MW, Mahdi SS (eds) Climate change: the principles and applications in horticultural science, vol I. CRC Press, Boca Raton, pp 1–396. isbn:9781771880312

Kavar T, Maras M, Kidriè M, Šuštar-Vozliè J, Megliè V (2011) The expression profiles of selected genes in different bean species (*Phaseolus* spp.) as response to water deficit. J Cent Eur Agric 12(4):557–576

Kavi Kishor PB, Hong Z, Miao GH, Hu CAA, Verma DPS (1995) Over expression of d-pyrroline-5-carboxylate synthetase increases proline production and confers osmotolerance in transgenic plants. Plant Physiol 25:1387–1394

Kawase M (1981) Anatomical and morphological adaptation of plants to water logging. Hortic Sci 16:30–34

Kerstiens G (1996) Cuticular water permeability and its physiological significance. J Exp Bot 47:1813–1832

Khah EM, Kakava E, Mavromatis A, Chachalis D, Goulas C (2006) Effect of grafting on growth and yield of tomato (*Lycopersicon esculentum* Mill.) in greenhouse and open-field. J Appl Hortic 8:3–7

Kim HL, Kim JH, Kim JJ, Sung CJ, Young H (2005) Molecular and functional characterization of *CaLEA6*, the gene for a hydrophobic *LEA* protein from *Capsicum annuum*. Gene 344:115–123

Kirnak H, Kaya C, Ismail TAS, Higgs D (2001) The influence of water deficit on vegetative growth, physiology, fruit yield and quality in eggplants. Bulg J Plant Physiol 27(3-4):34–46

Krammer PJ (1980) Drought resistance and the origin of adaptation. In: Turner NC, Krammer PJ (eds) Adaptation of plants to water and high temperature stress. Wiley-Interscience, New York, pp 7–20

Kremen C (2013) Integrated crop pollination for resilience against climate change and other problems. Presented at the California Department of Food and Agriculture Climate Change Adaptation Consortium, March, 20, American Canyon, CA

Krishna R, Karkute SG, Ansari WA, Jaiswal DK, Verma JP, Singh M (2019) Transgenic tomatoes for abiotic stress tolerance: status and way ahead. 3 Biotech 9(4):143. https://doi.org/10.1007/s13205-019-1665-0

Kumar RR, Goswami S (2012) Biochemical traits of crops for adaptation to climate change. Chapter 13, 149–172. In: Pathak H, Aggarwal PK, Singh SD (eds) Climate change impact, adaptation and mitigation in agriculture: methodology for assessment and applications. Indian Agricultural Research Institute, New Delhi, p xix + 302

Kumar R, Singh M (2006) Citation information genetic resources, chromosome engineering, and crop improvement vegetable crops (Singh RJ, ed) Vol 3. CRC Press, Boca Raton, pp 473–496

Kumar A, Verma JP (2018) Does plant—microbe interaction confer stress tolerance in plants: a review? Microbiol Res 207:41–52. https://doi.org/10.1016/j.micres.2017.11.004

Kumar R, Solankey SS, Singh M (2012) Breeding for drought tolerance in vegetables. Veg Sci 39(1):1–15

Kuo DG, Tsay JS, Chen BW, Lin PY (1982) Screening for flooding tolerance in the genus *Lycopersicon*. Hortic Sci 17(1):76–78

Kusvuran S (2012) Effects of drought and salt stresses on growth, stomatal conductance, leaf water and osmotic potentials of melon genotypes (*Cucumis melo* L.). Afr J Agric Res 7(5):775–781

Laporte MM, Shen B, Tarczynski MC (2002) Engineering for drought avoidance: expression of maize NADP-malic enzyme in tobacco results in altered stomatal function. J Exp Bot 53:699–705

Lawlor DW, Cornic G (2002) Photosynthetic carbon assimilation and associated metabolism in relation to water deficits in higher plants. Plant Cell Environ 25:275–294

Lazcano-Ferrat I, Louatt CJ (1999) Relationship between relative water content, nitrogen pools, and growth of *Phaseolus vulgaris* L. and *P. acutifolius* A. Gray during water deficit. Crop Sci 39:467–475

Leonardis AMD, Petrarulo M, Vita PD, Mastrangelo AM (2012) Genetic and molecular aspects of plant response to drought in annual crop species. In: Giuseppe M, Dichio B (eds) Advances in selected plant physiology aspects. InTech Publisher, Rijeka, pp 45–74

Levitt JB (1972) Responses of plants to environmental stresses. Academic, New York

Li Z, Palmer WM, Martin AP, Wang R, Rainsford F, Jin Y et al (2012) High invertase activity in tomato reproductive organs correlates with enhanced sucrose import into, and heat tolerance of, young fruit. J Exp Bot 63:1155–1166

Liao CT, Lin CH (1996) Photosynthetic response of grafted bitter melon seedling to flood stress. Environ Exp Bot 36:167–172

Linsley RK, Kohler MA, Paulhus JLH (1959) Applied hydrology. McGraw-Hill, New York

Lloyd J, Farquhar GD (2008) Effects of rising temperatures and [CO2] on the physiology of tropical forest trees. Philos Trans R Soc Biol Sci 363:1811–1817

Lobell DB, Schlenker W, Costa-Roberts J (2011) Climate trends and global crop production since 1980. Science 333:616–620

Lunin J, Gallatin MH, Batchelder AR (1963) Saline irrigation of several vegetable crops at various growth stages I. Effect on yields. Agron J 55:107–114

Lutts S, Kinet JM, Bouharmont J (1995) Changes in plant response to NaCl during development of rice (Oryza sativa L.) varieties differing in salinity resistance. J Exp Bot 46:1843–1852

Magan N, Medina A, Aldred D (2011) Possible climate-change effects on mycotoxin contamination of food crops pre- and postharvest. Plant Pathol 60:150–163

Maheswari M, Sarkar B, Vanaja M, Srinivasa Rao M, Srinivasa Rao C, Venkateswarlu B, Sikka AK (2015) Climate resilient crop varieties for sustainable food production under aberrant weather conditions. Central Research Institute for Dryland Agriculture (ICAR), Hyderabad, p 47

Maldonado C, Squeo FA, Ibacache E (2003) Phenotypic response of *Lycopersicon chilense* to water deficit. Revista Chilena Historia Natural 76:129–137

Martin B, Thorstenson YR (1988) Stabel carbon isotope composition (*delta 13C*), water use efficiency and biomass productivity of *Lycopersicon esculentum*, *Lycopersicon pennellii*, and the F_1 hybrid. Plant Physiol 88:213–217

Martin B, Tauer CG, Lin RK (1999) Carbon isotope discrimination as a tool to improve water-use efficiency in tomato. Crop Sci 39:1775–1783

Masle J, Gilmore SR, Farquhar GD (2005) The *ERECTA* gene regulates plant transpiration efficiency in *Arabidopsis*. Nature 436:866–870

Matsubara S (1989) Studies on salt tolerance of vegetables-3. Salt tolerance of rootstocks. Agric Bull, Okayama University 73:17–25

Mattos LM, Moretti CL, Jan S, Sargent SA, Lima CEP, Fontenelle MR (2014) Climate changes and potential impacts on quality of fruit and vegetable crops (Chapter 19). In: Ahmad P (ed)

Emerging technologies and management of crop stress tolerance, vol 1. Academic, San Diego, pp 467–486. https://doi.org/10.1016/B978-0-12-800876-8.00019-9
McKersie BD, Bowley SR, Harjanto E, Leprice O (1996) Water-deficit tolerance and field performance of transgenic alfalfa over-expressing superoxide dismutase. Plant Physiol 111:1177–1181
Medellı'n-Azuara J, Howitt RE, Duncan J, MacEwan, Lund JR (2011) Economic impacts of climaterelated changes to California agriculture. Climate Change 109:387–405
Midmore DJ, Roan YC, Wu DL (1997) Management practices to improve lowland subtropical summer tomato production: yields, economic returns and risk. Exp Agric 33:125–137
Mir RR, Zaman-Allah M, Sreenivasulu N, Trethowan R, Varshney RK (2012) Integrated genomics, physiology and breeding approaches for improving drought tolerance in crops. Theor Appl Genet. https://doi.org/10.1007/s00122-012-1904-9
Mitra J (2001) Genetics and genetic improvement of drought resistance in crop plants. Curr Sci 80(6):758–763
Mohammed M, Wilson LA, Gomes PI (1996) Influence of high temperature stress on postharvest quality of processing and non-processing tomato cultivars. J Food Qual 19:41–55
Moreno AA, Orellana A (2011) The physiological role of the unfolded protein response in plants. Biol Res 44:75–80
Moretti CL, Mattos LM, Calbo AG, Sargent SA (2010) Climate changes and potential impacts on postharvest quality of fruit and vegetable crops: a review. Food Res Int 43:1825–1832
Morgan JM (1980) Osmotic adjustment in the spikelet and leaves of wheat. J Exp Bot 31:655–665
Morgan JM (1983) Osmoregulatiom as selection criterion for drought tolerance in wheat. Aust J Agric Res 34:607–614
Muller F, Xu J, Kristensen L, Wolters-Arts M, de Groot PFM, Jansma SY, Mariani C, Park S, Rieu I (2016) High-temperature-induced defects in tomato (*Solanum lycopersicum*) anther and pollen development are associated with reduced expression of B-class floral patterning genes. PLoS One 11(12):e0167614. https://doi.org/10.1371/journal.pone.0167614
Munns R, Tester M (2008) Mechanisms of salinity tolerance. Annu Rev Plant Biol 59:651–681
Nahar K, Ullah SM (2011) Effect of water stress on moisture content distribution in soil and morphological characters of two tomato (*Lycopersicon esculentum* Mill) cultivars. J Sci Res 3(3):677–682
Nakagawa H, Horie T, Nakano HY, Kim K, Wada K, Kobayashi M (1993) Effects of elevated CO2 concentration and high temperature on the growth and development of rice. J Agri Meteorol 48:799–802
Narayan R (2009) Air pollution a threat in vegetable production. In: Sulladmath UV, Swamy KRM (eds) International conference on horticulture (ICH-2009): horticulture for livelihood security and economic growth, pp 158–159
Naveed A, Khan AA, Khan IA (2009) Generation mean analysis of water stress tolerance in okra (*Abelmoschous esculentus* L.). Pak J Bot 41(1):195–205
Ndunguru BJ, Ntare BR, Williams JH, Greenberg DC (1995) Assessment of groundnut cultivars for end of season drought tolerance in a Sahelian environment. J Agric Sci (Camb) 125:79–85
O'Connell MA, Medina AL, Sanchez Pena Pand Trevino MB (2007) Molecular genetics of drought resistance response in tomato and related species. In: Razdan MK, Mattoo AK (eds) Genetic improvement of Solanaceous crops, Tomato, vol 2. Science Publishers, Enfield, pp 261–283
O'Toole JC, Moya TB (1978) Genotypic variation in maintenance of leaf water potential in rice. Crop Sci 18:873–876
Obidiegwu JE, Bryan GJ, Jones HG, Prashar A (2015) Coping with drought: stress and adaptive responses in potato and perspectives for improvement. Front Plant Sci 6:542. https://doi.org/10.3389/fpls.2015.00542
Oh SJ, Song SI, Kim YS, Jang HJ, Kim SY, Kim M, Kim YK, Kim NYK, Nahm BH, Kim JK (2005) *Arabidopsis CBF3/DREB1A* and *ABF3* in transgenic rice increased tolerance to abiotic stress without stunting growth. Plant Physiol 138:341–351

Okada M, Ozawa K, Hamasaki T (1992) A use of TRC technique for an analysis of crop responses to temperature. Agric Meteorol Tohoku 37:23–25. [in Japanese with English abstract]

Opena RT, Lo SH (1981) Breeding for heat tolerance in heading Chinese cabbage. In: Talekar NS, Griggs TD (eds) Proceedings of the 1st international symposium on Chinese cabbage. AVRDC, Shanhua

Opena RT, Chen JT, Kuo CG, Chen HM (1992) Genetic and physiological aspects of tropical adaptation in tomato. In: Kuo CG (ed) Adaptation of food crops to temperature and water stress. AVRDC, Shanhua, pp 321–334

Palada MC, Wu DL (2008) Evaluation of chili rootstocks for grafted sweet pepper production during the hot-wet and hot-dry seasons in Taiwan. Acta Hortic 767:167–174

Pandey, S., Ansari, W.A., Jha, A., Bhatt, K.V. and Singh, B. (2011). Evaluations of melons and indigenous *Cucumis* spp. genotypes for drought tolerance, 2nd internatioanal symposium on underutilized plant species, 27th June–1st July, The Royal Chaulan Kuala Lumpur, Malaysis, (A-61), 95pp

Pandey S, Ansari WA, Atri N, Singh B (2016) Standardization of screening technique and evaluation of muskmelon genotypes for drought tolerance. Plant Genetic Resour. https://doi.org/10.1017/S1479262116000253

Pangga IB, Hannan J, Chakraborty S (2011) Pathogen dynamics in a crop canopy and their evolution under changing climate. Plant Pathol 60:70–81

Pani RK (2008) Climate change hits vegetable crop. Indian Express. Available from: http://www.expressbuzz.com

Park S, Li J, Pittman JK, Berkowitz GA, Yang H, Undurraga S, Morris J, Hirschi KD, Gaxiola RA (2005) Up-regulation of a H^{+-} pyrophosphatase (H^{+}- PPase) as a strategy to engineer drought resistant crop plants. Proc Natl Acad Sci USA 102:18830–18835

Peet MM, Wolf DW (2000) Crop ecosystem responses to climatic change vegetable crops. In: Reddy KR, Hodges HF (eds) Climate change and global crop productivity. CABI, Wallingford, pp 213–244

Peet MM, Willits DH, Gardner R (1997) Response of ovule development and post pollen production processes in male-sterile tomatoes to chronic, sub-acute high temperature stress. J Exp Bot 48(306):101–111

Peet MM, Sato S, Gardner R (1998) Comparing heat stress on male-fertile and male sterile tomatoes. Plant, Cell and Environment 21:225–231

Pope KS (2012) Climate change adaptation: temperate perennial crops. Presented at the California Department of Food and Agriculture Climate Change Adaptation Consortium, November 28, Modesto, CA

Porter JR, Semenov MA (2005) Crop responses to climatic variation. Philos Trans R Soc Biol Sci 360:2021–2035

Quizenberry JE (1982) Breeding for drought resistance and plant use efficiency. In: Christianses MN, Lewis CF (eds) Breeding plants for less favourable environments, pp 193–212

Rai N, Rai M (2006) Heterosis breeding in vegetable crops. In: Solanaceous crops, tomato. New India Pulishing Agency, New Delhi, pp 259–260

Rai N, Yadav DS (2005) Advances in vegetable production. Researchco Book Centre, New Delhi, India

Rai M, Pandey S, Kumar S (2008) Cucurbitaceae: proceedings of the IXth EUCARPIA meeting on genetics and breeding of Cucurbitaceae. In: Pitrat M (ed) INRA, Avignon (France), May 21–24th, 285–293pp

Rai N, Tiwari SK, KumarR, Singh M, Bharadwaj DR (2011) Genetic resources of solanaceous vegetables in India. National symposium on vegetable biodiversity. Jawaharlal Nehru Krishi Vishwa Vidyalaya, Jabalpur, M.P. April, 4–5, 91–103 pp

Rai GK, Rai NP, Rathaur S, Kumar S, Singh M (2013) Expression of rd29A::AtDREB1A/CBF3 in tomato alleviates drought-induced oxidative stress by regulating key enzymatic and non-enzymatic antioxidants. Plant Physiol Biochem 69:90–100. https://doi.org/10.1016/j.plaphy.2013.05.002

Rajcan I, Tollenaar M (1999) Source-sink ratio and leaf senescence in maize. I. Dry matter accumulation and partitioning during the grain-filling period. Field Crop Res 90:245–253
Ram HH (2005) Vegetable breeding: principles and practices. Kalyani Publishers, New Delhi
Ramanjulu S, Bartels D (2002) Drought and desiccation induced modulation of gene expression in plants. Plant Cell Environ 25:141–151
Ramos B, Miller FA, Brandão TRS, Teixeira P, Silva CLM (2013) Fresh fruits and vegetables – an overview on applied methodologies to improve its quality and safety. Innov Food Sci Emerg Technol 20:1–15
Rana MK, Kalloo G (1989) Morphological attributes associated with the adaptation under water deficit conditions in tomato (*L. esculentum* Mill.). 12th Eucarpia Congress, Vortrage Pflanzenzucht, 23–27 pp
Rao GGSN, Rao AVMS, Rao VUM (2009) Trends in rainfall and temperature in rainfed India in previous century. In: Aggarwal PK (ed) Global climate change and Indian Agriculture case studies from ICAR network project. ICAR Publication, New Delhi, pp 71–73
Rawson HM, Gifford RM, Condon BN (1995) Temperature gradient chambers for research on global environment change. I Portable chambers for research on short-stature vegetation. Plant Cell Environ 18(9):1048–1054
Razdan Maharaj K, Mattoo AK (2007) Genetic improvement of Solanaceous crops: tomato, vol 2. Science Publishers, p 47
Reddy AR, Chaitanya KV, Vivekanandan M (2004) Drought induced responses of photosynthesis and antioxidant metabolism in higher plants. J Plant Physiol 161:1189–1202
Richards RA (2006) Physiological traits used in the breeding of new cultivars for water-scarce environments. Agric Water Manage 80:197–211
Rick CM (1973) Potential genetic resources in tomato species: clues from observation in native habitats. In: Srb AM (ed) Genes, enzymes and populations. Plenum Press, New York, pp 255–269
Rivero RM, Ruiz JM, Romero L (2003a) Can grafting in tomato plants strengthen resistance to thermal stress? J Sci Food Agr 83:1315–1319
Rivero RM, Ruiz JM, Sanchez E, Romero L (2003b) Role of grafting in horticultural plants under stress conditions. Food Agric Environ 1:70–74
Rivero RM, Kojima M, Gepstein A, Sakakibara H, Mittler R, Gepstein S, Blumwald E (2007) Delayed leaf senescence induces extreme drought tolerance in a flowering plant. Proc Natl Acad Sci USA 104:19631–19636
Robertson BM, Hall AE, Foster KW (1985) A field technique for screening for genotypic differences in root growth. Crop Sci 25:1084–1090
Romero L, Belakbir A, Ragala L, Ruiz MJ (1997) Response of plant yield and leaf pigments to saline conditions effectiveness of different rootstocks in melon plant (*Cucumis melo* L). Soil Sci Plant Nutr 3:855–862
Ross H (1986) Potato breeding: problems and perspectives. J Plant Breed (Suppl.) 13:1–132. Berlin
Rouphael Y, Cardarelli M, Colla G, Rea E (2008) Yield, mineral composition, water relations, and water use efficiency of grafted mini-watermelon plants under deficit irrigation. HortScience 43(3):730–736
Sage RF, Kubien D (2007) The temperature response of C3 and C4 photosynthesis. Plant Cell Environ 30:1086–1106
Saijo Y, Hata S, Kyozuka J, Shimamoto K, Izui K (2000) Over-expression of a single Ca2+ – dependent protein kinase confers both cold and salt/ drought tolerance on rice plants. Plant J 23:319–327
Sánchez Peña P (1999) Leaf water potentials in tomato (*L. esculentum* Mill.) *L chilense* Dun. and their interspecific F_1. M. Sc. thesis, New Mexico State University, Las Cruces, NM, USA
Sánchez-Rodríguez E, Moreno DA, Ferreres F, Rubio-Wilhelmi M, Manuel RJ (2011) Differential responses of five cherry tomato varieties to water stress: changes on phenolic metabolites and related enzymes. Phytochemistry 72:723–729

Sato S, Kamiyama M, Iwata T, Makita N, Furukawa H, Ikeda H (2006) Moderate increase of mean daily temperature adversely affects fruit set of *Lycopersicon esculentum* by disrupting specific physiological processes in male reproductive development. Ann Bot 97:731–738

Schwarz D, Rouphael Y, Colla G, Venema JH (2010) Grafting as a tool to improve tolerance of vegetables to abiotic stresses: thermal stress, water stress and organic pollutants. Sci Hortic 127:162–171

Selvakumar R (2014) A text book of glaustas olericulture. New Vishal Publications, New Delhi, pp 1–1143

Serraj R, Sinclair TR (1996) Processes contributing to N_2-fixation insensitivity to drought in the soybean cultivar Jackson. Crop Sci 36:961–968

Shannon MC (1996) New insights in plant breeding efforts for improved salt tolerance. HortTechnology 6:96–99

Shannon MC, Grieve CM (1999) Tolerance of vegetable crops to salinity. Sci Hortic 78:5–38

Sharp RE, Poroyko V, Hejlek LG, Spollen WG, Springer GK, Bohnert HJ, Nguyen T (2004) Root growth maintenance during water deficits: physiology to functional genomics. J Exp Bot 55:2343–2351

Shibuya T, Tokuda A, Terakura R, Shimizu-Maruo K, Sugiwaki H, Kitaya Y, Kiyota M (2007) Short-term bottom-heat treatment during low-air-temperature storage improves rooting in squash (*Cucurbita moschata* Duch.) cuttings used for rootstock of cucumber (*Cucumis sativus* L.). J Japn Soc Hortic Sci 76(2):139–143

Shinozaki K, Yamaguchi-Shinozaki K (2007) Gene networks involved in drought stress response and tolerance. J Exp Bot 58:221–227

Shirsath PB, Aggarwal PK, Thornton PK, Dunnett A (2016) Prioritizing climate-smart agricultural land use options at a regional scale. Agric Syst. https://doi.org/10.1016/j.agsy.2016.09.018

Singh, H.P. (2010). Ongoing research in abiotic stress due to climate change in horticulture, curtain raiser meet on research needs arising due to abiotic stresses in agriculture management in India under global climate change scenario, Baramati, Maharashtra, October 29–30. 1–23 pp. http://www.niam.res.in/pdfs/DDG-Hort-lecture.pdf

Singh NN, Sarkar KR (1991) Physiological, genetical basis of drought tolerance in maize. In: Proceedings of the golden jubilee symposium. Genetic research and education. Indian Society of Genetics and Plant Breeding, New Delhi

Singh C.S., Anima, K., Kumar, B. and Gautam (2014). Disatrous challenge due to climate change in Bihar, developing state of India. Int J Sci Eng Res, 5 (6): 1038-1041

Slafer GA, Araus JL, Royo C, Del Moral LFG (2005) Promising ecophysiological traits for genetic improvement of cereal yields in Mediterranean environments. Ann Appl Biol 146:61–70

Solankey SS, Singh RK, Baranwal DK, Singh DK (2015) Genetic expression of tomato for heat and drought stress tolerance: an overview. Int J Veg Sci 21(5):496–515. https://doi.org/10.1080/19315260.2014.902414

Spaldon S, Samnotra RK, Chopra S (2015) Climate resilient technologies to meet the challenges in vegetable production. Int J Curr Res Acad Rev 3(2):28–47

Springate DA, Kover PX (2013) Plant responses to elevated temperatures: a field. Glob Chang Biol. https://doi.org/10.1111/gcb.12430

Stevens MA, Rudich J (1978) Genetic potential for overcoming physiological limitations on adaptability, yield, and quality in tomato. Hortic Sci 13:673–678

Stoner AK (1972) Merit, Red Rock and Potomac-tomato varieties adapted to mechanical harvesting. USDA Prod. Res. Rep

Swarup V (2006) Vegetable science and technology in India. Kalyani Publishers, New Delhi, pp 1–656

Symonds RC, Kadirvel P, Yen J, Lin J, Peña RDL (2010) Genetic, physiological and molecular approaches to improve drought tolerance in tropical tomato. Proceedings SOL2010, 42 pp.

Szabolcs I (1992) Salinisation of soils and water and its relation to desertification. In: Razdan MK, Mattoo AK (eds) Genetic improvement of Solanaceous crop, Beltsville, pp 521–590

Tirado MC, Clarke R, Jaykus LA, McQuatters-Gollop A, Frank JM (2010) Climate change and food safety: a review. Food Res Int 43(1745):65

Toppino L, Acciarri N, Mennella G, Lo Scalzo R, Rotino GL (2009) Introgression breeding in eggplant (*Solanum melongena* L.) by combining biotechnological and conventional approaches. Proceedings of the 53rd Italian Society of Agricultural Genetics Annual Congress Torino, Italy, 16/19 Sept

Aazami MA, Torabi M, Jalili E (2010) In vitro response of promising tomato genotypes for tolerance to osmotic stress. Afr J Biotechnol 9(26):4014–4017

Turner, N.C. (1979). Stress Physiology in Crop Plants. In: Mussell H, Staples RC (eds). Wiley, New York, 343–372 pp

Turner NC (1986) Crop water deficits: a decade of progress. Adv Agron 39:1051

Tyerman SD, Niemietz CM, Bramley H (2002) Plant aquaporins: multifunctional water and solute channels with expanding roles. Plant Cell Environ 25:173–194

Ugherughe PO (1986) Drought and tropical pasture management. J Agron Crop Sci-Zeitschrift Für Acker Und Pflanzenbau 157:13–23

Valliyodan B, Nguyen HT (2006) Understanding regulatory networks and engineering for enhanced drought tolerance in plants. Curr Opin Plant Biol 9:189–195

Wahid A, Gelani S, Ashraf M, Foolad MR (2007) Heat tolerance in plants: an overview. Environ Exp Bot 61:199–223

Walter A, Shurr U (2005) Dynamics of leaf and root growth: endogenous control versus environmental impact. Ann Bot 95:891–900

Wang Y, Ying J, Kuzma M, Chalifoux M, Sample A, McArthur C, Uchacz T, Sarvas C, Wan J, Tennis DT, McCourt P, Huang Y (2005) Molecular tailoring of farnesylation for plant drought tolerance and yield protection. Plant J 43:413–424

Wang S, Yang R, Cheng J, Zhao J (2007) Effect of rootstocks on the tolerance to high temperature of eggplants under solar greenhouse during summer season. Acta Hortic 761:357–360

Weiss J, Egea-Cortines M (2010) Transcriptomic analysis of cold response in tomato fruits identifies dehydrin as a marker of cold stress. J Appl Genet 50(4):311–319

Went FW (1953) The effect of temperature on plant growth. Annu Rev Plant Physiol 4:347–362

Wurr DCE, Fellows JR, Phelps K (1996) Investigating trends in vegetable crop response to increasing temperature associated with climate change. Sci Hortic 66:255–263

Xiao B, Huang Y, Tang N, Xiong L (2007) Over-expression of a *LEA* gene in rice improves drought resistance under the field conditions. Theor Appl Genet 115:35–46

Yan J, He C, Wang J, Mao Z, Holaday SA, Allen RD, Zhang H (2004) Overexpression of the *Arabidopsis 14-3-3* protein *GF14* lambda in cotton leads to a "stay-green" phenotype and improves stress tolerance under moderate drought conditions. Plant Cell Physiol 45:1007–1014

Yildirim E, Guvenc I (2006) Salt tolerance of pepper cultivars during germination and seedling growth. Turk J Agric For 30:347–353

Zhu BC, Su J, Chan MC, Verma DPS, Fan YL, Wu R (1998) Over expression of a d-pyrroline-5-carboxylate synthetase gene and analysis of tolerance to water-stress and salt stress in transgenic rice. Plant Sci 139:41–48

Zhu M, Chen G, Zhang J, Zhang Y, Xie Q, Zhao Z, Pan Y, Hu Z (2014) The abiotic stress-responsive NAC-type transcription factor SlNAC4 regulates salt and drought tolerance and stress-related genes in tomato (*Solanum lycopersicum*). Plant Cell Rep 33(11):1851–1863. https://doi.org/10.1007/s00299-014-1662-z

Źróbek-Sokolnik A (2012) Temperature stress and responses of plants. In: Ahmad P, Prasad MNV (eds) Environmental adaptations and stress tolerance of plants in the era of climate change. Springer, New York, pp 113–134

Genotypic Selection in Vegetables for Adaptation to Climate Change

Shirin Akhtar, Abhishek Naik, and Shashank Shekhar Solankey

1 Introduction

The major production constraint for any agricultural crop in today's world is the constantly changing climatic condition, impacting all spheres of life. The United Nations Framework Convention on Climate Change (UNFCCC) has defined climate change as: "a change of climate which is attributed directly or indirectly to human activity that alters the composition of the global atmosphere and which is in addition to natural climate variability observed over comparable time periods". The Intergovernmental Panel on Climate Change (IPCC), on the other hand, has defined climate change as "a change in the state of the climate that can be identified (e. g., by using statistical tests) by changes in the mean and/or the variability of its properties and that persists for an extended period, typically decades or longer". Thus climate change due to anthropological actions modifying the atmospheric constitution, and climate inconsistency having natural roots has been distinguished by UNFCCC, whereas the IPCC has not differentiated between them. Climate change encompasses extremities of temperature, erratic rainfall pattern, melting of the polar caps and glacier leading to rise in sea level. Every year different parts of the world are affected by flood, drought, frost, hailstorm, cyclones, etc. that are caused by climate change. The impact of climate change is most obvious through the general trend of

S. Akhtar (✉)
Department of Horticulture (Vegetable and Floriculture), Dr. Kalam Agricultural College, Kishanganj, Bihar Agricultural University, Sabour, Bhagalpur, Bihar, India

A. Naik
Marketing Department, Ajeet Seeds Pvt. Ltd. (Aurangabad, Maharashtra), Kolkata, West Bengal, India

S. S. Solankey
Department of Horticulture (Vegetable and Floriculture), Bihar Agricultural University, Sabour, Bhagalpur, Bihar, India

S. S. Solankey et al. (eds.), *Advances in Research on Vegetable Production Under a Changing Climate Vol. 1*, Advances in Olericulture,
https://doi.org/10.1007/978-3-030-63497-1_3

increasing temperature. There has been acceleration of global warming at an alarming rate in the last 6 decades. The average temperature over 2000–2010 has been higher than the 1961–1990 mean by 0.46 °C and 0.21 °C higher than the 1991–2000 mean (WMO 2011). There has been accelerated rise in the present decade and at the current rate of emission of greenhouse gases, the IPCC has projected that the global warming will reach 1.5 °C within 2030–2052 and more than 3–4 °C by 2100 resulting in loss of thousands of species, with the projection of 8% plants at 1.5 °C and 16% at 2 °C (Nullis 2018). The fourth report of IPCC has explained the impacts of projected climate change, both global and regional, on crop production, water resources, natural ecologies and food safety and assurance. The major effects of climate change are change in productivity and quality of crops, reduction of crop diversity, nitrogen loss through leaching, soil erosion and change in the pattern of water use and application of fertilizers, insecticides, and herbicides etc. South Asia has been depicted as one of the most vulnerable regions among the several highly populated regions of the world. Among the South Asian countries, India is one of the most vulnerable which is accountable to its huge population, dependence of major mass on agriculture, non-judicious use and minimum conservation of natural resources and least strategies for coping with the climatic vagaries. The agricultural production system of India has the havoc responsibility to feed 17.5% of the world's population, having only 2.4% of land and about 4% of the water resources at its disposal (NICRA policy paper 65). Under such circumstances, climate resilient agriculture seems to be the only viable option. It involves adaptation, mitigation and other strategies that enable the agricultural production system to bounce back quickly and resist the damage caused by the climatic vagaries (viz., heat/cold waves, drought, floods, erratic rainfall, dry spells, enhanced pest population, disease epidemics, etc.). Judicious handling of the natural resources, *viz.*, land, water, soil and other inputs along with genetic resources through best agricultural packages forms the backbone of climate resilient agriculture.

2 Vegetable Production Scenario

The global production of vegetable crops has escalated manifolds in the previous decades and at present the trade value of vegetables exceeds that of the cereal crops (De La Peña and Hughes 2007). The total production of vegetables in the world amounts to about 1169.45 million tonnes from an area of 61.22 million hectares (Anonymous 2017). The per annum growth in global production of vegetables has increased by 6.7% during 1990–2000 and 3.4% during 2000–2011 (FAOSTAT 2014). India ranks second next to China in production of melons, tomatoes, eggplant, cabbage, cauliflower, potatoes as well as overall vegetable production, while India tops in okra production in the world (Anonymous 2017). The growth trend of vegetables has increased in the last six decades by 2.99 folds from an area of 2.84 million ha, 8.88 folds in production from 16.5 million tonnes and 2.98 folds in productivity from an average of 5.8 t/ha in 1950–1951 (Vanitha et al. 2013). The most important vegetable crops grown throughout the world are tomatoes, potatoes, onion, cabbage, hot peppers and eggplant (Brown et al. 2005). Most vegetables prefer mild temperatures and frequent shallow irrigation for their optimum production and hence extremities of temperature (both high and low) and soil moisture (drought or waterlogging) in particular restrict vegetable production, which is enhanced in presence of other environmental constraints.

3 Climate Change and Vegetable Productivity

Environmental stress is the major reason behind global crop losses, resulting in 50% and even greater yield reduction in key crops (Bray et al. 2000; Boyer 1982). The primary constraints for sustainable vegetable production with enhanced productivity are temperature rise, increase in atmospheric carbon dioxide, scarcity of irrigation water, floods, salinity, which in turn will hamper the soil fertility and intensify soil erosion.

The severity of these environmental stresses on vegetable crops will be magnified by the changing climate. The influence of climate change on agriculture will be exhibited immensely through water availability (Mou 2011). There will be altered hydrological cycles, thus changing the precipitation pattern. There may be lesser rainfall, earlier melting of snow due to global warming leading to drought. There may be rainfall of higher quantities but lesser frequency which may result in floods, soil erosion and leaching of minerals, decrease oxygen level of crop fields, lead to soil compaction and even delay crop planting. Rise in temperature may even enhance transpiration rate in plants and pave way for faster evaporation of soil moisture. The duration of exposure to the severity of the stress and the plant developmental stage determines the response of the plant to the particular stress (Bray 2002). In vegetables, climatic vagaries affect the crop right from the seed germination stage to vegetative, flowering, fruitset and fruit development stages. There is hampering in the habitual pattern of crop growth and development and eventually the production and productivity are severely reduced (Spaldon et al. 2015). For example, when daily ozone concentration is greater than 5–15%, the vegetable yield may be reduced by 50 ppb (Raj Narayan 2009). There may be influence of environmental interactions that may result in harsher effect of climate change and make the plant response to stress more complex (Datta 2013). The morphological or biochemical mechanisms of the plant to avoid or tolerate one or many stresses may be similar (Capiati et al. 2006).

4 Impacts of Climate Change on Vegetable Production

4.1 High Temperature

Heat stress that results from high ambient temperatures is a serious threat to the crop production scenario of the world (Hall 2001). Heat stress has been defined by Wahid et al. (2007) as "the rise in temperature beyond a threshold level for a period of time sufficient to cause irreversible damage to plant growth and development". Generally, a temperature of 10–15 °C above the ambient temperature is regarded as heat shock or heat stress. It encompasses intensity (temperature in degrees), duration, and rate of upsurge in temperature. Its extent is determined by the probability and duration of diurnal and/or nocturnal high temperatures. There may be variations in plant growth, development, physiological processes, and yield as a result of the heat stress on plant (Hasanuzzaman et al. 2012, 2013).

The rise in temperature will lead to change in production timing, whereas not much alteration in photoperiod will ultimately lead to faster maturity of photosensitive vegetables. Higher temperatures will lead to reduction in tuber initiation in potato; tomato quality will be inferior, and pollination will be hampered in different crops. There will be bolting in cole crops, anthocyanin production in capsicum will be lesser, while tip burn and blossom end rot will be a very common phenomenon in tomatoes. Pollination will be adversely affected and frequent incidences of floral abortions flower and fruit drop will occur. When any vegetable crop is exposed to optimum temperature (21 °C night/28 °C day temperature), starch accumulation in pollen grains occur, which reaches its peak just before anthesis and thereafter decline gradually. Simultaneously, total soluble sugar concentration slowly increase in pollen grains, the highest being at anthesis. However, under high temperature conditions (27 °C night/37 °C day temperature) the transient enhancement of starch concentration will be checked and result in reduced soluble sugar concentration in pollen grains at anthesis, which will lead to reduced pollen viability, germinability and ultimately fruitset. The survival and incidence of pathogens, pests and weeds will be more at problems high temperature, and therefore application of more insecticides and pesticides will be required. On the other hand, the effectiveness of pesticides will decrease at higher temperature.

Developments of the modified vegetative parts that are economically used are very much influenced by the prevailing atmospheric temperature and have been discussed in the following Table 1.

Temperature also influences the flowering and fruit set in vegetable crops. The specific temperature requirement for fruit set in different vegetable crops has been shown in the following Table 2.

The quality of vegetables is also immensely affected by temperature. The following table shows the impact of temperature on the quality of vegetables (Table 3).

Vegetables have been classified as cool season and warm season crops. Cool season vegetables are tolerant to low temperature and sometimes even frost, such as onion, garlic, cabbage, Brussels sprout, knolkhol, turnip, radish, etc., besides being

Table 1 Temperature required for development of vegetative parts for economic use

Vegetable crop	Economic part	Temperature required for proper development (°C)	Impact of variation from optimum temperature
Cauliflower:	Curd		Early high temperature cultivars grown in low temperature result in quick curding without proper vegetative growth leading to small button-like curds. Late cultivars grown early result in high vegetative growth and lastly when curding occurs, the curds are small, button-like. High temperature makes curd yellow, hard, leafy and ricey.
Early		20–27	
Mid		20–25	
Mid late		16–19	
Late		10–12	
Cabbage:	Head		Heads are not compact
European		15–18	
Tropical Japanese		25–30	
Knolkhol	Knob	15–18	Poor quality of knob
Brussel's sprouts	Miniature heads	10–15	Head not formed
Sprouting broccoli	Flower heads	10–15	Bud clusters loose in high temperature
Chinese cabbage	Head	15.5–21	Seed stalks produced before proper head is produced if temperature goes below 15 °C
Onion	Bulb	19–22	Bulb development retarded above 40 °C
Garlic	Bulb	10–15	Bulb production hampered in high temperature
Potato	Tuber	18–20 (night temperature)	Above 21 °C night temperature, tuberization is inhibited
Radish	Root	20–25 initially, 10–18 at latter stage	Roots thin and improper at high temperature
Carrot	Root	16–21	Below 15 °C roots slender and colour development improper; at 24–25 °C roots shorter, thicker and colour is less
Beet	Root	15–21	Above 25 °C zoning in roots observed; at 10–15 °C seed stalks produced prematurely
Turnip	Root	15–20	Below 10 °C seed stalks emerge; at high temperature roots become tough and bitter in taste
Sweet potato	Root tuber	20–25	Tuberization hampered at high (35–40 °C) and low (15 °C) temperatures
Cassava	Root tuber	25–32	Tuberization hampered at high and low temperatures
Taro	Corm	25–30	Corms small and ill-developed
Elephant foot yam	Corm	20–25	Corms small and ill-developed
Yam	Stem tuber	25–30	Growth and tuberization affected
Yam bean	Root tuber	18–20	Growth and bulking of tuber affected

(continued)

Table 1 (continued)

Vegetable crop	Economic part	Temperature required for proper development (°C)	Impact of variation from optimum temperature
Asparagus	Spear	15–24	Spear production hampered
Palak	Leaves	16–21	Above 25 °C bolting occurs leading to reduction in leaf yield
Spinach	Leaves	16–21	Growth and leaf yield seriously hampered in high temperature
Lettuce	Leaves	13–18	Growth and leaf yield seriously hampered in high temperature
Celery	Leaves	15–18	At 10–15 °C bolting occurs; high temperature causes bitterness in leaves

Source: Hazra and Som (2006) and Akhtar et al. (2015)

sensitive to heat. On the other hand, warm season vegetables grow at relatively higher temperature, such as, tomato, brinjal, peppers, cucurbits, beans, etc.). Among them, tomato, brinjal, peppers, cucumber, pumpkin, etc. are warm loving crops, while winged bean, cluster bean, sweet potato, muskmelon, etc. tolerant of temperature upto 40 °C may be called heat resistant crops.

4.2 Low Temperature

Plants undergo low temperature stress when exposed to temperatures lower than 15 °C. Temperatures below this temperature but above 0 °C exert chilling stress, whereas when temperature goes below 0 °C plants experience freezing stress (Levitt 1980; Raison and Lyons 1986). The winter regime and chilling duration in temperate regions affect the normal growth of the temperate crops and their seed production. Besides, there is also decrease in temperatures in certain regions of the world in particular seasons and the mercury is dipping to record breaking temperatures. Frost or freezing temperature causes damage to the crops. Plants, particularly the ones native to warm habitat, exhibit injury symptoms when exposed to low temperatures, even when non-freezing (Jiang et al. 2002). These plants include vegetables like tomato, eggplant, peppers, gourds and pumpkin, which are sensitive to temperatures lesser than 10–15 °C and show signs of injury, which appear within 48–72 h of exposure to the cold stress depending upon the sensitivity of a plant. Plants exhibiting visual injuries at temperatures below 15 °C are referred to as "very sensitive to chilling" (Raison and Lyons 1986). Despite being cultivated in temperate regions for long time various tropical or subtropical crops, including vegetables such as tomato, cucumber, soybeans, etc., do not possess substantial resistance to chilling, (Wilson 1985). The effects of chilling temperatures in plants in temperate climates may be yield reduction or total crop failure due to either direct damage or

Table 2 Temperature required for flowering and fruit set in different vegetables

Vegetable crop	Temperature required for fruit set (°C)	Impact of variation from optimum temperature
Pea	10–18	Above 25 °C maturity hastened, pod set lowered, quality adversely affected
French bean	18–25	Flower number per plant increases at high temperature (27–32 °C) but pod set markedly decreases
Lima bean	26–30	Fruit set is drastically affected at high temperature (26–30 °C)
Tomato	15–21 (night temperature)	Fruit set does not occur if night temperature goes to 13 °C or if average maximum day/night temperature goes above 32 °C/21 °C
Brinjal	17–25	Above 35 °C fruit set and yield are adversely affected; below 10 °C splitting of ovary in bud stage and deformation of fruits occur
Hot pepper	18–20 (night temperature)	Flower drop occurs at high night temperature (24 °C) and also elevated day temperature (>30 °C) along with with low light intensity
Sweet pepper	20–25 (day temperature)	Flower drop occurs at high day temperature (30 °C and above); pollen viability is low resulting in parthenocarpic fruits at low night temperature (10–15 °C)
	15–16 (night temperature)	
Cucurbits:		High temperature induces more male flowers and thus adversely affects fruit set
Melons (watermelon, maskmelon); pumpkin, summer squash; bitter gourd; cucumber	12.3–18.3	
Bottle gourd, ridge gourd	24–28	

Source: Hazra and Som (2006) and Akhtar et al. (2015)

delayed maturation. There may be up to 50% decrease in productivity in chilling-sensitive plants due to even minute drop in temperature exhibiting visible damage to the crop (Jouyban et al. 2013).

Smaller leaf size, wilting, yellowing of leaves (chlorosis) are different phenotypic symptoms in response to chilling stress and it may even lead to necrosis (i.e., tissue death). Besides, the reproductive development of plants for is also severely hampered and chilling temperatures during anthesis often lead to sterility of flowers (Jiang et al. 2002).

The major symptoms of chilling injury in vegetable crops are surface lesions, water soaked tissues, breakdown of tissue, dehydration, shrivelling or desiccation, internal discolouration, slow, uneven or non ripening of fruits, enhanced ethylene production and quicker senescence, shorter shelf life and lesser keeping quality, changes in nutrient and biochemical composition, weakened capacity of sprouting or growth, wilting and leakage of plant metabolites leading to rot and decay, aided

Table 3 Impact of temperature on quality of vegetable

Vegetable crop	Impact of temperature on quality
Tomato	Optimum temperature for colour development is 10–25 °C; at temperature lesser than 10 °C neither red nor yellow colour develop; when temperature is more than 30 °C red colour development stops but it resumes when temperature reverts back to 30 °C and lower; lycopene pigment forming mechanism is destroyed at 40 °C resulting in no red colour development, and only development of yellow shoulders; scalding of fruits is also observed under high temperature
Hot pepper	Capsaicin content of fruit is elevated under high night temperature
Cauliflower	High temperature causes riceyness (formation of small buds over the curd); yellowing of curds besides loose and leafy curd is observed on exposure to high temperature
Cabbage	Compact heads are not formed at high temperature
Carrot	Carotene synthesis decreases at temperature below 15 °C or above 24 °C, thus reducing the colour intensity
Beet	Ideal temperature for roots with high sugar content and dark internal colour is 15–18 °C; above 25 °C alternate white and coloured rings are seen in roots when sliced (referred to as zoning)
Radish	High temperature makes root tough and pungent before it reaches edible size
Turnip	Above 25 °C roots become tough and bitter in taste
Pea	At 30 °C and above peas become tough, lose the tenderness; elevated temperature promotes enzymatic conversion of starch to sugar and hence sugar content is less thus reducing its quality
Palak	Above 25 °C early bolting occurs and leaf succulence decreases quickly
Spinach	At high temperature (above 25 °C) leaves become yellow, contain less sugar, succulence decreases quickly
Lettuce	Solid head, crispy leaves are produced under 18/20 °C night/day temperature; above 22 °C quality of head and leaves is reduced
Melons	High day temperature preceding cool night temperature enhances sugar content in fruits and thus increases sweetness

Source: Hazra and Som (2006) and Akhtar et al. (2015)

by secondary infection of micro organisms, especially fungal pathogen (Lukatkin et al. 2012 cross ref. Skog 1998).

The most common symptoms of chilling injury that we can see in vegetables are wilting of hypocotyls and leaves (Mitchell and Madore 1992; Frenkel and Erez 1996), followed by appearance of water soaked regions (McMahon et al. 1994; Sharom et al. 1994), thereafter appearance of pitted surface and big cavities (Dodds and Ludford 1990; Cabrera et al. 1992; Frenkel and Erez 1996), discoloured leaves and internal tissues (Sharom et al. 1994; Yoshida et al. 1996; Tsuda et al. 2003), quicker aging and senescence, rupture of tissues, and slow, partial or irregular ripening (Dodds and Ludford 1990), disintegration of the structure and loss of flavour (Harker and Maindonald 1994; Ventura and Mendlinger 1999); becoming more vulnerable to incidence of pathogen thus leading to rotting (Cabrera et al. 1992), withering of the margins or tips of leaf blades (Hahn and Walbot 1989). Prolonged chilling may even lead to leaf necrosis and death of the plant (Mitchell and Madore 1992; Frenkel and Erez 1996).

A typical effect of low temperature on sensitive plants is slow growth and is more distinct in susceptible species and varieties (Ting et al. 1991; Rab and Saltveit 1996; Venema et al. 1999). Besides, development is hindered and as a result the crop season is lengthened (Skrudlik and Koscielniak 1996). On the other hand, differentiation of the apical cone is late, thus the number and frequency of occurrence of the newly formed plant organs is decreased, besides alteration in the the root structure and flowering, fruitset and seed production are hampered (Buis et al. 1988, Barlow and Adam 1989; Rab and Saltveit 1996; Skrudlik and Koscielniak 1996; Lejeune and Bernier 1996) (Table 4).

4.3 Moisture Stress

Water is the basic requirement for life, be it plants or animals. Availability of this precious resource in our planet is limited, whereas the uses have no bounds. Therefore, it is not surprising that the amount of water for use in agriculture becomes extremely less. Vegetable crops typically require higher quantities of water than the agronomic crops. Hence for improvement of water use efficiency, reduction in irrigation cost, maintaining the acreage of the vegetable crops and most importantly to conserve the precious water, utilization of drought tolerant cultivars is the need of the hour (Mou 2011). On the other hand vegetables are also susceptible to

Table 4 Vegetables with their lowest safe temperatures and chilling injury symptoms

Vegetable crop	Lowest safe temperature (°C)	Symptoms of chilling injury
Tomato (ripe)	7–10	Water-soaked lesions, softening and rot, decay
Tomato (mature-green)	13	Improper colour development on ripening, fruit rot (particularly, *Alternaria*)
Eggplant	7	Surface scald, *Alternaria* rot, seed blackening
Pepper	7	Pitting of fruit surface, *Alternaria* rot, blackening of seeds
Potato	2	Browning, conversion of starch to sugar
Sweet potato	10	Pitting of surface, rotting, internal discoloration
Okra	7	Discoloration, water-soaked areas, pitting, decay
Bean (snap)	7	Pitting and russeting
Cucumber	7	Pitting of surface, formation of water-soaked lesions, fruit rot
Pumpkin	10	Rotting of fruits particularly due to *Alternaria*
Squash	10	Rotting of fruits particularly due to *Alternaria*
Asparagus	0–2	Dull gray-green coloured spears, tips are wilted

Source: Lukatkin et al. (2012)

waterlogging and excess moisture stress. Hence genotypes tolerant to high moisture in the soil has to be developed.

4.3.1 Water Deficit (Drought)

Deficit in precipitation for a long term is termed as meteorological drought while soil moisture deficit that leads to reduction in growth, development and yield of crops is referred to as agricultural drought. Yield loss up to 50% or more can be affected due to drought in the agricultural areas of the world and precipitation is inadequate for most agricultural uses in about 35% of the global land surface which are deemed as arid or semiarid (Bahadur et al. 2011). Vegetables, in general, are tender and succulent with a moisture content of 90% or higher. They require regular supply of water for their proper growth, development and quality. Deficit of water suppresses the growth and development of vegetables and markedly reduces the productivity and quality of the crops. However, plant reaction to the stress exerted by drought depends chiefly on the intensity and duration of the drought stress, the type of vegetable and the stage of the crop growth. When the stress occurs at the critical period of growth, it may radically shrink the productivity and diminish the quality of vegetables (Table 5).

4.3.2 Excess Moisture Stress (Waterlogging)

Waterlogging occurs due to heavy rainfall, inadequate drainage systems, natural flooding etc. About 10% of the total irrigated agricultural areas in the world suffers from regular waterlogging, leading to reduced productivity by about 20% (Biswas and Kalra 2018). The impacts of waterlogging are observed as growth reduction and chlorosis of older leaves (Ellington 1986) because of impaired physiological activities. Waterlogging checks gaseous exchange in the root zone that causes potential to damage roots, hampering the routine uptake of water and mineral. Improper development of roots lead to lesser uptake of nitrogen from the anaerobic soil thus resulting in chlorosis of older leaves (Belford et al. 1992). The harmful effect of waterlogging on crop depends on the growth stage of the crop, and the frequency and duration of the waterlogging event(s) (Johnston 1999). The greatest detrimental effect is observed on crop, particularly, yields when it occurs at germinating or early vegetative stages (Cannell et al. 1980). The prevalence of soil borne fungal diseases enhances under flooding conditions (Yanar et al. 1997). The fungal colonies attacks the germinating seeds more readily and wilting caused by *Phytophthora*, damping off caused by *Pythium* often takes place (Walker 1991), although doubts still exist whether the microbial infection or the direct effects of flooding are the major causes of the setback (Davison 1991). Oxygen levels diminish very rapidly in waterlogged root environment, creating anoxia condition which is detrimental to the growing root tips and enhanced by shortage of ATP since plants exhibit metabolic switch from aerobic respiration to anaerobic fermentation, resulting in reduced rate of

Table 5 Critical stages of water application in vegetables and its impact of water deficit

Vegetable crop	Critical stage of moisture requirement	Effect of water deficit
Tomato	Flowering and period of rapid fruit enlargement	Flower drop; reduced or no fertilization; small size of fruits; cracking of fruits; pave way for development of blossom end rot (BER)
Eggplant	Flowering and fruit development	Improper and faded colour development; lesser yield
Hot and sweet pepper	Flowering and fruit set	Flower and fruit drop; reduced nutrient uptake; lesser dry matter production
Potato	Tuberization and tuber enlargement	Improper tuber growth; tuber splitting; diminished yield
Carrot, radish and turnip	Root enlargement	Improper growth of roots; distorted and tough roots; accumulation of harmful nitrates in roots; development of strong and pungent odour in carrot
Onion	Bulb formation and enlargement	Splitting and doubling of bulbs; reduced storage life
Cabbage and cauliflower	Head/curd formation and enlargement	Tip burning and splitting of head in cabbage; browning and buttoning in cauliflower
Cucumber	Flowering as well as throughout fruit development	Distorted and non-viable pollen grains; deformed and bitter fruits
Melons	Flowering and evenly throughout fruit development	Diminished TSS, reducing sugars and ascorbic acid leading to reduced fruit quality in muskmelon; enhancement of nitrate content in watermelon
Okra	Flowering and pod development	Fibrous pods; reduced yield; increased infestation of mites
Pea	Flowering and pod filling	Decreased nodulation of roots; improper plant growth; improper grain filling
Lettuce	Consistently throughout development	Toughened leaves; stunted plant growth; tip burning
Leafy vegetables	Throughout growth and development of plant	Toughened leaves; improper leaf growth; nitrate accumulation

Source: Bahadur et al. (2011)

energy production by 65–97%. Several toxic secondary metabolites are produced from anaerobic metabolism of roots under waterlogged conditions, as has been observed in pea (Biswas and Kalra 2018). The flooding also hinders the diffusive escape and/or oxidative breakdown of gases such as ethylene or carbon dioxide resulting in its accumulation that hampers root growth and severely damages the roots of different species. This is combined with the oxidative damage of the reactive oxygen species. Deficiency of essential nutrients like nitrogen, magnesium, potassium, calcium is created by waterlogging that results in adverse effects on several physiological and biochemical processes of plants. Plants with surface-inhabiting root systems are comparatively tolerant to waterlogging for longer duration.

4.4 Salinity

About 20% of the total global irrigated lands (approximately, 45 million hectares) are distressed by soil salinity, which is a key adverse factor in crop production (Zhu 2001). Climate changes and the existing irrigation practices are likely to increase salinity over time (Rengasamy 2010; Munns and Tester 2008). Salinity is a major problem for sustainable agriculture. The evapotranspiration rates is enhanced by the ever rising temperature leading to increase in the need of the irrigation water and hence increase the amount of salts in the soil. This situation is particularly alarming in the arid and semi-arid regions of the world. Apart from this, there is often intrusion of saline sea water into farmlands around coastal areas which is expected to increase with the rising sea level, thus enhancing salinity problems. Growth in plants is reduced by salinity by processes that are either due to the accumulation of salt in the stem, or due to processes independent of this (Roy et al. 2014). The effects are measured immediately (within minutes to a few days) upon addition of salt (without providing any time for salt to accumulate in the shoot) or much later (several days to weeks), providing sufficient time to salt for accumulation in the shoot and affecting shoot growth. Soon after plants are affected by salt there are rapid response of the plants, stomatal closure and associated leaf temperature enhancement (Sirault et al. 2009), and inhibition of shoot elongation (Munns and Passioura 1984; Rajendran et al. 2009), being two of the best documented consequences. Hence, new leaves production is highly reduced causing significant reduction in shoot growth. In the second phase of plant responses to salinity, slower onset of growth inhibition occurs that encompasses several days to weeks, due to salt accumulation over time, particularly in the older leaves, resulting in premature senescence of those older leaves. This is referred to as the 'ionic phase' of salt toxicity. It is the resultant of both accumulation of salts, and the inability of the shoot to tolerate the higher concentration of salt accumulated to the level of toxicity (Munns and Tester 2008). Growers require salinity tolerant genotypes to cope up and sustain vegetable production in these areas.

5 Utilization of Diverse Genotypes

Adaptation to the climate change is necessity to meet the food and nutritional demand of the ever growing population. Sustainable vegetable production under such circumstances requires multi-disciplinary research efforts. Studies of stress physiology determine the mechanisms of tolerance and identify traits for screening for tolerance thus leading to the approaches and methodology to be adopted. Studies of genomics and molecular biology is necessary for understanding the structural organization and functional properties of these stress linked traits, their variation at genetic level, selection based on molecular markers and high throughput genotyping. Plant breeders utilize these findings for developing stress tolerant varieties using conventional and biotechnological methods.

The diminishing genetic diversity of vegetable crops, narrow genetic base of different crops and monoculture are the limiting factors for breeding genotypes for adapting to future environments (Mou 2011). A collection of diverse germplasm is the basis of any vegetable improvement programme. Collection, characterization, management, maintenance and preservation of the germplasm and utilization for recombination of genes for broadening of the genetic base is necessary for combating climate change.

5.1 Development of Tolerant/Resistant Genotypes

5.1.1 Conventional Breeding for Development of Tolerance

Combating climate change will be possible by developing genotypes that will be able to tolerate the vagaries of the climate. Breeders aim at developing high yielding genotypes having tolerance to extreme climatic conditions. Low heritability of different traits under selection in vegetables makes it a herculean task to develop genotypes resilient to climate change (Byerlee and Moya 1993; Trethowan et al. 2002). Novel genes for abiotic and biotic stress tolerance may be identified which would lead to elite varieties adapted to wider range of climatic conditions. Hybridization of tolerant lines may result superior combinations of alleles at multiple loci which may be identified and with proper selection, elite lines may be developed which may be used as varieties or pre-breeding lines for the targeted stress in crop improvement programmes. Often the source of these traits are wild relatives that can survive in stress conditions which is detrimental for growth of the popular high yielding cultivated varieties, being exposed to variable climate in their natural habitat, and through proper techniques the genes responsible for these tolerance traits may be identified and further utilized. Through conventional breeding few high yielding genotypes with stress tolerance have been developed, but the ultimate genetic and physiological bases behind these are still not clear. Phenotypic or biochemical or molecular markers would make the selection process easier but marker development based on the available information is challenging. Very narrow genetic base of vegetable crops is another limitation for developing tolerant genotype (Ladizinsky 1985; Paran and Van Der Knaap 2007). Diverse breeding approaches may be used to widen the genetic base of the crops exploiting the biodiversity and different gene pools.

The environmental condition of the location of the breeding programme should not be different from that for which the genotype is being bred and the selection carried out in the similar climatic condition will be effective (Mickelbart et al. 2015). Considering the fact that that cultivars from warmer regions are often more heat-tolerant than those from cooler regions, this option seems to be the more feasible one (Smillie and Nott 1979; Momonoki and Momonoki 1993; Tonsor et al.

2008; Yamamoto et al. 2011; Kugblenu et al. 2013). Breeders should focus on traits related to the specific stress tolerance and such passive selection would result in high yielding tolerant variety.

5.1.1.1 Traits Targeted for Heat Tolerance

- Extensive root system for efficient uptake of water and nutrients from the soil.
- Brief span of life cycle of crop that enables minimising the temperature effect on plant.
- Leaf and stem pubescence that repels direct sunrays and which creates partial shady condition for cell wall and cell membrane.
- Diminished size of leaf where number of stomata is reduced leading to lesser evapotranspiration.
- Orientation of leaf to increase photosynthetic activity for heat tolerance.
- Glossy and waxy leaves for repelling sunlight.
- Yield *per se*

5.1.1.2 Traits Targeted for Chilling Tolerance

- Thin highly pubescent stem.
- Small and narrow leaflets, highly pubescent leaves and also fruits.
- Deep and extensive root system for coping up with water stress
- High photosynthetic rate and
- Capability of seedling to survive at zero and sub-zero temperatures.
- Yield *per se*

5.1.1.3 Traits Targeted for Drought Stress

Drought escape

- Earliness

Drought Avoidance Characteristics

- Reduced transpiration
- Stomatal sensitivity
- Osmotic adjustment
- Cuticular wax
- Abscisic acid
- Leaf pubescence
- Leaf angle
- Leaf movement
- Leaf rolling

- Increased water uptake
- Deep root system
- Root length density
- Root hydraulic resistance

Drought Tolerance Characteristics

- Maintenance of membrane integrity
- Seedling survival
- Seedling rooting pattern
- Seed germination
- Translocation of stem reserves
- Proline accumulation
- Yield *per se*

5.1.1.4 Traits Targeted for Waterlogging Stress

- Plant growth
- Root, shoot and total biomass
- Root porosity
- Seedling survival
- Yield *per se*

5.1.1.5 Traits Targeted for Salinity Stress

- Cell survival
- Seed germination
- Dry matter accumulation
- Senescence and number of dead leaves
- Leaf ion content
- Leaf necrosis
- Root growth
- Osmoregulation
- Yield *per se*

5.2 Identification of Tolerant Lines for Use in Breeding Programmes

Screening of genotypes for stress tolerance and climate resilience is the first step towards development of varieties. Several varieties have been identified and developed which can act as climate resilient ones and tolerate the vagaries of nature.

Besides, traits have to be identified that may help in tolerance of any particular stress and selection based on these traits should be done.

5.2.1 Heat and Cold Tolerant Genotypes

Numerous temperature tolerant genotypes have been developed in vegetables, like tomato, peppers, cole crops, etc. However, the limitation with the stress tolerant varieties is their low yield. High yields with tolerance in cultivars can be achieved through the broadening of the genetic base through crosses between tolerant tropical lines and disease-resistant temperate varieties (Opena and Lo 1981). Word Vegetable Centre, Taiwan, has made significant contributions in this direction and particularly high temperature tolerant tomato and Chinese cabbage lines adapted to hot and humid climate have been developed. But the heat tolerant genotypes, in general, have lower yield potential which is a limitation, and the breeding goal should be developing tolerant varieties with yield potential at par with high yielding popular varieties. The biodiversity of the crops must be explored for identification of new tolerant sources and use in the breeding work (Metwally et al. 1996). Heat tolerant landraces of USA and Philippines have been used for developing heat tolerant breeding lines (Opena et al. 1992), while primitive cultivars of Turkey have been used as sources for cold resistance. Different accessions of wild related species have been used as sources of tolerance to extremities in temperature, for example LA-1777, LA3921 and LA3925, all accessions belonging to *Solanum habrochaites* being sources of chilling tolerance.

Early maturation in broccoli is targeted for improved head quality since it results in escape of high temperature later in season which causes flower initiation (Farnham and Bjorkman 2011). Higher temperature and longer days during reproductive phase led to higher average grain yield in new varieties of cowpea (Ehlers and Hall 1998). In potato reports of genetic gain for heat tolerance and increase in yield up to 37.8% after three cycles of recurrent selection could be attained (Benites and Pinto 2011).

The screening for high or low temperature stress may be carried out under natural field environment, in field condition in offseason to expose the plants to a climatic condition adverse from its normal regime, or in controlled environment in greenhouses, or even *in vitro* for certain assays (Singh 2015). For heat stress, seed germination, growth of the plant, pollen fertility, and flower, fruit, seed, etc. formation, membrane stability, photosynthetic sensitivity, and yield should be the selection criteria. However, during heat stress screening care should be taken that drought is not imposed and there should be regular supply of moisture to the plants. For chilling stress, the selection criteria are somewhat similar with addition of chlorophyll loss under stress conditions that may be observed through the leaf or seedling colour. When the stress goes to the level of freezing, survival of plants during freezing, recovery and regrowth after freezing and osmoregulation following freeze-hardening measured through sugar content may be emphasized upon (Table 6).

Table 6 Various sources of stress tolerance in vegetable crops

Crop	Stress	Tolerant sources	References
Tomato	Heat	VC11-3-1-8, VC11-2-5, Divisoria-2, (landraces from the Philippines); Tamu Chico III, PI289309 (landraces from United States)	Opena et al. (1992)
Tomato	Heat	CL5915	Metwally et al. (1996)
Tomato	Cold	PI-120256, a primitive tomato from Turkey	
Tomato	Cold	LA-1777 (*Solanum habrochaites*) from AVRDC, Taiwan	
Tomato	Cold	*Lycopersicon hirsutum* LA3921 and LA3925, both *Solanum habrochaites* from AVRDC, Taiwan	
Tomato	Both high and low temperature	EC-520061 (*Solanum habrochaites*)	
Tomato	Drought	*S. chilense*	Rick (1973) and Maldonado et al. (2003)
			Sánchez Pena (1999)
Tomato	Drought	*S. pennellii*	O'Connell et al. (2007)
			Rick (1973)
Tomato	Drought	*S. esculentum* var. *cerasiforme, S. cheesmanii, S. habrochaites, S. chilense, S. hirsutum, S. sitiens, S peruvianum, S. habrochaites* (EC-520061), *S. pennelli* (IIHR 14-1, IIHR 146-2, IIHR 383, IIHR 553, IIHR 555, K-14, EC-130042, EC-104395, Sel-28), *S. pimpinellifoloium* (PI-205009, EC-65992, Pan American)	Rai et al. (2011)
Tomato	Drought	Arka Vikas, RF-4A	Singh (2010)
Tomato	Drought	*L. pennellii* (LA0716), *L. chilense* (LA1958, LA1959, LA1972), *S. sitiens* (LA1974, LA2876, LA2877, LA2878, LA2885), *S. pimpinellifolium* (LA1579)	Razdan and Mattoo (2007)
Eggplant	Drought	*S. macrocarpon, S. gilo S. macrosperma, S. integrifolium,* Bundelkhand Deshi	Rai et al. (2011)
Eggplant	Drought	SM-1, SM-19, SM-30, VioletteRound, Supreme	Kumar and Singh (2006)
Eggplant	Drought	*S. sodomaeum (syn. S. linneanum)*	Toppino et al. (2009)
Peppers	Drought	*C. chinense*, *C. baccatum* var. *pendulum*, *C. eximium* Arka Lohit, IIHR – Sel.-132	Singh (2010)
Potato	Drought	*S. acaule*, *S. demissum* and *S. stenotonum,* Alpha, Bintje	Arvin and Donnelly (2008)
Potato	Drought	*S.ajanhuiri, S.curtilobum, S.xjuzepczukii*	Ross (1986)

(continued)

Table 6 (continued)

Crop	Stress	Tolerant sources	References
Potato	Drought	Kufri Sheetman, *Solanum chacoense,* Kufri Sindhuri	Pandey et al. (2007)
Okra	Drought	*A. caillei, A. rugosus, A. tuberosus*	Charrler (1984)
Onion	Drought	*Allium fistulosum, A. munzii,* Arka Kalyan, MST 42, MST 46	Singh (2010)
French bean	Drought	*P. acutifolius*	Kavar et al. (2011)
Water melon	Drought	*Citrullus colocynthis (L.) Schrad*	Dane et al. (2007)
Cucumber	Drought	INGR-98018 (AHC-13)	Rai et al. (2008)
Winter squash	Drought	*Cucurbita maxima*	Chigumira and Grubben (2004)
Cucumis Spp.	Drought	*Cucumis melo var. momordica* VRSM-58, INGR-98015 (AHS-10), INGR-98016 (AHS-82), CU 159, CU 196 Cucumis pubescens, INGR-98013 (AHK-119) *Cucumis melo* var. *callosus,* AHK-200, SKY/DR/RS-101 *Cucumis melo var. chat,* Arya *Cucumis melo,* SC-15	Rai et al. (2008), Kusvuran (2012), Pandey et al. (2011)
Cassava	Drought	CE-54, CE-534, CI-260, CI-308, CI-848, 129, 7, 16, TP White, Narukku-3, Ci-4, Ci-60, Ci-17, Ci-80	Singh (2010)
Sweet potato	Drought	VLS6, IGSP 10, IGSP 14, Sree Bhadra	Singh (2010)
Tomato	Salinity	*S. esculentum* accession (PI174263)	Foolad and Jones (1991)
Tomato	Salinity	LA1579 and LA1606, (*S. pimpinellifolium*) and LA4133 (*S. lycopersicum* var. *cerasiforme*)	Flowers (2004)
Tomato	Salinity	*S. cheesmani, S. peruvianum, S. pennellii, S. pimpinellifolium,* and *S. habrochaites*	Flowers (2004), Foolad (2004), Cuartero et al. (2006)
Pepper	Salinity	Beldi (Tunisian cultivar); Demre, Ilica 250, 11-B-14, Bagci Carliston, Mini Aci Sivri, Yalova Carliston, and Yaglik 28	Yildirim and Guvenc (2006)

Source: Modified from Kumar et al. (2012)

5.2.2 Drought Tolerance

Vegetables are highly drought sensitive crops; however few crops such as eggplant, cowpea, amaranthus, and tomato can tolerate drought to some level. Cultivated vegetable varieties exhibit very less genetic variability for drought tolerance. Wild species are the best sources of resistance to drought tolerance. In tomato, various accessions of wild relatives, *viz. S. chilense*, *S. pennellii*, both of Eriopersicon group having indeterminate growth habit and small sized green coloured fruits, and *S. pimpinellifolium* possess drought tolerance. *S. chilense* possesses longer primary root and more extensive secondary root system than cultivated tomato and finely serrated leaves (Sánchez Pena 1999) and can grow even where other vegetation does not grow and adapted to desert regions also (Rick 1973; Maldonado et al. 2003). On the other hand, *S. pennellii* possesses thick, round, waxy leaves, and produces acylsugars in its trichomes, while its leaves are able to take up dew (Rick 1973). *S. chilense* has been found to be five-times more tolerant to wilting than cultivated tomato, while *S. pennellii* increases its water-use efficiency under drought conditions (O'Connell et al. 2007). In eggplant, *S. gilo*, *S. integrifolium*, *S. macrocarpon*, *S. macrosperma* (Rai et al. 2011); in peppers *C. chinense*, *C. eximum*, *C. baccatum* var. pendulum (Singh 2010); in potato *S. acaule*, *S. demissum*, *S. stenotomum*, *S. curtilobum*, *S. ajanhuiri*, *S. juzepzuckii* (Arvin and Donnelly 2008; Ross 1986); in okra *A. Callei* and *A. rugosus* (Charrler 1984), etc. have been found to be sources of drought tolerance. Transfer of genes from these drought-tolerant lines can improve the tolerance level in the targeted cultivars. However, not all wild species are cross-compatible in nature, therefore intervention of somatic hybridization and embryo rescue techniques should be used.

The screening environment for drought should be very precise having full control not only over water regime, but also all variables viz., homogeneity of site, nutrition availability to crop, disease pest as well as competition of weeds. The best screening may be done under greenhouse condition which is more controlled for natural precipitation. Field environment having low precipitation may also be used for the drought stress condition and the moisture stress may be regulated by supplemented irrigation. The selection criteria for drought avoidance may be leaf rolling, canopy temperature, leaf pubescence and waxiness, leaf water retention and root characteristics. On the other hand, for drought tolerance, seed germination, seedling growth, plant growth, plant phenology in terms of pollen shed, pollen fertility and fruitset, cell membrane stability, water use efficiency and yield should be the selection criteria.

5.2.3 Salt Tolerance

Conventional breeding has been very less successful in the improvement of salt tolerance due to the genetic and physiological complexity of the targeted trait (Flowers 2004). For a successful breeding programme for salt tolerance most important are presence of sufficient genetic variability, possibility of transfer of targeted genes from the donor to the recipient, proper selection pressure and efficient screening techniques. The screening techniques should be sound and screening in field conditions should not be carried out due to variable levels of salinity in soils under natural field conditions. Soil-less culture with nutrient solutions of known salt concentrations should be utilized for the purpose (Cuartero and Fernandez-Munoz 1999). The screening may be carried out in non-saline field with saline condition created with saline irrigation having controlled level of salinity, in microplots, or in controlled techniques like on filter papers, antibiotic agar, pots filled with soil or sand and gravel and in hydroponics. Apart from yield under stress, several other factors such as germination, survival of cell, accumulation of dry matter, senescence, leaf necrosis, ion content of leaf, growth and development of root and osmoregulation measured through accumulation of proline or carbohydrate should be the emphasized on as criteria for selection.

Most of the vegetables are sensitive to salinity conditions, except a few like beet, palak, asparagus, kale, turnip, ash gourd, bitter gourd and lettuce that can tolerate salt to some extent (Hazra and Som 2006). Genetic variation for salt tolerance in various vegetable crops has been identified, but cultivated and commercial varieties offer limited variation and wild species often are found to be sources of salinity tolerance. Wild species have been used to develop mapping populations and also transfer quantitative trait loci (QTLs) have been done along with development of pre-breeding lines that have also been used to interpret the genetic details of salt tolerance. Explanation of the mechanism of salt tolerance at different growth stages of the crop and the introgression of salinity tolerance genes into vegetables would enable development of varieties with high or variable levels of tolerance to salinity and compatible with different production environments.

6 Molecular and Biotechnology Assisted Breeding for Abiotic Stress Tolerance

6.1 Quantitative Trait Loci (QTLs) for Tolerance

The tolerance to the different abiotic stresses (heat, cold, drought, waterlogging, etc.) seems to be governed by polygenes mostly, and thus their genetics is so poorly understood (Wahid et al. 2007; Collins et al. 2008; Ainsworth and Ort 2010). Identification of quantitative trait loci (QTL) in segregating mapping populations are aimed by the researchers for knowledge of genetic governance of the tolerance. Jha et al. (2014) recently listed QTLs associated with heat tolerance in various plants, including vegetables like tomato, potato and cowpea and reported the influence of diverse factors on heat tolerance and variation in QTL between crops. Various markers linked with traits influencing heat tolerance have been reported in different crops that help in selection (Driedonks et al. 2016). QTLs for different traits for heat tolerance *viz.*, higher chlorophyll fluorescence, canopy temperature at different stages of crop growth, etc. have been discovered (Pinto et al. 2010; Lopes et al. 2013; Vijayalakshmi et al. 2010). Heat-tolerant photosynthesis is depicted by high chlorophyll fluorescence, whereas efficient water uptake may be depicted by lower canopy temperature that is correlated with extensive root system (Pinto and Reynolds 2015).

In pea, 161 QTLs located on seven linkage groups have been identified for different traits such as plant height at harvest, number of branches per plant, pod number per plant, seed number per plant, seed number per pod, seed weight per plant, 1000-seed weight, straw dry weight per plant, biomass dry weight per plant, seed protein content, and harvest index under six environments for developing frost tolerant variety (Klein et al. 2014). 10 QTLs, located on linkage groups 1, 3, and 4 were identified by Iglesias-García et al. (2015), which may help in selection of drought-tolerant genotypes in pea.

14 QTLs were detected for drought stress in common bean and mapped on chromosomes 1, 3, 4, 7, 8, and 9 and may be utilized for developing drought-tolerant cultivars (Mukeshimana et al. 2014). In lettuce, a major QTL for seed germination under high temperature, Htg6.1, is linked with a temperature-sensitive gene encoding an abscisic acid biosynthesis enzyme (LsNCED4) (Argyris et al. 2008, 2011). Nine QTLs for internal heat necrosis in tubers were detected in potato that (McCord et al. 2011). In tomato, six QTLs for fruitsetting under high temperature condition were identified (Ventura et al. 2007). Multi-parent advanced generation inter-cross (MAGIC) populations may be even more efficiently utilized to create genetic variation and widen the genetic for heat tolerance related traits (Ye et al. 2015).

With domestication of the crops which may be considered as the first stage of plant breeding, the plant architecture was severely modified (Meyer and Purugganan 2013; Gross and Olsen 2010), and in the process the genetic base of the crops were narrowed down and cultivated varieties retained minimum genes and alleles from its wild progenitor and wild relatives (Godfray et al. 2010; Ladizinsky 1985; Olsen and Wendel 2013). The wild relatives bring rich in genes related to abiotic stress tolerance, the domesticated crop mostly lost these traits (Dolferus 2014; Maduraimuthu and Prasad 2014). Hence, identification and breeding for superior alleles for tolerance became the chief aim of plant breeders (Tanksley and McCouch 1997; Grandillo et al. 2007; Lippman et al. 2007; Feuillet et al. 2008). However, utilization of the wild relatives as a source of novel alleles has ill effects of linked, undesirable traits, and availability of markers for these traits is also scarce (Dolferus 2014). Complete genomes of very limited crops and their wild relatives have been sequenced *viz*., tomato (Aflitos et al. 2014), cucumber (Qi et al. 2013), sorghum (Mace et al. 2013), soybean (Li et al. 2013) among the vegetable crops. Breeders still face hindrance when QTLs crossed in a particular genetic background and show very little or no effect at all. Epistatic interactions often make it difficult or even impossible to predict whether a QTL can be transferred to elite backgrounds or not (Podlich et al. 2004; Collins et al. 2008). Molecular markers based on linked flanking polymorphisms of the QTL can enable successful transfer of a QTL into a germplasm. Therefore, it may be said that QTL analysis and subsequent fine mapping and cloning may be used to identify genes and loci for abiotic stress tolerance. Various candidate genes for heat, chilling, drought, salinity and waterlogging have been proposed, but characterizing the exact causal gene underlying the tolerance QTL still remains challenging.

6.2 Use of Transgenics in Stress Management

A recent approach towards developing abiotic stress tolerance is by creating transgenic plants. Overexpression of osmotin gene in tomato has shown resistance to salinity and drought in controlled conditions showing higher germination, increased proline and relative water capacity (Goel et al. 2010). CaXTH3, a hot pepper xyloglucan endotransglucosylase/hydrolase has been transformed and expressed in tomato showing drought and salinity tolerance in controlled conditions and the seedlings have maintained sufficient chlorophyll even at 100 mM of NaCl (Choi et al. 2011). Chloroplastic BADH gene from *Spinacia oleracea* (SoBADH) has been transformed and expressed in sweet potato that have shown tolerance to low temperature stress, oxidative stress and salinity under growth chamber as evident through maintained cell membrane integrity, photosynthetic activity and Glycine betaine accumulation, (Fan et al. 2012). Transgenic potato with Nucleoside diphosphate kinase 2 (AtNDPK2) gene has been developed which shows salinity and drought tolerance in growth chamber (Tang et al. 2008).

The commercialization of transgenics in vegetables, however, has been minimum, due to direct and fresh consumption of the produce that has led to lesser consumer acceptance. Besides, the transformation and transgene expression varies from crop to crop, which has held back the success in vegetables. There are other issues like the intellectual property rights, market regulation, etc. The researchers are aiming at assessment of transgenics on biodiversity, environment and human health, since transgenics provide an alternate and viable option to cope up with plant stress factors, particularly, abiotic stress.

7 Conclusion

Climate change is a continuous process in today's world. It cannot be stopped but may be deterred. The challenge is production of crops under the changing climate scenario. Apart from different management aspects, identifying traits that confer resistance to climatic vagaries and exploiting genetic variability of the vegetable crops having better performance in this scenario can help in combating the climate change and not compromise the production. The resistant sources being mostly wild species lack the desirable quality of the vegetables and require long time for conventional breeding, hence use of *cisgenics* may be encouraged to improve the resistance traits in crop plants.

References

Aflitos S, Schijlen E, de Jong H, de Ridder D, Smit S, Finkers R, Wang J, Zhang G, Li N, Mao L, Bakker F, Dirks R, Breit T, Gravendeel B, Huits H, Struss D, Swanson-Wagner R, van Leeuwen H, van Ham RC, Fito L, Guignier L, Sevilla M, Ellul P, Ganko E, Kapur A, Reclus E, de Geus B, van de Geest H, Te L, Hekkert B, van Haarst J, Smits L, Koops A, Sanchez-Perez G, van Heusden AW, Visser R, Quan Z, Min J, Liao L, Wang X, Wang G, Yue Z, Yang X, Xu N, Schranz E, Smets E, Vos R, Rauwerda J, Ursem R, Schuit C, Kerns M, van den Berg J, Vriezen W, Janssen A, Datema E, Jahrman T, Moquet F, Bonnet J, Peters S (2014) Exploring genetic variation in the tomato (*Solanum* section *lycopersicon*) clade by whole-genome sequencing. Plant J 80:136–148. https://doi.org/10.1111/tpj.12616

Ainsworth EA, Ort DR (2010) How do we improve crop production in a warming world? Plant Physiol 154:526–530. https://doi.org/10.1104/pp.110.161349

Akhtar S, Naik A, Hazra P (2015) Harnessing heat stress in vegetable crops towards mitigating impacts of climate change (Chapter 10), 173–200. In: Chaudhary ML, Patel VB, Siddiqui MW, Mahdi SS (eds) Climate dynamics in horticultural science, Principles and applications, vol I. Apple Academic Press, Burlington, p 416

Anonymous (2017) Horticultural Statistics at a Glance, Horticultural Statistics Division, Department of Agriculture, Cooperation & Farmers Welfare Ministry of Agriculture & Farmers Welfare Government of India, pp xxix + 481

Argyris J, Dahal P, Hayashi E, Still DW, Bradford KJ (2008) Genetic variation for lettuce seed thermoinhibition is associated with temperature-sensitive expression of abscisic acid, gibberellin, and ethylene biosynthesis, metabolism, and response genes. Plant Physiol 148:926–947. https://doi.org/10.1104/pp.108.125807

Argyris J, Truco MJ, Ochoa O, Mc Hale L, Dahal P, Van Deynze A, Michelmore RW, Bradford KJ (2011) A gene encoding an abscisic acid biosynthetic enzyme (LsNCED4) collocates with the high temperature germination locus Htg 6.1 in lettuce (*Lactuca spp.*). Theor Appl Genet 122:95–108. https://doi.org/10.1007/s00122-010-1425-3

Arvin MJ, Donnelly DJ (2008) Screening potato cultivars and wild species to abiotic, stresses using an electrolyte leakage bioassay. J Agric Sci Technol 10:33–42

Bahadur A, Chatterjee A, Kumar R, Singh M, Naik PS (2011) Physiological and biochemical basis of drought tolerance in vegetables. Vegetable Sci 38(1):1–16

Barlow PW, Adam JS (1989) Anatomical disturbances in primary roots of *Zea mays* following periods of cool temperature. Environ Exp Bot 29(3):323–336

Belford RK, Dracup M, Tennant D (1992) Limitations to growth and yield of cereal and lupin crops on duplex soils. Aust J Exp Agric 32(7):929–945

Benites FRG, Pinto CABP (2011) Genetic gains for heat tolerance in potato in three cycles of recurrent selection. Crop Breed Appl Biotechnol 11:133–140. https://doi.org/10.1590/S1984-70332011000200005

Biswas JC, Kalra N (2018) Effect of waterlogging and submergence on crop physiology and growth of different crops and its remedies: Bangladesh perspectives. Saudi J Eng Technol 3(6):315–329. https://doi.org/10.21276/sjeat.2018.3.6.1

Boyer JS (1982) Plant productivity and environment. Science 218:443–448

Bray EA (2002) Abscisic acid regulation of gene expression during water-deficit stress in the era of the Arabidopsis genome. Plant Cell Environ 25:153–161

Bray EA, Bailey-Serres J, Weretilnyk, E (2000) Responses to abiotic stresses. In: Gruissem W, Buchannan B, Jones R, (eds), Biochemistry and Molecular Biology of Plants. American Society of Plant Physiologists, Rockville, pp 1158–1249

Brown P, Lumpkin T, Barber S, Hardie E, Kraft K, Luedeling E, Rosenstock T, Tabaj K, Clay D, Luther G, Marcotte P, Paul R, Weller S, Youssefi F, Demment M (2005) Global horticulture assessment. ISHS, Gent-Oostakker. 135 pp

Buis R, Barthou H, Roux B (1988) Effect of temporary chilling on foliar and caulinary growth and productivity in soybean (Glycinemax). Ann Bot 61(6):705–715

Byerlee D, Moya P (1993) Impacts of International Wheat Breeding Research in the Developing World. CIMMYT, Mexico, D.F, pp 1966–1990

Cabrera RM, Saltveit ME, Owens K (1992) Cucumber cultivars differ in their response to chilling temperatures. J Am Soc Hortic Sci 117(5):802–807

Cannell RQ, Belford RK, Gales K, Dennis CW, Prew RD (1980) Effects of waterlogging at different stages of development on the growth and yield of winter wheat. J Sci Food Agric 31(2):117–132

Capiati DA, Pais SM, Tellez-Inon MT (2006) Wounding increases salt tolerance in tomato plants: evidence on the participation of calmodulin-like activities in cross-tolerance signaling. J Exp Bot 57:2391–2400. https://doi.org/10.1093/jxb/erj212.

Charrler A (1984) Genetic Resources of the Genus Abelmoschus Med. (Okra). International Board for Plant Genetic Resources; IBPGR Secretariat-Rome. Available online at: http://pdf.usaid.gov/pdf_docs/PNAAT275.pdf.

Chigumira NF, Grubben GJH (2004) *Cucurbita maxima* Duchesne. In: Grubben GJH, Denton OA (eds) PROTA2: vegetables/legumes. PROTA, Wageningen

Choi JY, Seo YS, Kim SJ, Kim WT, Shin JS (2011) Constitutive expression of CaXTH3, a hot pepper xyloglucan endotransglucosylase/hydrolase, enhanced tolerance to salt and drought stresses without phenotypic defects in tomato plants (Solanum lycopersicum cv. Dotaerang). Plant Cell Rep 30(5):879–881

Collins NC, Tardieu F, Tuberosa R (2008) Quantitative trait loci and crop performance under abiotic stress: where do we stand? Plant Physiol 147:469–486. https://doi.org/10.1104/pp.108.118117

Cuartero J, Fernandez-Munoz R (1999) Tomato and salinity. Sci Hortic 78(83):125

Cuartero J, Bolarin MC, Asins MJ, Moreno V (2006) Increasing salt tolerance in tomato. J Exp Bot 57(1045):1058

Dane F, Liu J, Zhang C (2007) Phylogeography of the bitter apple, Citrullus colocynthis. Genet Resour Crop Evol 54:327–336

Datta S (2013) Impact of climate change in Indian horticulture – a review. Int J Scie Environ Technol 2(4):661–671

Davison IR (1991) Environmental effects on photosynthesis: I. Temperature. J Phycol 27:2–8

De La Peña R, Hughes J (2007) Improving vegetable productivity in a variable and changing climate. SAT e-J 4(1):1–22

Dodds GT, Ludford PM (1990) Surface topology of chilling injury of tomato fruit. HortScience 25(11):1416–1419

Dolferus R (2014) Plant science to grow or not to grow: a stressful decision for plants. Plant Sci 229:247–261. https://doi.org/10.1016/j.plantsci.2014.10.002

Driedonks N, Rieu I, Vriezen WH (2016) Breeding for plant heat tolerance at vegetative and reproductive stages. Plant Rep 29:67–79. https://doi.org/10.1007/s00497-016-0275-9

Ehlers JD, Hall AE (1998) Heat tolerance of contrasting cowpea lines in short and long days. Field Crop Res 55:11–21. https://doi.org/10.1016/S0378-4290(97)00055-5

Ellington A (1986) Effects of deep ripping, direct drilling, gypsum and lime on soils, wheat growth and yield. Soil Tillage Res 8:29–49

Fan W, Zhang M, Zhang H, Zhang P (2012) Improved tolerance to various abiotic stresses in transgenic sweet potato (*Ipomoea batatas*) expressing spinach betaine aldehyde dehydrogenase. PLoS One 7:e37344

FAOSTAT (2014) FAO statistical yearbook 2014: Asia and the Pacific Food and Agriculture, Food and Agriculture Organization of the United Nations Regional Office for Asia and the Pacific, Bangkok

Farnham MW, Bjorkman T (2011) Breeding vegetables adapted to high temperatures: a case study with broccoli. HortScience 46:1093–1097

Feuillet C, Langridge P, Waugh R (2008) Cereal breeding takes a walk on the wild side. Trends Genet 24:24–32. https://doi.org/10.1016/j.tig.2007.11.001

Flowers TJ (2004) Improving crop salt tolerance. J Exp Bot 55(307):319

Foolad MR (2004) Recent advances in genetics of salt tolerance in tomato. Plant Cell Tissue Org Cult 76:101–119

Foolad MR, Jones RA (1991) Genetic analysis of salt tolerance during germination in *Lycopersicon*. Theor Appl Genet 81(321):326

Frenkel C, Erez A (1996) Induction of chilling tolerance in cucumber (*Cucumis sativus*) seedlings by endogenous and applied ethanol. Physiol Plant 96(4):593–600

Godfray C, Beddington J, Crute I, Haddad L, Lawrence D, Muir JF, Pretty J, Robinson S, Thomas SM, Toulmin C (2010) Food security: the challenge of feeding 9 billion people. Science 327:812–818. https://doi.org/10.1126/science.1185383

Goel D, Singh AK, Yadav V, Babbar SB, Bansal KC (2010) Overexpression of osmotin gene confers tolerance to salt and drought stresses in transgenic tomato (*Solanum lycopersicum* L.). Protoplasma 245:133–141

Grandillo S, Tanksley SD, Zamir D (2007) Exploitation of natural biodiversity through genomics. In: Genomics-assisted crop improvement, Genomics approaches and platforms, vol 1. Kluwer Academic Press, Dordrecht, pp 121–150. https://doi.org/10.1007/978-1-4020-6295-7_6

Gross BL, Olsen KM (2010) Genetic perspectives on crop domestication. Trends Plant Sci 15:529–537. https://doi.org/10.1016/j.tplants.2010.05.008

Hahn M, Walbot V (1989) Effect of cold-treatment on protein synthesis and mRNA levels in rice leaves. Plant Physiol 91(3):930–938

Hall A (2001) Crop developmental responses to temperature, photoperiod, and light quality. In: Anthony E. Hall (ed) Crop Response to Environment, CRC, Boca Raton, pp 83–87

Harker FR, Maindonald JH (1994) Ripening of nectarine fruit. Changes in the cell wall, vacuole, and membranes detected using electrical impedance measurements. Plant Physiol 106(l):165–171

Hasanuzzaman M, Hossain MA, da Silva JAT, Fujita M (2012) Plant responses and tolerance to abiotic oxidative stress: antioxidant defense is a key factor. In: eds V. Bandi, A. K. Shanker, C. Shanker, and M. Mandapaka (eds) Crop stress and its management: perspectives and strategies, Springer, Berlin, pp 261–316

Hasanuzzaman M, Nahar K, Alam MM, Roychowdhury R, Fujita M (2013) Physiological, biochemical, and molecular mechanisms of heat stress tolerance in plants. Int J Mol Sci 14:9643–9684. https://doi.org/10.3390/ijms14059643

Hazra P, Som MG (2006) Vegetable science. Kalyani Publishers, New Delhi, India.

Iglesias-García R, Prats E, Fondevilla S, Satovic Z, Rubiales D (2015) Quantitative trait loci associated to drought adaptation in pea (*Pisum sativum* L.). Plant Mol Biol Rep 33:1768. https://doi.org/10.1007/s11105-015-0872-z

Jha UC, Bohra A, Singh NP (2014) Heat stress in crop plants: its nature, impacts and integrated breeding strategies to improve heat tolerance. Plant Breed 133:679–701. https://doi.org/10.1111/pbr.1221

Jiang QW, Kiyoharu O, Ryozo I (2002) Two novel mitogen-activated protein signaling components, OsMEK1 and OsMAP1, are involved in a moderate low-temperature signaling pathway in Rice1. Plant Physiol 129:1880–1891

Johnston (1999) Subsoil drainage for increased productivity on duplex soils. Final report for the Grains Research and Development Corporation. Department of Natural Resources and Environment, Agriculture Victoria, Rutherglen

Jouyban Z, Hasanzade R, Sharafi S (2013) Chilling stress in plants. Int J Agricult Crop Sci 5(24):2961–2968

Kavar T, Maras M, Meglie V (2011) The expression profiles of selected genes in different bean species as response to water deficit. J Central European Agric 12:557–576

Klein A, Houtin H, Rond C, Marget P, Jacquin F, Boucherot K, Huart M, Rivière N, Boutet G, Lejeune-Hénaut I, Burstin J (2014) QTL analysis of frost damage in pea suggests different mechanisms involved in frost tolerance. Theor Appl Genet 127:1319–1330

Kugblenu YO, Oppong Danso E, Ofori K, Andersen MN, Abenney-Mickson S, Sabi E, Plauborg F, Abekoe MK, Ortiz R, Jørgensen ST (2013) Screening tomato genotypes in Ghana for adaptation to high temperature. Acta Agricult Scand Sect B Soil Plant Sci 63:516–522. https://doi.org/10.1080/09064710.2013.813062

Kumar R, Singh M (2006) Citation information genetic resources. In: Singh RJ (ed) Chromosome engineering, and crop improvement vegetable crops, vol 3. CRC Press, Boca Raton, pp 473–496

Kumar R, Solankey SS, Singh M (2012) Breeding for drought tolerance in vegetables. Vegetable Sci 39(1):1–15

Kusvuran S (2012) Effects of drought and salt stresses on growth, stomatal conductance, leaf water and osmotic potentials of melon genotypes (*Cucumis melo* L.). Afr J Agric Res 7(5):775–781

Ladizinsky G (1985) Founder effect in crop-plant evolution. Econ Bot 39:191–199

Lejeune P, Bernier G (1996) Effect of environment on the early steps of ear initiation in maize (*Zea mays* L.). Plant Cell Environ 19(2):217–224

Levitt J (1980) Responses of plants to environmental stresses, Chilling, freezing and high temperatures stresses, vol 1. Academic, New York. 426 p

Li Y, Zhao S, Ma J, Li D, Yan L, Li J, Qi X, Guo X, Zhang L, He W, Chang R, Liang Q, Guo Y, Ye C, Wang X, Tao Y, Guan R, Wang J, Liu Y, Jin L, Zhang X, Liu Z, Zhang L, Chen J, Wang K, Nielsen R, Li R, Chen P, Li W, Reif JC, Purugganan M, Wang J, Zhang M, Wang J, Qiu L (2013) Molecular footprints of domestication and improvement in soybean revealed by whole genome re-sequencing. BMC Genomics 14:579. https://doi.org/10.1186/1471-2164-14-579

Lippman ZB, Semel Y, Zamir D (2007) An integrated view of quantitative trait variation using tomato inter specific introgression lines. Curr Opin Genet Dev 17:545–552. https://doi.org/10.1016/j.gde.2007.07.007ent

Lopes MS, Reynolds MP, McIntyre CL, Mathews KL, Jalal Kamali MR, Mossad M, Feltaous Y, Tahir ISA, Chatrath R, Ogbonnaya F et al (2013) QTL for yield and associated traits in the Seri/Babax population grown across several environments in Mexico, in the West Asia, North Africa, and South Asia regions. Theor Appl Genet 126:971–984. https://doi.org/10.1007/s00122-012-2030-4

Lukatkin AS, Brazaitytė A, Bobinas C, Duchovskis P (2012) Chilling injury in chilling-sensitive plants: a review. Žemdirbystė=Agriculture 99(2):111–124

Mace ES, Tai S, Gilding EK, Li Y, Prentis PJ, Bian L, Campbell BC, Hu W, Innes DJ, Han X, Cruickshank A, Dai C, Frère C, Zhang H, Hunt CH, Wang X, Shatte T, Wang M, Su Z, Li J, Lin X, Godwin ID, Jordan DR, Wang J (2013) Whole-genome sequencing reveals untapped genetic potential in Africa's indigenous cereal crop sorghum. Nat Commun 4:2320. https://doi.org/10.1038/ncomms3320

Maduraimuthu D, Prasad PVV (2014) High temperature stress. In: Jackson M, Ford-Lloyd BV, Perry ML (eds) Plant genetic resources and climate change. CABI, Wallingford, pp 201–220

Maldonado C, Squeo FA, Ibacache E (2003) Phenotypic response of *Lycopersicon chilense* to water deficit. Rev Chil Hist Nat 76(129):137

McCord PH, Sosinski BR, Haynes KG, Clough ME, Yencho GC (2011) QTL mapping of internal heat necrosis in tetraploid potato. Theor Appl Genet 122:129–142. https://doi.org/10.1007/s00122-010-1429-z

McMahon MJ, Permit AJ, Arnold JE (1994) Effects of chilling on Episcia and dieffenbachia. J Am Soc Hortic Sci 119(1):80–83

Metwally E, El-Zawily A, Hassan N, Zanata O (1996) Inheritance of fruit set and yields of tomato under high temperature conditions in Egypt. First Egypt-Hungary Horticult Conf I:112–122

Meyer RS, Purugganan MD (2013) Evolution of crop species: genetics of domestication and diversification. Nat Rev Genet 14:840–852. https://doi.org/10.1038/nrg3605

Mickelbart MV, Hasegawa PM, Bailey-Serres J (2015) Genetic mechanisms of abiotic stress tolerance that translate to crop yield stability. Nat Rev Genet 16:237–251. https://doi.org/10.1038/nrg3901

Mitchell DE, Madore MA (1992) Patterns of assimilate production and translocation in muskmelon (*Cucumis melo* L.). 2. Low temperature effects. Plant Physiol 99(3):966–971

Momonoki YS, Momonoki T (1993) Changes in acetylcholinehydrolyzing activity in heat-stressed plant cultivars. Biosci Biotechnol Biochem 62:438–446. https://doi.org/10.1248/cpb.37.3229

Mou B (2011) Improvement of horticultural crops for abiotic stress tolerance: an introduction. HortScience 46(8):1068–1069

Mukeshimana G, Butare L, Cregan PB, Blair MW, Kelly JD (2014) Quantitative trait loci associated with drought tolerance in common bean. Crop Sci 54:923–938. https://doi.org/10.2135/cropsci2013.06.0427

Munns R, Passioura JB (1984) Hydraulic resistance of plants: effects of NaCl in barley and lupin. Aust J Plant Physiol 11:351–359

Munns R, Tester M (2008) Mechanisms of salinity tolerance. Annu Rev Plant Biol 59:651–681

Nullis C (2018) IPCC issues special report on global warming of 1.5 °C every bit of warming matters. WMO Bull 67(2):4–7

O'Connell MA, Medina AL, Sanchez Pena P, Trevino MB (2007) Molecular genetics of drought resistance response in tomato and related species. In: Razdan MK, Mattoo AK (eds) Genetic improvement of solanaceous crops, Tomato, vol 2. Science Publishers, Enfield, pp 261–283

Olsen KM, Wendel JF (2013) A bountiful harvest: genomic insights into crop domestication phenotypes. Annu Rev Plant Biol 64:47–70. https://doi.org/10.1146/annurev-arplant-050312-120048

Opena RT, Lo SH (1981) Breeding for heat tolerance in heading Chinese cabbage. In: Talekar NS, Griggs TD (eds) Proceedings of the 1st international symposium on Chinese cabbage. AVRDC, Shanhua

Opena RT, Chen JT, Kuo CG, Chen HM (1992) Genetic and physiological aspects of tropical adaptation in tomato. In: Adaptation of food crops to temperature and water stress. AVRDC, Shanhua, pp 321–334

Pandey SK, Naik PS, Sud KC, Chakrabarti SK (2007) CPRI – perspective plan vision 2025. Central Potato Research Institute, Shimla, pp 1–70

Pandey S, Ansari WA, Jha A, Bhatt KV, Singh B (2011) Evaluations of melons and indigenous *Cucumis spp.* genotypes for drought tolerance. In: 2nd international symposium on underutilized plant species, 27th June–1st July. The Royal Chaulan, Kuala Lumpur. 95 pp

Paran I, Van Der Knaap E (2007) Genetic and molecular regulation of fruit and plant domestication traits in tomato and pepper. J Exp Bot 58:3841–3852. https://doi.org/10.1093/jxb/erm257

Pinto RS, Reynolds MP (2015) Common genetic basis for canopy temperature depression under heat and drought stress associated with optimized root distribution in bread wheat. Theor Appl Genet 128:575–585. https://doi.org/10.1007/s00122-015-2453-9

Pinto RS, Reynolds MP, Mathews KL, McIntyre CL, Olivares-Villegas JJ, Chapman SC (2010) Heat and drought adaptive QTL in a wheat population designed to minimize confounding agronomic effects. Theor Appl Genet 121:1001–1021. https://doi.org/10.1007/s00122-010-1351-4

Podlich DW, Winkler CR, Cooper M (2004) Mapping as you go. An effective approach for marker assisted selection of complex traits. Crop Sci 44:1560–1571

Qi J, Liu X, Shen D, Miao H, Xie B, Li X, Zeng P, Wang S, Shang Y, Gu X, Du Y, Li Y, Lin T, Yuan J, Yang X, Chen J, Chen H, Xiong X, Huang K, Fei Z, Mao L, Tian L, Stadler T, Renner SS, Kamoun S, Lucas WJ, Zhang Z, Huang S (2013) A genomic variation map provides insights into the genetic basis of cucumber domestication and diversity. Nat Genet 45:1510–1515. https://doi.org/10.1038/ng.2801

Rab A, Saltveit ME (1996) Sensitivity of seedling radicles to chilling and heat shock-induced chilling tolerance. J Am Soc Hortic Sci 121(4):711–715

Rai M, Pandey S, Kumar S (2008) In: Pitrat M (ed) Cucurbitaceae: proceedings of the IXth EUCARPIA meeting on genetics and breeding of Cucurbitaceae, INRA, Avignon (France), May 21–24, pp 285–293

Rai N, Tiwari SK, Kumar R, Singh M, Bharadwaj DR (2011) Genetic resources of Solanaceous vegetables in India. In: National symposium on vegetable biodiversity. Jawaharlal Nehru Krishi Vishwa Vidyalaya, Jabalpur, pp 91–103

Raison JK, Lyons JM (1986) Chilling injury: a plea for uniform terminology. Plant Cell Environ 9:685–686

Raj Narayan (2009) Air pollution-a threat to vegetable production. International Conference on Horticulture (ICH-2009): horticulture for livelihood security and economic growth, Bangalore, Karnataka, November, 2009, pp 158–159
Rajendran K, Tester M, Roy SJ (2009) Quantifying the three main components of salinity tolerance in cereals. Plant Cell Environ 32:237–249
Razdan MK, Mattoo AK (2007) Genetic improvement of solanaceous crops: tomato, vol 2. Science Publishers, Oakville, p 47
Rengasamy P (2010) Soil processes affecting crop production in salt-affected soils. Funct Plant Biol 37:613–620
Rick CM (1973) Potential genetic resources in tomato species: clues from observation in native habitats. In: Srb AM (ed) Genes, enzymes and populations. Plenum Press, New York, pp 255–269
Ross H (1986) Potato breeding: problems and perspectives. J Plant Breeding, Supplement 13:1–132
Roy SJ, Negra S, Tester M (2014) Salt resistant crop plants. Curr Opin Biotechnol 26:115–124. https://doi.org/10.1016/j.copbio.2013.12.004
Sánchez Pena P (1999) Leaf water potentials in tomato (*L. esculentum* Mill.) *L chilense* Dun. and their interspecific F1. M. Sc. Thesis, New Mexico State University, Las Cruces, NM, USA
Sharom M, Willemot C, Thompson JE (1994) Chilling injury induces lipidphase changes in membranes of tomato fruit. Plant Physiol 105(1):305–308
Singh HP (2010) Ongoing research in abiotic stress due to climate change in horticulture, Curtain Raiser Meet on research needs arising due to abiotic stresses in agriculture management in India under global climate change scenario, Baramati, Maharashtra, October 29–30, 1–23pp. http://www.niam.res.in/pdfs/DDG_Hort_lecture.pdf
Singh BD (2015) Plant breeding: Principles and methods. Kalyani Publishers, New Delhi, India
Sirault XRR, James RA, Furbank RT (2009) A new screening method for osmotic component of salinity tolerance in cereals using infrared thermography. Funct Plant Biol 36:970–977
Skog LJ (1998) Chilling injury of horticultural crops. Ontario Ministry of Agriculture, Food and Rural Affairs Factsheet. http://www.omafra.gov.on.ca/english/crops/facts/98–021.htm. Accessed 11 May 2011
Skrudlik G, Koscielniak J (1996) Effects of low-temperature treatment at seedling stage on soybean growth, development and final yield. J Agron Crop Sci 176(2):111–117
Smillie RM, Nott R (1979) Heat injury in leaves of alpine, temperate and tropical plants. Funct Plant Biol 6:135–141
Spaldon S, Samnotra RK, Chopra S (2015) Climate resilient technologies to meet the challenges in vegetable production. Int J Current Research Academic Rev 3:28–47
Tang L, Kim MD, Yang KS, Kwon SY, Kim SH, Kim JS, Yun DJ, Kwak SS, Lee HS (2008) Enhanced tolerance of transgenic potato plants overexpressing nucleoside diphosphate kinase 2 against multiple environmental stresses. Transgenic Res 17:705–715
Tanksley SD, McCouch SR (1997) Seed banks and molecular maps: unlocking genetic potential from the wild. Science 277:1063–1066. https://doi.org/10.1126/science.277.5329.1063
Ting CS, Owens TG, Wolfe DW (1991) Seedling growth and chilling stress effect on photosynthesis in chilling-sensitive and chilling tolerant cultivars of *Zea mays*. J Plant Physiol 137(5):559–564
Tonsor SJ, Scott C, Boumaza I, Liss TR, Brodsky JL, Vierling E (2008) Heat shock protein 101 effects in A. thaliana: genetic variation, fitness and pleiotropy in controlled temperature conditions. Mol Ecol 17:1614–1626. https://doi.org/10.1111/j.1365-294X.2008.03690.x
Toppino L, Acciarri N, Mennella G, Lo Scalzo R, Rotino GL (2009) Introgression breeding in eggplant (*Solanum melongena* L.) by combining biotechnological and conventional approaches. In: Proceedings of the 53rd Italian Society of Agricultural Genetics Annual Congress Torino, Italy, 16/19 Sept
Trethowan RM, van Ginkel M, Rajaram S (2002) Progress in breeding wheat for yield and adaptation in global drought affected environments. Crop Sci 42:1441–1446
Tsuda H, Niimura Y, Katoh T (2003) Chill injury in Saintpaulia leaf with special reference to leaf spot formation. J Agricult Sci Tokyo Nogyo Daigaku 47(4):283–289

Vanitha SM, Chaurasia SNS, Singh PM, Naik PS (2013) Vegetable ststistics. Technical bulletin no. 51. ICAR-Indian Institute of Vegetable Research, Varanasi, p 2

Venema JH, Posthumus F, De Vries M, Van Hasselt PR (1999) Differential response of domestic and wild *Lycopersicon species* to chilling under low light: growth, carbohydrate content, photosynthesis and the xanthophyll cycle. Physiol Plant 105(1):81–88

Ventura Y, Mendlinger S (1999) Effects of suboptimal low temperature on yield, fruit appearance and quality in muskmelon (*Cucumis melo* L.) cultivars. J Hortic Sci Biotechnol 74(5):602–607

Ventura G, Grilli G, Braz LT, Gertrudes E, Lemos M (2007) QTL identification for tolerance to fruit set in tomato by fAFLP markers. Crop Breeding Appl Biotechnol 7:234–241

Vijayalakshmi K, Fritz AK, Paulsen GM, Bai G, Pandravada S, Gill BS (2010) Modeling and mapping QTL for senescence-related traits in winter wheat under high temperature. Mol Breed 26:163–175. https://doi.org/10.1007/s11032-009-9366-8

Wahid A, Gelani S, Ashraf M, Foolad M (2007) Heat tolerance in plants: an overview. Environ Exp Bot 61:199–223. https://doi.org/10.1016/j.envexpbot.2007.05.011

Walker GE (1991) Chemical, physical and biological control of carrot seedling diseases. Plant Soil 136(1):31–39

Wilson JM (1985) The economic importance of chilling injury. Outlook on Agric 14:197–203. https://doi.org/10.1177/003072708501400407

WMO (2011) World Meteorological Organization statement on the status of the global climate in 2010, WMO Publication No. 1074. WMO, Geneva

Yamamoto K, Sakamoto H, Momonoki YS (2011) Maize acetylcholinesterase is a positive regulator of heat tolerance in plants. J Plant Physiol 168:1987–1992. https://doi.org/10.1016/j.jplph.2011.06.001

Yanar Y, Lipps PE, Deep IW (1997) Effect of soil saturation duration and soil water content on root rot of maize caused by Pythium arrhenomanes. Plant Dis 81(5):475–480

Ye C, Tenorio FA, Argayoso MA, Laza MA, Koh HJ, Redona ED, Jagadish KS, Gregorio GB (2015) Identifying and confirming quantitative trait loci associated with heat tolerance at flowering stage in different rice populations. BMC Genet 16:41. https://doi.org/10.1186/s12863-015-0199-7

Yildirim E, Guvenc I (2006) Salt tolerance of pepper cultivars during germination and seedling growth. Turk J Agric For 30(347):353

Yoshida R, Kanno A, Sato T, Kameya T (1996) Cool temperature induced chlorosis in rice plants. I. Relationship between the induction and a disturbance of etioplast development. Plant Physiol 110:997–1005

Zhu JK (2001) Plant salt tolerance. Trends Plant Sci 6:66–71

Response of Solanaceous Vegetables to Increasing Temperature and Atmospheric CO_2

Durga Prasad Moharana, Ramesh Kumar Singh, Sarvesh Pratap Kashyap, Nagendra Rai, D. R. Bhardwaj, and Anand Kumar Singh

1 Introduction

The climate change has made its strong footprint in the world to the scientific consensus. The foremost reason for this havoc is the emission of greenhouse gases mainly through the anthropogenic (human-caused) activities (Cook et al. 2016). The climate system is on the verge of facing a long-term irreversible impact in all of its components. Global warming, certainly, is one of the serious threats to our ecosystems (IPCC 2014). There will be further warming of the globe with the continued greenhouse gas emissions. The last few decades have witnessed a slow but steady escalation in the global temperature (Fig. 1) as well as atmospheric CO_2 concentrations (Fig. 2). In this chapter, the consequences of rising temperature and elevated atmospheric CO_2 (eCO_2) on plants in general and Solanaceous vegetables in particular are elaborately discussed.

D. P. Moharana
Department of Horticulture, Institute of Agricultural Sciences, Banaras Hindu University, Varanasi, Uttar Pradesh, India

Division of Crop Improvement, ICAR-Indian Institute of Vegetable Research, Varanasi, Uttar Pradesh, India

R. K. Singh (✉) · S. P. Kashyap · N. Rai · D. R. Bhardwaj
Division of Crop Improvement, ICAR-Indian Institute of Vegetable Research, Varanasi, Uttar Pradesh, India

A. K. Singh
Department of Horticulture, Institute of Agricultural Sciences, Banaras Hindu University, Varanasi, Uttar Pradesh, India

S. S. Solankey et al. (eds.), *Advances in Research on Vegetable Production Under a Changing Climate Vol. 1*, Advances in Olericulture,
https://doi.org/10.1007/978-3-030-63497-1_4

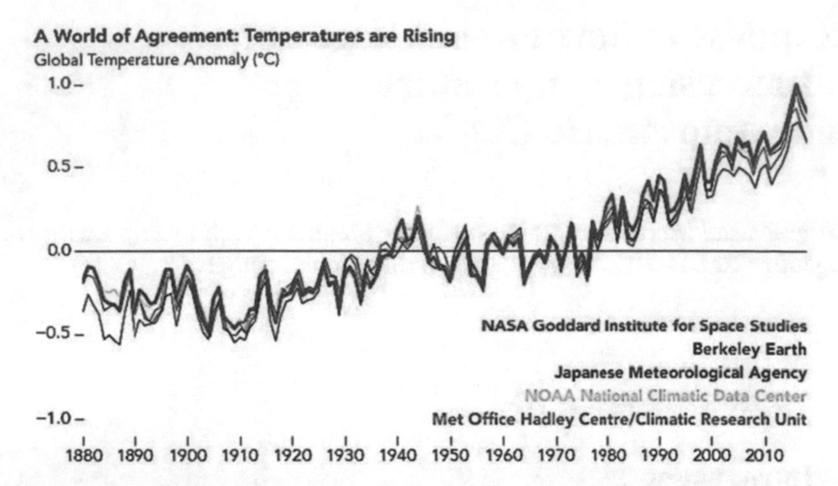

Fig. 1 Temperature data showing rapid warming of the globe. (Source: https://climate.nasa.gov/scientific-consensus/)

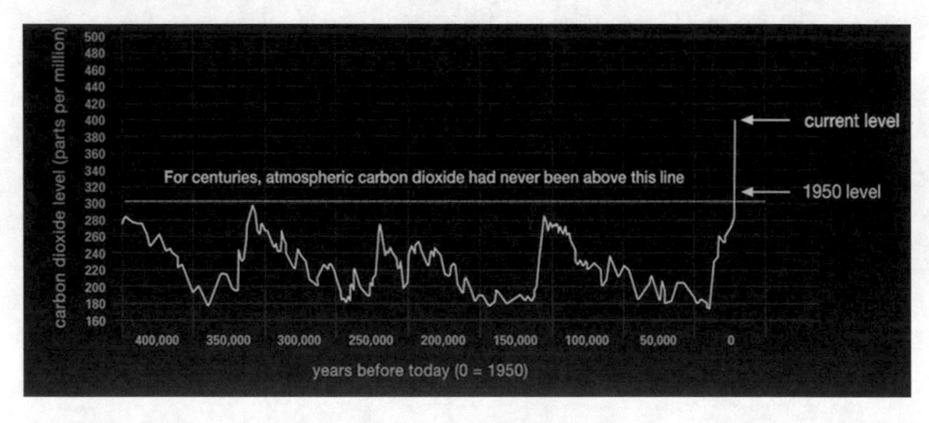

Fig. 2 The relentless rise of atmospheric carbon dioxide. (Source: https://climate.nasa.gov/evidence/)

2 Impact of Rising Temperature on Plants

According to IPCC (2007), about 2–3 °C change in temperature is predicted over the coming 30–50 years. Temperature plays a pivotal role in plant growth and development. Each plant species requires a defined range of temperature for completion of its life cycle. With the rise in temperature to the optimum level, the vegetative development is higher as compared to reproductive development. During the reproductive phase of the plant, the extremely high temperature influences the fertilization, pollen viability, and fruit/grain production (Hatfield et al. 2011). Yield potential can be significantly hampered if there are extreme temperatures during the pollination, fruit set, or initial fruit growth stage as well as the reproductive stage. Plant's cardinal temperature requirements decide its yield response towards the extreme temperatures. The rise in temperature changes the vapour pressure deficit at the leaf surface resulting in increased transpiration. Hastening of foliage aging and shortening of plant growing season (e.g., shorter grain-filling period) are the major detrimental effects of higher canopy temperature (Van de Geijn and Goudriaan 1996). Enzymes are highly influenced by temperature and play a significant role in carrying out different biochemical reactions inside the plant cells. The plant may not function properly or even dies if any one of the essential enzymes fails. For this reason, most of the plant species can survive the high temperature up to a relatively narrow range, i.e., 40–45 °C (Senioniti et al. 1986). The temperature, causing the inhibition of various cellular functions in C_3 species in the cool season, may not apparently influence the warm-season C_3 species like rice, etc. and C_4 species, viz., sugarcane, maize, sorghum, etc. (Abrol and Ingram 1996). Abrupt exposure to high temperature may lead to membrane injury, loss of cellular contents, disruption of cellular functioning, or even death of plant (Ahrens and Ingram 1988). In higher plant species, heat stress results in the reduced synthesis of normal proteins while increased production of a new set of proteins, viz., heat shock proteins (HSPs) is accompanied (Wang et al. 2014). Phenological development of photosensitive crops like soybean is also likely to be prominently disturbed due to the high temperature. Throughout the twenty-first century, about 2.5–10% of yield loss may occur across numerous crop species (Hatfield et al. 2011). As compared to annual crops, perennial crops possess a more complicated relationship with temperature. Exposing the apple plants to the high temperature stress (>22 °C) improved the fruit size and soluble solids but the firmness, a desirable quality, reduced (Warrington et al. 1999).

3 Elevated Atmospheric CO_2 Levels and Plant Responses

The concentration of atmospheric CO_2 is increasing at an alarming rate. The average atmospheric level of CO_2 persistently rose from 315 parts per million (ppm) in 1959 to about 409.78 ppm in 2019 (https://www.esrl.noaa.gov/gmd/ccgg/trends/global.html). Such a higher level of CO_2 is not only likely to have a profound effect on the global climate system but also imparts a substantial direct effect on growth, development, and various physio-chemical processes of plants (Ziska 2008).

CO_2 is the base of all life forms on the earth. Plants utilize CO_2 as the basic raw material for building up their tissues, which consequently become the ultimate source of food for all animals, including humans. A number of research outcomes have established the fact that elevated CO_2 level in the atmosphere leads to better plant growth (Singer and Idso 2009). Long back in 1804, de Saussure, for the first time demonstrated better growth in pea plants that are exposed to the increased CO_2 concentrations as compared to the control plants in ambient air. An upsurge of about 33% in agricultural yield with CO_2 enrichment has been reported (Kimball 1983). There is an upsurge in the photosynthetic carbon fixation rate by leaves in response to the elevated CO_2 concentrations in the atmosphere. Towards the rising CO_2

concentrations, plants experience and response differently in an open-field condition as compared to the closed chamber system. Keeping this in mind, free-air carbon dioxide enrichment (FACE) technique was developed in the late 1980s. This method allows the researchers to raise the CO_2 concentrations in an isolated area and measure the resultant response of the plants.

Ainsworth and Rogers (2007) reported about 40% increase in the leaf photosynthetic rates in a range of plant species at an increased CO_2 level of 475–600 ppm. Over the last 35 years, a substantial amount of greenery has been noticed in a quarter to half of the global vegetated lands due to the escalation in the level of atmospheric CO_2 that ultimately led to the increase in leaves on the plants causing greening (Zhu et al. 2016).

3.1 CO_2 Fertilization Effect

It is a well-believed perception that the C_4 plant species (maize, sugar cane, sorghum, millets, etc.) are likely to be less responsive towards elevated CO_2 than the plant species following the C_3 photosynthetic pathway (potatoes, rice, cotton, wheat, barley, etc.). But, in a twenty-year field experiment, this C_3-C_4 elevated CO_2 paradigm is surprisingly reversed as during the last 8 years of the experiment, a significant enhance in the biomass was observed in C_4 instead of the C_3 plants (Reich et al. 2018). This contradictory finding may be because of the availability of nitrogen to the C_3 plants in a lesser amount than that of the C_4 plants with passing time. So, not only the plants but also the soil chemistry and microbes play an important role in getting this astonishing outcome.

3.2 CO_2 Anti-Transpirant Effect

The atmospheric CO_2 also influences the crop plant in another imperative manner. Higher levels of atmospheric CO_2 lead to the decrease of water lost through transpiration in the plants, thus increasing the water-use efficiency. This is due to the contraction and/or decrease in the number of tiny pores, i.e., stomata present in the leaves through which plants transpire (Wolfe and Erickson 1993; Kimball 2011; Deryng et al. 2016).

4 Solanaceous Vegetables

Solanaceae or the nightshade family consists of about 98 genera and 2700 species (Olmstead and Bohs 2006). The economically important edible members of this family are tomato (*Solanum lycopersicum* L.), potato (*Solanum tuberosum* L.), eggplant (*Solanum melongena* L.), and pepper (*Capsicum annuum* L.). This family also includes a wide array of plant species like tobacco, cape gooseberry, henbane, climbing nightshade, belladonna, mandrake, Jimson weed, petunia, etc. belonging to the diverse groups based on their nature (Rubatzky and Yamaguchi 1997).

5 Response of Solanaceous Vegetables Towards Rising Temperature and Elevated Atmospheric CO_2

Climate change, caused mainly by the anthropogenic greenhouse gas emissions, has disrupted the ecosystem and these greenhouse gases (CFCs, N_2O, CH_4, CO_2, etc.) are the major culprits in the depletion of ozone layer. Vegetable crops under the family Solanaceae have occupied a prime position in the world of vegetables. With the changing scenario vis-à-vis climate systems, along with other agricultural crops, these crops are also significantly influenced. Solanaceous vegetables are basically warm season crops requiring an optimum temperature of 20–27 °C for better growth and development except for potato which requires a cooler climate. The consequences of high temperature (Table 1) and elevated CO_2 are briefly discussed with special reference to important members of the Solanaceous vegetable group.

5.1 Tomato

At the optimum temperature range, *viz.*, 21–24 °C, the tomato plant growth achieves the peak of the sigmoid curve. Deviation from this range will impart a detrimental influence on the growth and development of plant. The exposure of tomato plants to short periods of high temperature affect more severely if coincide with the critical plant growth phase (Geisenberg and Stewart 1986; Haque et al. 1999; Araki et al. 2000). Both day and night temperature play significant role in fruit yield and quality

Table 1 High-temperature injury symptoms in solanaceous vegetables

Vegetable	Injury symptoms
Tomato	Flower drop and underdevelopment of ovaries, no fruit setting beyond 35 °C day temperature, interruption of lycopene synthesis in fruits; sunscald and blotchy ripening in the affected tissues
Potato	Reduction or complete inhibition of tuberization, reducing sugar contents, physiological weight loss of tuber, bacterial wilt, black scurf/canker and black heart disorder
Hot and sweet pepper	Decrease in pollen production, reduced fruit set, reduced seed set, smaller fruit size, premature loss of fruits with sunburn necrosis, blossom end rot on the fruits, poor ripening, and color development of fruits.
Brinjal	Poor fruit and seed set, distorted floral buds and fruits, decrease in pollen production, yellow, bronze, or brown spot on the fruit due to sunburn.

(Iwahori and Takahashi 1964; Abdalla and Verkerk 1968; Kuo et al. 1979; Hann and Hernandez 1982). However, within a certain range, tomato plants have the potential to integrate temperature. Tomato plants exposed to a constant temperature regime may suffer a yield loss whereas a fluctuating temperature regime often does not impart any significant effect on yield (Adams et al. 2001; de Koning 1988, 1990). Both the vegetative and reproductive phases are considerably influenced by high temperature which ultimately affects the fruit yield and quality (Figs. 3, 4 and 5). The number of days for seed germination in tomato is decreased in response to the elevated temperature. Temperature does not have any pivotal impact on dry matter partitioning (Heuvelink 1995) while the fruit ripening can be hindered due to the extreme temperatures (Lurie et al. 1996). In the experiments conducted by Hurd and Graves (1984, 1985), it is demonstrated that there is a decrease in the time taken for fruit maturity in the initial part of the season. The reason may be due to the higher mean temperature in the early phase of the season. There is a positive correlation between fruit temperature of 10–30 °C and fruit growth rate, with an increase of 5 μm h^{-1} $°C^{-1}$ in the fruit diameter (Pearce et al. 1993). A shorter crop production time is the outcome of higher temperature during the plant growth, but with lower yield and small-sized fruits (Rylski 1979a, b; Sawhney and Polowick 1985).

The plant developmental rate and timing of first flowering are affected by the temperature differences during vegetative phase. The fruit firmness, development time, and yield are also significantly influenced by the timing, duration, and magnitude of short-term temperature pulses (Adams and Valdés 2002; Mulholland et al. 2003). For successful tomato production, an average daily temperature of 29 °C during the two-week period up to the opening of flowers has been considered as the critical temperature (Deuter et al. 2012). The constant air temperatures of ≥30 °C can hinder the normal ripening and softening of detached mature green tomato fruits (Mitcham and McDonald 1992). If the temperature surpasses 32.2 °C during

Fig. 3 Effect of high temperature on tomato plants (**a**) susceptible (pot 1) vs. heat-tolerant (pot 2) tomato plant, and (**b**) Field view of susceptible (yellow encircled) and heat-tolerant (white encircled) tomato plants. (Source: NICRA project, ICAR-Indian Institute of Vegetable Research, Varanasi)

Fig. 4 Response of tomato varieties (pot 1: Punjab Chhuhara, pot 2: CLN-1621, and pot 3: H-88-78-1) to high temperature; exposure period (**a**) 0 h (**b**) 16 h (**c**) 32 h, and (**d**) 48 h. (Source: ICAR-Indian Institute of Vegetable Research, Varanasi)

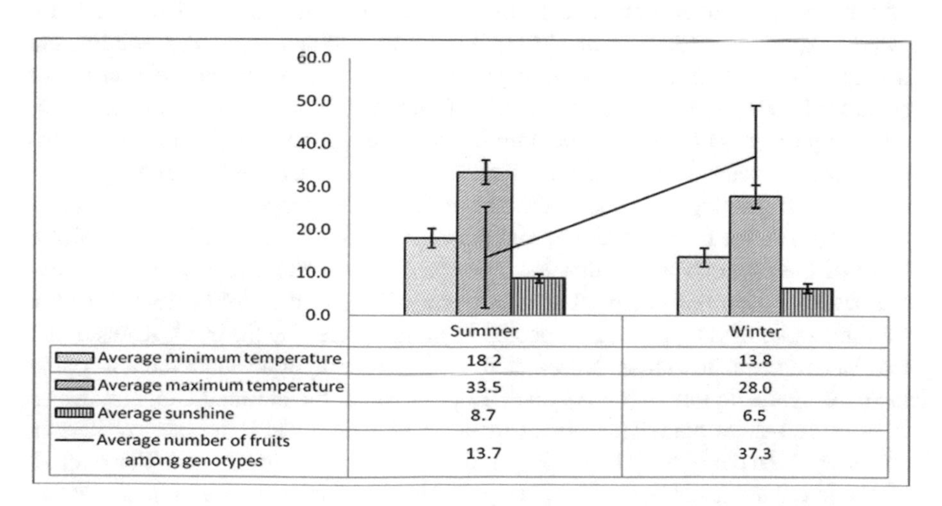

	Summer	Winter
Average minimum temperature	18.2	13.8
Average maximum temperature	33.5	28.0
Average sunshine	8.7	6.5
Average number of fruits among genotypes	13.7	37.3

Fig. 5 Effect of temperature on fruit set in tomato during winter and summer seasons. (Source: Kumar et al. 2017)

the critical stages of flowering and pollination, the yield is severely affected as the fruit set is reduced and the fruits become smaller in size and of poor quality (Sato et al. 2001). At elevated temperatures, the flower clusters emerge faster (Adams et al. 2001), and consequently, more number of fruits per plant appear initially (Fig. 3). Vegetative growth is penalised due to the growth of these fruits, however a delay in growth and development of newly set fruits, or flower or fruit abortion may occur (De Koning 1989; Kumar et al. 2017). A single factor cannot be the sole reason for poor fruit set in tomato at elevated temperature as it is considered as a complex trait (Rudich et al. 1977; Prendergast 1983).

Temperature alone or with other environmental parameters influences both the vegetative and reproductive phases of tomato. Several consequences of high temperatures on tomato plants include unnatural flower development, bud drop, persistent flower and calyx, anther splitting, poor anther dehiscence, degeneration of embryo sac and endosperm, scanty pollen production, reduced stigma receptivity and fertilization, low pollen viability, ovule abortion, decreased carbohydrate availability and protein content, reduction in number of seeds per fruit and fruit size, and other developmental abnormalities (Hazra et al. 2007). Sato et al. (2002) discussed the developmental modifications in anthers, especially anomalies in the endothecium and epidermis, poor pollen formation, and the problem of strontium opening are the resultant of elevated temperature during the pre-anthesis stage. There are also reports of splitting of stigma, antheridial cone, and stylar exsertion (Rudich et al. 1977; Levy et al. 1978; El Ahmadi and Stevens 1979). Peet et al. (1998) observed the harmful effect of heat stress on development of ovule and embryo, and ovule viability. The pollination is restricted as the stigma dehydrated due to high temperature. A 2–4 °C increase from the optimal temperature imparted detrimental effect on gamete development and supressed the capability of pollinated flowers to become seeded fruits (Peet et al. 1997; Sato et al. 2001; Firon et al. 2006). Seven to fifteen days prior to anthesis is the critical phase of sensitivity to moderately elevated temperatures (Sato et al. 2002). Under moderately high temperature, the decrease in the fruit set is not primarily due to scanty pollen production rather because of poor pollen release and viability (Sato et al. 2006). The high temperature-tolerant genotypes produced higher number of pollen grains than the sensitive genotypes (Abdelmageed et al. 2003). So, this criterion can be useful in picking heat-tolerant genotypes. In cherry tomato, reduction in lycopene and starch content is observed as the fruit temperature increased by about 1 °C (Gautier et al. 2005). Most of the research works that have targeted the thermo-effect on fruit quality attributes were primarily on postharvest ripening (Dalal et al. 1968; Lurie et al. 1996).

Tomato, being a C_3 plant, is expected to be influenced by the eCO_2 levels under the changing climate system. Nilsen et al. (1983) have reported about the upsurge in the photosynthetic rate in tomato cv. Virosa, grown at the elevated CO_2 concentrations of 500–2000 ppm. There is an increase of early yield in tomato by 15% in response to the elevated CO_2 of 900 $\mu mol\ mol^{-1}$ with additional light (Fierro et al. 1994). In tomato (cv. Arka Ashish), a significantly higher plant height, leaf area per plant, and stem dry mass were recorded when the plants are exposed to 550 ppm of CO_2 whereas for the parameters like number of branches, leaves per plant, leaf dry mass, and total dry mass, 700 ppm CO_2 was found to be superior. Moreover, CO_2 at 700 ppm concentrations have demonstrated highest photosynthetic rate, number of flowers, fruits per plant, fruit set (%), and fruit yield per plant as compared to CO_2 at 500 ppm and the control (Mamatha et al. 2014). Nutritional quality of tomato is also significantly influenced by increased CO_2 levels. According to Wei et al. (2018), eCO_2 has resulted in a substantial increase in nitrate content in tomato fruits. The total antioxidant capacity, phenols, and flavonoids have decreased to a greater degree at 550 μmol mol^{-1} CO_2. However, ascorbic acid concentration increased notably at 700 μmol mol^{-1} (Mamatha et al. 2014). At early fruiting phase, the sucrose content of fruits has increased greatly due to the elevated CO_2 as compared to the later stage of fruiting (Islam et al. 1996). There has been a mismatch between fruit colour and maturity as eCO_2 exposure of plants leads to the increase in the synthesis of colour pigments in fruits but up to a lower degree than the total solids and soluble sugar synthesis (Zhang et al. 2014). As the sugar concentration in tomato fruits gradually increases from green to red stage, the elevated CO_2 thus resulted in higher accumulation of soluble sugar (Winsor et al. 1962). Khan et al. (2013) demonstrated the acceleration of maturity along with the promotion of fiber and soluble sugar accumulation in *cv.* Eureka. In normal nitrogen availability condition, eCO_2 had negatively affected the lycopene, soluble sugar, and soluble solids content in the fruits whereas in higher nitrogen availability, their concentrations are promoted (Helyes et al. 2012). Under salt-stress conditions (7 dS m^{-1}), there was an upsurge in the fruit yield due to the elevated CO_2 levels while the other quality attributes like acidity, total soluble solids, and total soluble sugar remain constant (Li et al. 1999). The effect of eCO_2 on lycopene content is inconsistent, possibly due to the thermo-sensitivity of lycopene (Krumbein et al. 2012), so no significant impact has been established (Dong et al. 2018).

5.2 *Potato*

The potato has occupied the third most important food crop position after rice and wheat, and is considered as the most important non-grain crop in the world. Potato performs well under cool climatic conditions and is highly affected by elevated temperatures at different stages of its life cycle (Levy and Veilleux 2007). For net photosynthesis, the lowest, optimum, and highest temperatures reported are 0–7 °C, 16–25 °C, and 40 °C, respectively (Kooman and Haverkort 1995). The establishment stage is affected by the temperature, particularly soil temperature. There are alterations in the morphological attributes of the plant like smaller size of compound leaves and leaflets leading to the reduction in leaf area index (Ewing 1997; Fleisher et al. 2006). A linear correlation is observed between the leaf appearance rate and temperature range (9–25 °C), and beyond 25 °C, no subsequent increase is detected (Kirk and Marshall 1992). Moreover, Benoit et al. (1983) also reported 25 °C to be the optimum temperature for leaf expansion. Light interception reduces as the leaves cannot undergo full expansion in response to the increasing temperature. A reduced specific leaf area is reported in the varieties grown in hot climatic conditions (Midmore and Prange 1991). Therefore, there is a linear rise in the leaf expansion up to 24 °C, but at 35 °C, a linear decline is observed (Kooman and Haverkort 1995). Similarly, a linear relationship between temperature and stem elongation is noted up to 35 °C (Manrique 1990) and is accelerated by low night and high day temperatures (Moreno 1985). As yield is a resultant of light use efficiency and intercepted radiation, yield is penalized at elevated temperatures owing to the declined ground-cover duration that is positively interlinked with yield (Vander Zaag and Demagante 1987). High temperatures strongly suppress the tuber formation, reduce the fitness of seed tubers, lessen the shelf life of potato tubers, lower the amount of assimilated carbon partitioned to tuber starch, and hastens the leaf senescence (Menzel 1985; Ewing 1981; Fahem and Haverkort 1988; Wolf et al. 1991; Hancock et al. 2014; Sonnewald et al. 2015). Warmer temperature below 21 °C was found to accelerate the tuber initiation (Kooman et al. 1996). Moderately elevated temperature can also lead to severe tuber yield reduction without any considerable effect on total biomass and photosynthesis (Peet and Wolfe 2000). Moreover, at higher temperature, the translocation of biomass production is also restricted. Above two-thirds of the total photosynthates were translocated to the tubers at 18 °C, however hardly 50% translocated at 28 °C (Randeni and Caesar 1986) which suggests that the shoot portion of the plant is more favored for assimilates over the tubers resulting higher growth of haulm and restricted tuber production. High temperature also significantly restricts the tuber bulking rate in potato (Struik 2007) which may be due to the relationship with the hindered sugar to starch conversion (Krauss and Marschner 1984). The sprout growth is also inhibited by temperatures above the optimum (Midmore 1984) as a result of subapical necrosis (McGee et al. 1986). Kim et al. (2017) revealed a contrasting result of substantial decrease of 11% of tuber yield per degree of temperature rise in the range of 19.1–27.7 °C than the reported value of about 3–4.6% reduction per 1 °C temperature increase in the range of 13.81–25.45 °C (Peltonen-Sainio et al. 2010; Fleisher et al. 2017). At elevated temperatures, the marketable tubers become smaller in size due to the decreased sink strengths of tubers (Geigenberger 2003; Baroja-Fernández et al. 2009). At elevated temperature, there is a higher utilization of assimilated carbohydrate for respiration, thus resulting in reduced tuber formation (Hijmans 2003). During a study on potato cvs. Kufri Surya (heat tolerant) and Kufri Chandramukhi (heat susceptible), a significantly higher rate of transpiration, stomatal conductance, and photosynthesis was observed in Kufri Surya at the higher temperature. Moreover, the rise in temperature led to an upsurge in chlorophyll content in both the cultivars whereas biosynthesis of gibberellic acid was restricted in cv. Kufri Surya (Singh et al. 2015). The CO_2 compensation point and dark respiration rates increased at high temperatures. The net photosynthesis rate has revealed a decline at the high temperature of 40–42 °C or after shifting the plants from the temperature regimes (daytime) of 22–32 °C (Wolf et al. 1990). The combined effect of heat and drought stress that sustained for 14 days has reduced the yield of the tolerant cultivars by about 25% and by over 50% in susceptible cultivars of potato (Rykaczewska 2013). A rise in temperature during the later stages of plant development had a detrimental effect on the sprouting of tubers in the soil prior to harvest. The growth stage of the plant largely influences the thermo-response of the potato cultivars. The earlier the growth stage, higher will be the damage severity with respect to the plant growth and total tuber yield. The physiological defects of tubers and secondary tuberization should also be taken into consideration along with total tuber yield regarding the thermo-tolerance in potato (Rykaczewska 2015). High temperature also affected the infection rate of various diseases of potato. Chung et al. (2006) have reported the highest number of plants infected with potato virus Y-O and potato virus A at 20 °C; and potato leafroll virus at 25 °C. The infestation of *Myzus persicae* (potato peach aphid) is advanced by 2 weeks for each 1 °C increase in the mean temperature. Furthermore, there is a positive correlation between aphid population upsurge; and minimum relative humidity and maximum temperature (Dias et al. 1980; Biswas et al. 2004). Most of the Indian potato varieties are furnished with single specific trait of interest, viz., high yielding, early maturing, high biotic/abiotic stress tolerance/resistance, etc. (Table 2).

Potato plants with short-term exposure to increased CO_2 have demonstrated an increase in the photosynthetic rates (Donnelly et al. 2001a; Vandermeiren et al. 2002). Sicher and Bunce (1999) proposed that the acclamatory reaction to enhanced CO_2 is the resultant of the reduced RuBisCO (Ribulose bisphosphate carboxylase/oxygenase) activity rather than any decline in the leaf content of this protein. Contrarily, Schapendonk et al. (2000) reported that the acclimation is a complicated mechanism caused due to the negative response of sink-source balance induced by high temperature and irradiance. A positive correlation is found between the CO_2 assimilation and concentration. An increase in total biomass by 27–66% is observed by doubling the ambient CO_2 level (Collins 1976; Wheeler et al. 1991; Van De Geijn and Dijkstra 1995; Donnelly et al. 2001a; Olivo et al. 2002; Heagle et al. 2003). At elevated (up to 700 μmol mol^{-1}) and super-elevated (1000–10,000 μmol mol^{-1}) CO_2

Table 2 Trait-specific performance of Indian potato varieties

Variety name	Yield potential (t/ha)	Crop maturity period[a]	Heat tolerance	Drought tolerance	Late blight resistance
Kufri Sindhuri	30–35	Late	High	Medium	Sensitive
Kufri Arun	30–35	Medium	Sensitive	High	High
Kufri Chandramukhi	20–25	Early	Sensitive	High	Sensitive
Kufri Chipsona	30–35	Medium	Sensitive	Medium	High
Kufri Bahar	30–35	Medium	Sensitive	Medium	Sensitive
Kufri Kanchan	25–30	Medium	Sensitive	Medium	Medium
Kufri Surya	25–30	Early	High	Medium	Sensitive
Kufri Megha	25–30	Medium	Sensitive	Medium	High
Kufri Jyoti	25–30	Medium	Sensitive	Medium	Medium
Kufri Khyati	25–30	Early	Sensitive	High	High
Kufri Pukhraj	35–40	Early	Sensitive	High	Medium

Source: Gatto et al. (2016, 2018)
[a]Early: 70–90 days, Medium: 90–100 days, and Late: >110 days

levels, a decrease in the stomatal conductance of potato leaves has been reported (Sicher and Bunce 1999; Lawson et al. 2001; Finnan et al. 2002). It is expected that the reduction in the stomatal conductance enhances the water use efficiency of potato. Olivo et al. (2002) have reported a reduction of 16% in the transpiration rate and increase in the instantaneous transpiration efficiency by 80%. In potato, the leaf chlorophyll content pattern fluctuates in correspondence with the developmental stages of the plants (Finnan et al. 2005). During the later stage of plant growth (after tuber initiation), increased CO_2 level negatively affected the leaf chlorophyll content (Lawson et al. 2001; Bindi et al. 2002). In an open top chamber experiment, an increased CO_2 concentration of 680 ppm resulted in a 40% rise in the light-saturated photosynthetic rate of completely expanded leaves in the upper canopy of cv. Bintje during tuber initiation phase due to the cumulative influence of decrease in the photosynthetic ability and a 12% decline in the stomatal conductance (Vandermeiren et al. 2002). With the exposure to elevated CO_2, the tuber yield is stimulated and the extent of tuber yield is highly dependent on numerous additional factors like growing conditions, agronomy, and cultivar. The starch and dry matter content in potato tubers enhances while the glycoalkaloid and nitrogen content in tubers reduces in response to enriched CO_2 (Finnan et al. 2005). Miglietta et al. (1998) reported about 10% increase rate in the tuber yield for each 100 ppm rise in CO_2 level. The response of potato plant towards the elevated CO_2 varies on the basis of variety and nutrition (Olivo et al. 2002). With optimum supply of nutrients, dry matter yield is significantly influenced by doubling the ambient CO_2 level, whereas, in nitrogen deficit condition, a minor negative response to CO_2 enrichment is noticed (Goudriaan and De Ruiter 1983). Under increased CO_2 concentrations, there are no alterations in the number of tubers; however, improvement was seen in the tuber weight primarily due to the rise in the cell number in tubers without affecting the cell volume (Collins 1976; Donnelly et al. 2001b; Chen and Setter 2003). Contrarily, Miglietta et al. (1998) and Craigon et al. (2002) have reported an increase in the number of tubers. The intensification of starch and soluble sugars in the tubers in response to the elevated CO_2 has resulted in enhanced browning and acrylamide synthesis upon frying (Donnelly et al. 2001b; Kumari and Agrawal 2014). Högy and Fangmeier (2009) revealed a varying effect of eCO_2 on the processing and nutritional quality of potato. Fangmeier et al. (2002) emphasized the need for necessary alterations in fertilizer practices in the upcoming CO_2-rich global climate system with special significance on the quality of potato tubers as the concentration of crude protein was influenced by CO_2 along with O_3. At different water-stress levels, consistency was observed in potato yield enhancement under elevated CO_2 concentrations (Fleisher et al. 2008).

A significant effect on potato production in India is projected in response to the cumulative effect of eCO_2 and high temperature (Table 3).

Table 3 Per cent change in potato production in India as influenced by increased temperature and elevated CO_2 (without adaptations)

	Increase in temperature (°C)					
Atmospheric CO_2 conc. (ppm)	Nil (2009)	1 (2020)	2	3 (2050)	4	5 (2090)
369 (2009)	0.0	−6.27	−17.09	−28.10	−42.55	−60.55
400 (2020)	3.40	−3.16	−14.57	−25.54	−58.63	−58.63
550 (2050)	18.65	11.12	−1.25	−13.72	−30.25	−49.94

Values in parentheses are likely years for associated temperature increase and CO_2 concentrations (Source: Singh et al. 2009)

5.3 Pepper

Temperature possesses a pivotal role in proper growth, flowering, and fruit set in sweet pepper (Rylski and Spigelman 1982; Polowick and Sawhney 1985). Even a short-term exposure (20 min) of pepper plants to high temperature, viz., >40 °C can be detrimental to net photosynthesis rate (Hanying et al. 2001).

The pre-anthesis period in sweet pepper is not sensitive to elevated temperature and there is no effect on stamen or pistil viability. However, at later stages of the plant's life cycle, fertilization is affected by high temperature and a reduction in fruit setting is observed (Erickson and Markhart 2002). Fierro et al. (1994) reported a yield increment of 11% in pepper under the influence of enhanced CO_2 (900 μmol mol^{-1}) with supplementary light (ambient + 100 μmol m^{-2} PAR). Elevated temperatures at the time of flowering are responsible for improper pollen tube growth, faulty germination and fertilization leading to flower abscission and decrease in fruit setting (Usman et al. 1999; Aloni et al. 2001).

The productivity of greenhouse pepper and other C_3 plants increases up to 50% or higher with CO_2 enrichment (Nederhoff 1994; Akilli et al. 2000). With the increase in application duration, the efficiency of CO_2 enrichment has improved. Vafiadis et al. (2012) suggested that the application of CO_2 enrichment is feasible under elevated temperatures with a positive response in terms of yield in pepper. The increased CO_2 level has a significant effect on components of free amino acid in sweet pepper (Piñero et al. 2017). A reduction in nitrogen level in leaves by 10% was observed in comparison with the reference leaves after 58 days under variable CO_2 enrichment conditions (Porras et al. 2017). Under Mediterranean conditions, variable CO_2 enrichment has demonstrated an enhanced production of sweet pepper (Alonso et al. 2010).

5.4 Eggplant

Flowering and fruit setting is eggplant is highly thermo-responsive (Nothmann et al. 1979). Among the Solanaceous vegetables, eggplant is the most thermophilic one (Abak et al. 1996). An increase in stem diameter and plant height in response to enhanced temperatures was observed in eggplant (Pearson 1992; Uzun 1996; Cemek 2002). Cemek et al. (2005) observed a higher plant height in eggplants raised in double polyethylene-cladded greenhouses (having a higher temperature) than that of the single polythene-cladded greenhouses. Fruit set in eggplant reduced as low as 10% in warm Mediterranean regions in response to the cumulative effect of low humidity and high temperatures (Passam and Bolmatis 1997). When the maximum temperature enhanced by 1 °C over the range of 28–34 °C during the first five days of flowering, the rate of fruit setting reduced by 0.83% (Sun et al. 1990).

Fruit yield in the eggplant cv. Cava increased by 13%, 28%, and 18% with concentrations of 0.1, 0.2, and 0.3 g l^{-1} CO_2 in irrigation water, respectively as compared to the control (Aguilera et al. 2000). They also reported the highest fruit weight and fruit yield in 0.3 and 0.2 g l^{-1} dose.

6 Conclusion

Vegetable crops are highly vulnerable towards the climatic vagaries and respond in terms of reduction in production, productivity, and quality. Environmental constraints like increasing temperature and elevated CO_2 have significantly affected the Solanaceous vegetable crops influencing their yield and quality. Thus, the need of the hour is to undertake more research works to identify and/or develop climate-resilient genotypes and advanced technologies to maintain a sustainable production system and safeguard the food as well as nutritional security under the inconsistent global climate system.

References

Abak K, Dasgan HY, Ikiz Ö, Uygun N, Sayalan M, Kaftanoglu O, Yeninar H (1996) Pollen production and quality of pepper grown in unheated greenhouses during winter and the effects of bumblebees (*Bombus terrestris*) pollination on fruit yield and quality. VII International Symposium on Pollination, Cape Town, pp 303–308

Abdalla AA, Verkerk K (1968) Growth, flowering and fruit-set of the tomato at high temperature. Neth J Agric Sci 16:71–76

Abdelmageed AH, Gruda N, Geyer B (2003) Effect of high temperature and heat shock on tomato (*Lycopersicon esculentum* Mill.) genotypes under controlled conditions. Conference on International Agricultural Research for Development, Montpellier, pp 1064–1076

Abrol YP, Ingram KT (1996) Effects of higher day and night temperatures on growth and yields of some crop plants. In: Global climate change and agricultural production: direct and indirect effects of changing hydrological, pedological and plant physiological processes. FAO, Rome, pp 123–140

Adams SR, Valdés VM (2002) The effect of periods of high temperature and manipulating fruit load on the pattern of tomato yields. J Hortic Sci Biotechnol 77:461–466

Adams SR, Cockshull KE, Cave CRJ (2001) Effect of temperature on the growth and development of tomato fruits. Ann Bot 88:869–877

Aguilera C, Murcia D, Ruiz A (2000) Effects of carbon dioxide enriched irrigation on yield of eggplant (*Solanum melongena*) production under greenhouse conditions. In: V International Symposium on Protected Cultivation in Mild Winter Climates: Current Trends for Sustainable Technologies, pp 223–228

Ahrens MJ, Ingram DL (1988) Heat tolerance of citrus leaves. HortScience 23:747–748

Ainsworth EA, Rogers A (2007) The response of photosynthesis and stomatal conductance to rising CO_2: mechanisms and environmental interactions. Plant Cell Environ 30:258–270

Akilli M, Özmerzi A, Ercan N (2000) Effect of CO_2 enrichment on yield of some vegetables grown in greenhouses. In: International conference and British-Israeli Workshop on Greenhouse Techniques towards the 3rd Millennium, pp 231–234

Aloni B, Peet M, Pharr M, Karni L (2001) The effect of high temperature and high atmospheric CO_2 on carbohydrate changes in bell pepper (*Capsicum annuum*) pollen in relation to its germination. Physiol Plant 112:505–512

Alonso FJ, Lorenzo P, Medrano E, Sánchez-Guerrero MC (2010) Greenhouse sweet pepper productive response to carbon dioxide enrichment and crop pruning. In: XXVIII International Horticultural Congress on Science and Horticulture for People, pp 345–351

Araki T, Kitano M, Equchi H (2000) Dynamics of fruit growth and photoassimilation translocation in tomato plant under controlled environment. Acta Hortic 534:85–92

Baroja-Fernández E, Muñoz FJ, Montero M, Etxeberria E, Sesma MT, Ovecka M, Bahaji A, Ezquer I, Li J, Prat S, Pozueta-Romero J (2009) Enhancing sucrose synthase activity in transgenic potato (*Solanum tuberosum* L.) tubers results in increased levels of starch, ADPglucose and UDPglucose and total yield. Plant Cell Physiol 50:1651–1662

Benoit LF, Skelly JM, Moore LD, Dochinger LS (1983) The influence of ozone on Pinus strobus L. pollen germination. Can J For Res 1. https://doi.org/10.1139/x83-025

Bindi M, Hacour A, Vandermeiren K, Craigon J, Ojanperä K, Selldén G, Hogy P, Fibbi L (2002) Chlorophyll concentration of potatoes grown under elevated carbon dioxide and/or ozone concentrations. Eur J Agron 17:319–335

Biswas MK, De BK, Nath PS, Mohasin M (2004) Influence of different weather factors on the population of vectors of potato virus. Ann Plant Protect Sci 12:352–355

Cemek B (2002) Effects of different covering materials on growth, development and yield of crop and environmental conditions inside greenhouses. Unpublished Ph.D. Thesis, Ondokuz Mayis University, Samsum, Turkey

Cemek B, Demir Y, Uzun S (2005) Effects of greenhouse covers on growth and yield of aubergine. Eur J Hortic Sci 70:16–22

Chen CT, Setter TL (2003) Response of potato tuber cell division and growth to shade and elevated CO_2. Ann Bot 91:373–381

Chung BN, Canto T, Tenllado F, San Choi K, Joa JH, Ahn JJ, Kim CH, Do KS (2006) The effects of high temperature on infection by potato virus Y, potato virus A, and potato leaf roll virus. Plant Pathol J 32:321–328

Collins WB (1976) Effect of carbon dioxide enrichment on growth of the potato plant. HortScience 11:467

Cook J, Oreskes N, Doran PT, Anderegg WR, Verheggen B, Maibach EW, Carlton JS, Lewandowsky S, Skuce AG, Green SA, Nuccitelli D (2016) Consensus on consensus: a synthesis of consensus estimates on human-caused global warming. Environ Res Lett 11:048002

Craigon J, Fangmeier A, Jones M, Donnelly A, Bindi M, De Temmerman L, Persson K, Ojanpera K (2002) Growth and marketable-yield responses of potato to increased CO_2 and ozone. Eur J Agron 17:273–289

Dalal KB, Salunkhe DK, Olson LE, Do JY, Yu MH (1968) Volatile components of developing tomato fruit grown under field and greenhouse conditions. Plant Cell Physiol 9:389–400

de Koning ANM (1988) The effect of different day/night temperature regimes on growth, development and yield of glasshouse tomatoes. J Horticult Sci 63:465–471

De Koning ANM (1989) The effect of temperature on fruit growth and fruit load of tomato. International symposium on models for plant growth, environmental control and farm

management in protected cultivation. ISHS Acta Horticulturae 248. https://doi.org/10.17660/Actahortic.1989.248.40
de Koning ANM (1990) Long-term temperature integration of tomato. Growth and development under alternating temperature regimes. Sci Hortic 45:117–127
Deryng D, Elliott J, Folberth C, Müller C, Pugh TA, Boote KJ, Conway D, Ruane AC, Gerten D, Jones JW, Khabarov N (2016) Regional disparities in the beneficial effects of rising CO_2 concentrations on crop water productivity. Nat Clim Chang 6:786–793
Deuter P, White N, Putland D (2012) Critical temperature thresholds case study: tomato. Agriscience, Queensland
Dias JAC, Yuki VA, Costa AS, Teixeira PRM (1980) Study of the spread of potato virus diseases in a warm climate as compared to a cold climate, with a view to obtaining seed potatoes with a low rate of virus diseases. Summa Phytopathol 6:24–59
Dong J, Gruda N, Lam SK, Li X, Duan Z (2018) Effects of elevated CO_2 on nutritional quality of vegetables: a review. Front Plant Sci 9:1–11
Donnelly A, Lawson T, Craigon J, Black CR, Colls JJ, Landon G (2001a) Effects of elevated CO_2 and O_3 on tuber quality in potato (*Solanum tuberosum* L.). Agric Ecosyst Environ 87:273–285
Donnelly A, Craigon J, Black CR, Colls JJ, Landon G (2001b) Elevated CO_2 increases biomass and tuber yield in potato even at high ozone concentrations. New Phytol 149:265–274
El Ahmadi AB, Stevens MA (1979) Reproductive responses of heat-tolerant tomatoes to high temperatures. J Am Soc Hortic Sci 104:686–691
Erickson AN, Markhart AH (2002) Flower developmental stage and organ sensitivity of bell pepper (*Capsicum annuum* L.) to elevated temperature. Plant Cell Environ 25:123–130
Ewing EE (1981) Heat stress and the tuberization stimulus. Am J Potato Res 58:31–49
Ewing EE (1997) Potato. In: Wien HC (ed) The physiology of vegetable crops. CAB International, Wallingford, p 662
Fahem M, Haverkort AJ (1988) Comparison of the growth of potato crops grown in autumn and spring in North Africa. Potato Res 31:557–568
Fangmeier A, De Temmerman L, Black C, Persson K, Vorne V (2002) Effects of elevated CO_2 and/or ozone on nutrient concentrations and nutrient uptake of potatoes. Eur J Agron 17:353–368
Fierro A, Gosselin A, Tremblay N (1994) Supplemental carbon dioxide and light improved tomato and pepper seedling growth and yield. HortScience 29:152–154
Finnan JM, Donnelly A, Burke JI, Jones MB (2002) The effects of elevated concentrations of carbon dioxide and ozone on potato (*Solanum tuberosum* L.) yield. Agric Ecosyst Environ 88:11–22
Finnan JM, Donnelly A, Jones MB, Burke JI (2005) The effect of elevated levels of carbon dioxide on potato crops: a review. J Crop Improv 13:91–111
Firon N, Shaked R, Peet MM, Phari DM, Zamski E, Rosenfeld K, Althan L, Pressman NE (2006) Pollen grains of heat tolerant tomato cultivars retain higher carbohydrate concentration under heat stress conditions. Sci Hortic 109:212–217
Fleisher DH, Timlin DJ, Reddy VR (2006) Temperature influence on potato leaf and branch distribution and on canopy photosynthetic rate. Agron J 98:1442–1452
Fleisher DH, Timlin DJ, Reddy VR (2008) Interactive effects of CO2 and water stress on potato canopy growth and development. Agron J 100:711–719
Fleisher DH, Condori B, Quiroz R, Alva A, Asseng S, Barreda C, Bindi M, Boote KJ, Ferrise R, Franke AC, Govindakrishnan PM (2017) A potato model intercomparison across varying climates and productivity levels. Glob Chang Biol 23:1258–1281
Gatto M, Pradel W, Suarez V, Qin J, Hareau G, Bhardwaj V, Pandey SK (2016) In: Dataset for: modern potato varietal release information in selected countries in Southeast and South Asia. International Potato Center Dataverse, V1, https://doi.org/10.21223/P3/2UOG9I
Gatto M, Hareau G, Pradel W, Suarez V, Qin J (2018) Release and adoption of improved potato varieties in Southeast and South Asia, Social sciences working paper no. 2018-2. International Potato Center (CIP), Lima, pp 1–42

Gautier H, Rocci A, Buret M, Grasselly D, Causse M (2005) Fruit load or fruit position alters response to temperature and subsequently cherry tomato quality. J Sci Food Agric 85:1009–1016

Geigenberger P (2003) Regulation of sucrose to starch conversion in growing potato tubers. J Exp Bot 54:457–465

Geisenberg C, Stewart K (1986) Field crop management. The tomato crop. Chapman & Hall, London, pp 511–557

Goudriaan J, De Ruiter HE (1983) Plant growth in response to CO_2 enrichment, at two levels of nitrogen and phosphorus supply. 1. Dry matter, leaf area and development. Neth J Agric Sci 31:157–169

Hancock RD, Morris WL, Ducreux LJ, Morris JA, Usman M, Verrall SR, Fuller J, Simpson CG, Zhang R, Hedley PE, Taylor MA (2014) Physiological, biochemical and molecular responses of the potato (*Solanum tuberosum* L.) plant to moderately elevated temperature. Plant Cell Environ 37:439–450

Hann YH, Hernandez TP (1982) Response of six tomato genotypes under the summer and spring weather conditions in Louisiana. HortScience 17:758–759

Hanying W, Shenyan S, Zhujun Z, Xinting Y (2001) Effects of high temperature stress on photosynthesis and chlorophyll fluorescence in sweet pepper (*Capsicum frutescens* L.). Acta Horticult Sinica 28:517–521

Haque MA, Hossain AKMA, Ahmed KU (1999) A comparative study on the performance of different varieties of tomato. II. Varietal response of different seasons and temperature in respect of yield and yield components. Bangl Horticult 26:39–45

Hatfield JL, Boote KJ, Kimball BA, Ziska LH, Izaurralde RC, Ort D, Thomson AM, Wolfe D (2011) Climate impacts on agriculture: implications for crop production. Agron J 103:351–370

Hazra P, Samsul HA, Sikder D, Peter KV (2007) Breeding tomato (*Lycopersicon esculentum* Mill.) resistant to high temperature stress. Int J Plant Breed 1:31–40

Heagle AS, Miller JE, Pursley WA (2003) Growth and yield responses of potato to mixtures of carbon dioxide and ozone. J Environ Qual 32:1603–1610

Helyes L, Lugasi A, Neményi A, Pék Z (2012) The simultaneous effect of elevated CO_2-level and nitrogen-supply on the fruit components of tomato. Acta Aliment 41:265–271

Heuvelink E (1995) Effect of temperature on biomass allocation in tomato (*Lycopersicon esculentum*). Physiol Plant 94:447–452

Hijmans RJ (2003) The effect of climate change on global potato production. Am J Potato Res 80:271–279

Högy P, Fangmeier A (2009) Atmospheric CO_2 enrichment affects potatoes: 2. Tuber quality traits. Eur J Agron 30:85–94

Hurd RG, Graves CJ (1984) The influence of different temperature patterns having the same integral on the earliness and yield of tomatoes. Acta Hortic 148:547–554

Hurd RG, Graves CJ (1985) Some effects of air and root temperatures on the yield and quality of glasshouse tomatoes. J Horticult Sci 60:359–371

IPCC (2007) Impacts, adaptation and vulnerability: contribution of working group II to the fourth assessment report of the Intergovernmental Panel on Climate Change. Cambridge University Press, Cambridge/New York

IPCC (2014) Synthesis report. In: Core Writing Team, Pachauri RK, Meyer LA (eds) Geneva: contribution of working groups I, II and III to the fifth assessment report of the Intergovernmental Panel on Climate Change. IPCC, Geneva

Islam MS, Matsui T, Yoshida Y (1996) Effect of carbon dioxide enrichment on physico-chemical and enzymatic changes in tomato fruits at various stages of maturity. Sci Hortic 65:137–149

Iwahori S, Takahashi K (1964) High temperature injuries in tomato. III. Effects of high temperature on flower buds and flowers of different stages of development. J Jpn Soc Horticult Sci 33:67–74

Khan I, Azam A, Mahmood A (2013) The impact of enhanced atmospheric carbon dioxide on yield, proximate composition, elemental concentration, fatty acid and vitamin c contents of tomato (*Lycopersicon esculentum*). Environ Monit Assess 185:205–214

Kim YU, Seo BS, Choi DH, Ban HY, Lee BW (2017) Impact of high temperatures on the marketable tuber yield and related traits of potato. Eur J Agron 89:46–52

Kimball BA (1983) Carbon dioxide and agricultural yield: an assemblage and analysis of 430 prior observations. Agron J 75:779–788

Kimball BA (2011) Handbook of climate change and agroecosystems, impacts, adaptation, and mitigation, vol 1. Indian Agricultural Research Institute, New Delhi, pp 87–107

Kirk WW, Marshall B (1992) The influence of temperature on leaf development and growth in potatoes in controlled environments. Ann Appl Biol 120:511–525

Kooman PL, Haverkort AJ (1995) Modelling development and growth of the potato crop influenced by temperature and daylength: LINTUL-POTATO. In: Potato ecology and modelling of crops under conditions limiting growth. Springer, Dordrecht, pp 41–59

Kooman PL, Fahem M, Tegera P, Haverkort AJ (1996) Effects of climate on different potato genotypes 2. Dry matter allocation and duration of the growth cycle. Eur J Agron 5:207–217

Krauss A, Marschner H (1984) Growth rate and carbohydrate metabolism of potato tubers exposed to high temperatures. Potato Res 27:297–303

Krumbein A, Schwarz D, Kläring HP (2012) Effects of environmental factors on carotenoid content in tomato (*Lycopersicon esculentum* (L.) Mill.) grown in a greenhouse. J Appl Bot Food Qual 80:160–164

Kumar R, Kumar R, Singh RK, Rai VP, Singh M, Singh PK (2017) Selection of tomato (*Solanum lycopersicum*) genotypes for heat stress and analyzing stability in their physio-morphological traits under different seasons. Curr Horticult 5(1):30–39

Kumari S, Agrawal M (2014) Growth, yield and quality attributes of a tropical potato variety (*Solanum tuberosum* L. cv. Kufri Chandramukhi) under ambient and elevated carbon dioxide and ozone and their interactions. Ecotoxicol Environ Saf 101:146–156

Kuo CG, Chen BW, Chou MH, Tsai CL, Tsay TS (1979) Tomato fruit-set at high temperatures. In: First international symposium on tropical tomato. Asian Vegetable Research Development Centre, Shanhua, pp 94–109

Lawson T, Craigon J, Tulloch AM, Black CR, Colls JJ, Landon G (2001) Photosynthetic responses to elevated CO_2 and O_3 in field-grown potato (*Solanum tuberosum*). J Plant Physiol 158:309–323

Levy D, Veilleux RE (2007) Adaptation of potato to high temperatures and salinity: a review. Am J Potato Res 84:487–506

Levy A, Rabinowitch HD, Kedar N (1978) Morphological and physiological characters affecting flower drop and fruit set of tomatoes at high temperatures. Euphytica 27:211–218

Li JH, Sagi M, Gale J, Volokita M, Novoplansky A (1999) Response of tomato plants to saline water as affected by carbon dioxide supplementation. I. Growth, yield and fruit quality. J Hortic Sci Biotechnol 74:232–237

Lurie S, Handros A, Fallik E, Shapira R (1996) Reversible inhibition of tomato fruit gene expression at high temperature (Effects on tomato fruit ripening). Plant Physiol 110:1207–1214

Mamatha H, Rao NS, Laxman RH, Shivashankara KS, Bhatt RM, Pavithra KC (2014) Impact of elevated CO_2 on growth, physiology, yield, and quality of tomato (*Lycopersicon esculentum* Mill) cv. Arka Ashish. Photosynthetica 52:519–528

Manrique LA (1990) Growth and yield of potato grown in the greenhouse during summer and winter in Hawaii. Commun Soil Sci Plant Anal 21:237–249

McGee E, Jarvis MC, Duncan HJ (1986) The relationship between temperature and sprout growth in stored seed potatoes. Potato Res 29:521–524

Menzel CM (1985) Tuberization in potato at high temperatures: interaction between temperature and irradiance. Ann Bot 55:35–39

Midmore DJ (1984) Potato (*Solanum* spp.) in the hot tropics I. Soil temperature effects on emergence, plant development and yield. Field Crop Res 8:255–271

Midmore DJ, Prange RK (1991) Sources of heat tolerance amongst potato cultivars, breeding lines, and *Solanum* species. Euphytica 55(3):235–245

Miglietta F, Magliulo V, Bindi M, Cerio L, Vaccari FP, Loduca V, Peressotti A (1998) Free air CO_2 enrichment of potato (*Solanum tuberosum* L.): development, growth and yield. Glob Chang Biol 4:163–172

Mitcham EJ, McDonald RE (1992) Effect of high temperature on cell wall modifications associated with tomato fruit ripening. Postharvest Biol Technol 1:257–264

Moreno U (1985) Environmental effects on growth and development of potato plants. In: Potato physiology. Academic Press Inc, London, pp 481–501

Mulholland BJ, Edmondson RN, Fussell M, Basham J, Ho LC (2003) Effects of high temperature on tomato summer fruit quality. J Hortic Sci Biotechnol 78:365–374

Nederhoff EM (1994) Effects of CO_2 concentration on photosynthesis, transpiration and production of greenhouse fruit vegetable crops. Dissertation. Agricultural University, Wageningen

Nilsen S, Hovland K, Dons C, Sletten SP (1983) Effect of CO_2 enrichment on photosynthesis, growth and yield of tomato. Sci Hortic 20:1–14

Nothmann J, Rylski I, Spigelman M (1979) Flowering-pattern, fruit growth and color development of eggplant during the cool season in a subtropical climate. Sci Hortic 11:217–222

Olivo N, Martinez CA, Oliva MA (2002) The photosynthetic response to elevated CO_2 in high altitude potato species (*Solanum curtilobum*). Photosynthetica 40:309–313

Olmstead RG, Bohs L (2006) A summary of molecular systematic research in Solanaceae: 1982–2006. In: VI International Solanaceae Conference: Genomics Meets Biodiversity, pp. 255–268

Passam HC, Bolmatis A (1997) The influence of style length on the fruit set, fruit size and seed content of aubergines cultivated under high ambient temperature. Trop Sci 37:221–227

Pearce BD, Grange RI, Hardwick K (1993) The growth of young tomato fruit. II. Environmental influences on glasshouse crops grown in rockwool or nutrient film. J Horticult Sci 68:13–23

Pearson S (1992) Modelling the effects of temperature on the growth and development of horticultural crops. Doctoral dissertation, University of Reading

Peet MM, Wolfe DW (2000) Climate change and global productivity. Crop ecosystem responses to climate change: vegetable crops. CABI Publishing, New York/Wallingford, pp 213–243

Peet MM, Willits DH, Gardner RG (1997) Responses of ovule development and post pollen production processes in male-sterile tomatoes to chronic, sub-acute high temperature stress. J Exp Bot 48:101–111

Peet MM, Sato S, Gardner RG (1998) Comparing heat stress effects on male fertile and male sterile tomatoes. Plant Cell Environ 21:225–231

Peltonen-Sainio P, Jauhiainen L, Trnka M, Olesen JE, Calanca P, Eckersten H, Eitzinger J, Gobin A, Kersebaum KC, Kozyra J, Kumar S (2010) Coincidence of variation in yield and climate in Europe. Agric Ecosyst Environ 139:483–489

Piñero MC, Otálora G, Porras ME, Sánchez-Guerrero MC, Lorenzo P, Medrano E, del Amor FM (2017) The form in which nitrogen is supplied affects the polyamines, amino acids, and mineral composition of sweet pepper fruit under an elevated CO_2 concentration. J Agric Food Chem 65:711–717

Polowick PL, Sawhney VK (1985) Temperature effects on male fertility and flower and fruit development in *Capsicum annuum* L. Sci Hortic 25:117–127

Porras ME, Lorenzo P, Medrano E, Sánchez-González MJ, Otálora-Alcón G, Piñero MC, del Amor FM, Sánchez-Guerrero MC (2017) Photosynthetic acclimation to elevated CO_2 concentration in a sweet pepper (*Capsicum annuum*) crop under Mediterranean greenhouse conditions: influence of the nitrogen source and salinity. Funct Plant Biol 44:573–586

Prendergast JD (1983) Carbon assimilation and partitioning in heat tolerant tomato genotypes. Dissert Abstract Int 43:2109B

Randeni G, Caesar K (1986) Effect of soil temperature on the carbohydrate status in the potato plant (*S. tuberosum* L.). J Agron Crop Sci 56:217–224

Reich PB, Hobbie SE, Lee TD, Pastore MA (2018) Unexpected reversal of C_3 versus C_4 grass response to elevated CO_2 during a 20-year field experiment. Science 360:317–320

Rubatzky VE, Yamaguchi M (1997) Tomatoes, peppers, eggplants, and other solanaceous vegetables. World vegetables. Springer, Boston, pp 532–576
Rudich J, Zamski E, Regev Y (1977) Genotype variation for sensitivity to high temperature in the tomato: pollination and fruit set. Bot Gaz 138:448–452
Rykaczewska K (2013) The impact of high temperature during growing season on potato cultivars with different response to environmental stresses. Am J Plant Sci 4:2386–2393
Rykaczewska K (2015) The effect of high temperature occurring in subsequent stages of plant development on potato yield and tuber physiological defects. Am J Potato Res 92:339–349
Rylski I (1979a) Effect of temperatures and growth regulators on fruit malformation in tomato. Sci Hortic 10:27–35
Rylski I (1979b) Fruit set and development of seeded and seedless tomato fruits under diverse regimes of temperature and pollination. J Am Soc Hortic Sci 104:835–838
Rylski I, Spigelman M (1982) Effects of different diurnal temperature combinations on fruit set of sweet pepper. Sci Hortic 17:101–106
Sato S, Peet MM, Gardner RG (2001) Formation of parthenocarpic fruit, undeveloped flowers and aborted flowers in tomato under moderately elevated temperatures. Sci Hortic 90:243–254
Sato S, Peet MM, Thomas JF (2002) Determining critical pre- and post-anthesis periods and physiological processes in *Lycopersicon esculentum* Mill. Exposed to moderately elevated temperatures. J Exp Bot 53:1187–1195
Sato S, Kamiyama M, Iwata T, Makita N, Furukawa H, Ikeda H (2006) Moderate increase of mean daily temperature adversely affects fruit set of *Lycopersicon esculentum* by disrupting specific physiological processes in male reproductive development. Ann Bot 97:731–738
Sawhney VK, Polowick PL (1985) Fruit development in tomato: the role of temperature. Can J Bot 63:1031–1034
Schapendonk AH, van Oijen M, Dijkstra P, Pot CS, Jordi WJ, Stoopen GM (2000) Effects of elevated CO_2 concentration on photosynthetic acclimation and productivity of two potato cultivars grown in open-top chambers. Funct Plant Biol 27:1119–1130
Senioniti E, Manetos Y, Gavales NA (1986) Co-operative effects of light and temperature on the activity of phosphoenolpyruvate carboxylase from *Amaranthus paniculatus*. Plant Physiol 82:518–522
Sicher RC, Bunce JA (1999) Photosynthetic enhancement and conductance to water vapor of field-grown *Solanum tuberosum* (L.) in response to CO_2 enrichment. Photosynth Res 62:155–163
Singer SF, Idso C (2009) Climate change reconsidered: the report of the Nongovernmental International Panel on Climate Change (NIPCC). The Heartland Institute, Chicago
Singh JP, Lal SS, Pandey SK (2009) Effect of climate change on potato production in India. Central Potato Research Institute, Shimla, Newsl 40:17–18
Singh A, Siddappa S, Bhardwaj V, Singh B, Kumar D, Singh BP (2015) Expression profiling of potato cultivars with contrasting tuberization at elevated temperature using microarray analysis. Plant Physiol Biochem 97:108–116
Sonnewald S, van Harsselaar J, Ott K, Lorenz J, Sonnewald U (2015) How potato plants take the heat? Procedia Environ Sci 29:97
Struik PC (2007) Responses of the potato plant to temperature. In: Potato biology and biotechnology. Elsevier Science BV, Amsterdam, pp 367–393
Sun W, Wang D, Wu Z, Zhi J (1990) Seasonal change of fruit setting in eggplants (*Solanum melongena* L.) caused by different climatic conditions. Sci Horic 44:55–59
Usman IS, Mamat AS, Mohd HSZS, Aishah HS, Anuar AR (1999) The non-impairment of pollination and fertilization in the abscission of chilli (*Capsicum annuum* L. var. Kulai) flowers under high temperature and humid conditions. Sci Hortic 79:1–11
Uzun S (1996) The quantitative effects of temperature and light environment on the growth, development and yield of tomato and aubergine. Unpublished PhD Thesis, The University of Reading, England, pp 45–48

Vafiadis DT, Papamanthos C, Ntinas GK, Nikita-Martzopoulou C (2012) Influence of CO_2 enrichment in greenhouses on pepper plant (*Capsicum annuum* L.) yield under high temperature conditions. Acta Hortic 952:749–754

Van De Geijn SC, Dijkstra P (1995) Physiological effects of changes in atmospheric carbon dioxide concentration and temperature on growth and water relations of crop plants. In: Potato ecology and modelling of crops under conditions limiting growth. Springer, Dordrecht, pp 89–99

Van de Geijn SC, Goudriaan J (1996) The effects of elevated CO_2 and temperature change on transpiration and crop water use. In: Global climate change and agricultural production. FAO/John Wiley & Sons, New York, pp 101–122

Vander Zaag P, Demagante AL (1987) Potato (*Solanum* spp.) in an isohyperthermic environment. I. Agronomic management. Field Crop Res 17:199–217

Vandermeiren K, Black C, Lawson T, Casanova MA, Ojanperä K (2002) Photosynthetic and stomatal responses of potatoes grown under elevated CO_2 and/or O_3– results from the European CHIP-programme. Eur J Agron 17:337–352

Wang K, Zhang X, Goatley M, Ervin E (2014) Heat shock proteins in relation to heat stress tolerance of creeping bentgrass at different N levels. PLoS One 9:e102914

Warrington IJ, Fulton, TA, Halligan EA, deSilva HN (1999) Apple fruit growth and maturity are affected by early season temperatures. J Am Soc Hortic Sci 124:468–477

Wei Z, Du T, Li X, Fang L, Liu F (2018) Interactive effects of elevated CO_2 and N fertilization on yield and quality of tomato grown under reduced irrigation regimes. Front Plant Sci 9:1–10

Wheeler RM, Tibbitts TW, Fitzpatrick AH (1991) Carbon dioxide effects on potato growth under different photoperiods and irradiance. Crop Sci 31:1209–1213

Winsor GW, Davies JN, Massey DM (1962) Composition of tomato fruit. IV. Changes in some constituents of the fruit walls during ripening. J Sci Food Agric 13:141–145

Wolf S, Olesinski AA, Rudich J, Marani A (1990) Effect of high temperature on photosynthesis in potatoes. Ann Bot 65:179–185

Wolf S, Marani A, Rudich J (1991) Effect of temperature on carbohydrate metabolism in potato plants. J Exp Bot 42:619–625

Wolfe DW, Erickson JD (1993) Carbon dioxide effects on plants: uncertainties and implications. In: Agricultural dimensions of global climate change. IPCC, Geneva, pp 153–178

Zhang Z, Liu L, Zhang M, Zhang Y, Wang Q (2014) Effect of carbon dioxide enrichment on health-promoting compounds and organoleptic properties of tomato fruits grown in greenhouse. Food Chem 153:157–163

Zhu Z, Piao S, Myneni RB, Huang M, Zeng Z, Canadell JG, Ciais P, Sitch S, Friedlingstein P, Arneth A, Cao C, Cheng L, Kato E, Koven C, Li Y, Lian X, Liu Y, Liu R, Mao J, Pan Y, Peng S, Peñuelas J, Poulter B, Pugh TAM, Stocker BD, Viovy N, Wang X, Wang Y, Xiao Z, Yang H, Zaehle S, Zeng N (2016) Greening of the Earth and its drivers. Nat Clim Chang 6:791–795

Ziska LH (2008) Rising atmospheric carbon dioxide and plant biology: the overlooked paradigm. DNA Cell Biol 27:165–172

Climate Change Impact on Cole Crops and Mitigation Strategies

D. P. Singh, Meenakshi Kumari, and H. G. Prakash

1 Introduction

The change in the climate is a statistical variation in the properties of various climatic factors *viz.* temperature, light intensity, rainfall, precipitation and atmospheric gases composition, etc. The productivity of agriculture is mainly dependent upon the climatic factors and land resources and impact of climate change influences crop productivity. Different stages of plant growth influenced by the weather parameters and thereby affect the crop productivity. For maximizing the production of any crop, knowledge of relationships between growth stages of crop and weather parameters is very essential (Ray and Mishra 2017). Most of the vegetables are highly affected by environmental extremes. For vegetable production, increasing rate of CO_2 is beneficial but quality of internal product may alter or result in photosynthetic down regulation (Bisbis et al. 2018). Moreover, increasing temperature, heat stress, flooding, rainfall and salinity are the major limiting factors reduces both quality and fruit set in fruiting vegetables and also shorten time for photo-assimilation, altered metabolism and enzymatic activity as they greatly influence several biochemical and physiological processes. Climate change influence pest and disease occurrences which results in crop failures, yield shortage and reduction in quality thus render unprofitable vegetable production (Abewoy 2018). Different species of *Brassica oleraceae* (cabbage, cauliflower, broccoli, knol khol, Brussels sprout) comes under cole crops family, Brassicaceae.

D. P. Singh (✉) · H. G. Prakash
Directorate of Research, Chandra Shekhar Azad University of Agriculture and Technology, Kanpur, Uttar Pradesh, India

M. Kumari
Department of Vegetable Science, Chandra Shekhar Azad University of Agriculture & Technology, Kanpur, Uttar Pradesh, India

S. S. Solankey et al. (eds.), *Advances in Research on Vegetable Production Under a Changing Climate Vol. 1*, Advances in Olericulture,
https://doi.org/10.1007/978-3-030-63497-1_5

The Brassicas vegetables are also known as "cool season", crop because of their resistancy towards frost and light freeze. Several species of Brassica not only contain high amount of protein and fat but also various nutrients such as vitamins, minerals, carbohydrates, phenols, different group of phytochemicals (brassinin, spirobrassinin, brassilexin, camalexin) and glucosinolates (mainly glucoiberin, glucoraphanin, glucoalyssin, gluconapin, lucobrassicanapin, glucobrassicin, gluconasturtiin and neoglucobrassicin), which provide protection against DNA damage and cancer due to fight against free radical damage (Jahangir ct al. 2009). Thus, cole crops are recognized as best resource for overcoming nutrient deficiencies and it's production also provide higher income to small holders farmers. Most of the cole crops grown well with optimum temperature of 60–65 °F (Maynard and Hochmuth 1997). As compare to warm season crops, cole crops are more adversely affected by temperature extreme. In cauliflower, delay in head formation and insufficient vernalization is due to higher temperature (Bisbis et al. 2018). All cole crops are closely related hence, require similar climatic pattern for their growth and development and also infested by similar disease and pests. High level of antioxidants sulphoraphanes are found in broccoli, which are anticarcinogenic compounds provide protection against damage caused by UV light that can lead to mascular degeneration (Xiangqun and Talalay 2004) and also against high blood pressure, heart disease, and stroke (Lingyun et al. 2004). Continous weather shifting through high and low temperature, variability in rainfall threatened agriculture productivity is a result of climate change.

The production, flavor and nutritional quality of any crop can be affected by linkage between environmental factors and metabolic processes of plant itself (Hochmuth 2003). Higher temperature affects the antioxidant activities of fruit and vegetable (Reblova 2012) (Fig. 1, Table 1).

2 Effects of Temperature

Many important plant activities like physiological, bio-chemical and metabolic process are temperature dependent which play an important role in plant growth and development. In all the plant species, pollination is one of the most sensitive phenological stage, which is affected by high temperature and during pollination temperature extremes would greatly affect production (Hatfield and Prueger 2015). Generally, for proper growth most of the cole crops require maximum temperature of 25 °C above this temperature maturity of cabbage and cauliflower head can be delayed (Nieuwhof 1969). Due to high temperature in case of cabbage and broccoli, heading and flower initiation may be delayed or prevented, respectively (Wein 1997) and incidence of diseases may be increased (Saure 1998). High temperature also affects flavor and other quality parameters, like in case of cabbage more than 30 °C air temperature during head development leads to development of undesirable "hot" cabbage (Radovich et al. 2004). Growing of Chinese cabbage under high temperature stress, slightly increased cell membrane permeability and heat tolerance

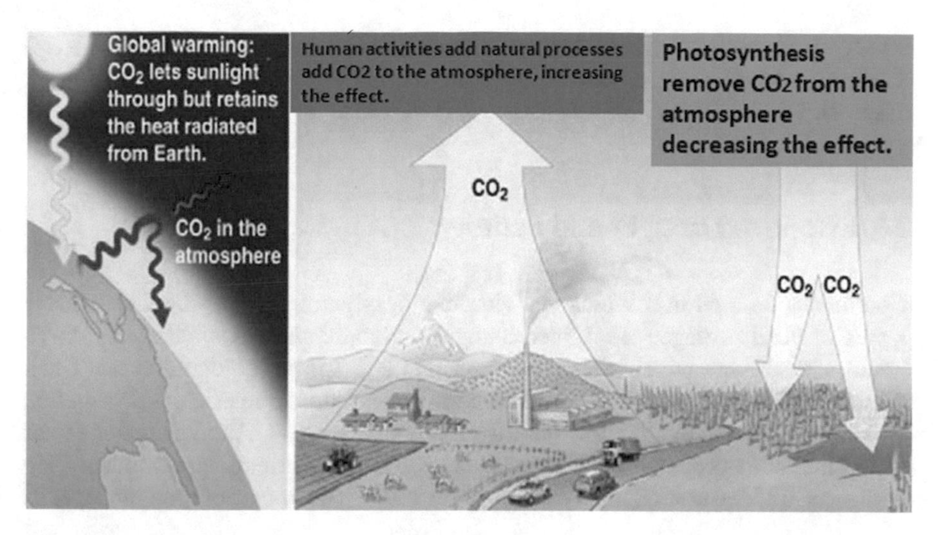

Fig. 1 Esion of Greenhouse gases in the atmosphere (CO_2). (Source: Abou-Hussein (2012))

Table 1 Important cole crops, botanical name and its edible part

Sl. no.	Common name	Scientific name	Edible part
1	Cauliflower	*Brassica oleracea* var. *botrytis*	Immature flower stalk
2	Cabbage	*Brassica oleracea* var. *capitata*	Head or leaf
3	Brussels sprout	*Brassica oleracea* var. *gemmifera*	Axillary bud
4	Sprouting broccoli	*Brassica oleracea* var. *italic*	Immature flower stalk
5	Knolkhol or kohlrabi	*Brassica oleracea* var. *gongylodes*	Enlarged stem
6	Kale	*Brassica oleracea* var. *alboglabra*	Leaf, flower stalk
7	Savoy cabbage	*Brassica oleracea* var.*sabuda*	Leaf
8	Pak choi	*Brassica rapa* var.*chinensis*	Leaf
9	Portuguese cabbage	*Brassica oleracea* var.*costata*	Leaf and inflorescence
10	Rutabaga	*Brassica napus* var.*napobrassica*	Root, leaf
11	Chinese cabbage	*Brassica rapa* var.*pekinensis*	Leaf

Source: Maynard and Hochmuth (1997)

index (Data 2013). Exposure of the plant to chilling injury (0 °C–10 ° C) or freezing injury (less than 0 °C) cause damage. Several researchers reported that excepted loss in broccoli is found due to increase in temperature (United States Global Change Program Research 2000). Yield and number of growing days at temperature above 30 °C are negatively correlated in cold adopted *Brassica* species including cabbage, cauliflower, and other vegetables (Warland et al. 2006). Cauliflower curds matured earlier when grown at higher temperature than grown under lower temperature (Wurr et al. 1996). The initiation of flower buds often depends on temperature. High temperature cause bolting in cole crops like in broccoli and cabbage. Elongation of stem and flower stalk initiation depends on the difference between day and night temperature (Agrawal et al. 1993). Due to high temperature, flowering and bolting occurs in cole crops which are highly undesirable (Peirce

1987). In case of winter cabbage (*Brassica oleracea* L., *capitata* Group) accumulation of sucrose was temperature dependent (Nilsson 1988) (Fig. 2, Table 2).

3 Effects of Drought and Salinity

Continuous drought in combination with heat or other stresses results in extensive losses of yield of vegetable. Under changing climatic situations, drought is most important extremes which affect yield losses in agriculture (Potopova et al. 2015). The availability of water is expected to be decreases due to inefficient distribution systems in developing countries and it affects productivity. Another major factor which limits the productivity of crop is salinity. Salinity affects approximately 7% of the world's land area (Rozema and Flowers 2008). The impact of drought depends

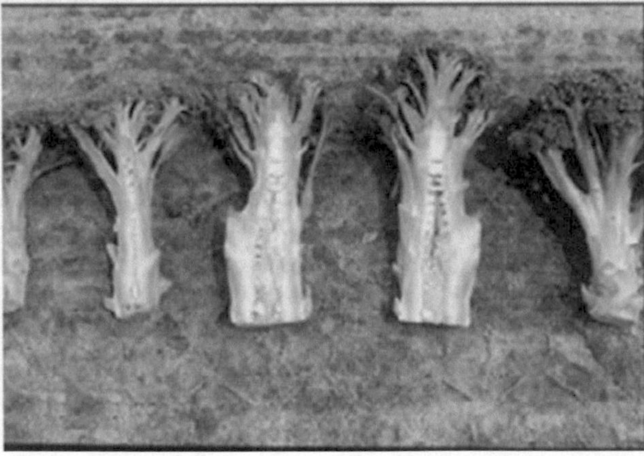

Fig. 2 High temperature cause bolting in cabbage and broccoli and hollow stem in broccoli. (Source: Abou-Hussein (2012))

Table 2 Symptoms of frost damage on some important cole crops

S. no.	Crop	Symptoms
1.	Cauliflower	Cooked curd turn brown and have a strong off-odour.
2.	Cabbage	Water soaked area develop on the leaves and become translucent and limp.
3.	Sprouting broccoli	Center curd (young florets) turns brown and gave off strong odour.

Source: Caplan (1988)

on the plant physiology and biochemical processes of plant itself (Massonnet et al. 2007).

The concentration of the salt in soil increases due to drought is a serious problem that reduces growth and production of vegetable crops. Vegetables are sensitive to salt throughout its life cycle. Several physiological and biochemical processes of plants such as photosynthesis, respiration *etc.*, are affected by salinity hence reduces yield (Pena and Hughes 2007). Excess salt stress causes death of plants due to loss of turgor, reduced growth, wilting, leaf abscission, decreased respiration and photosynthesis (Cheeseman 2008). Salinity causes reduction in germination rate, root and shoot length and fresh root and shoots weight in cabbage (Jamil and Rha 2004). Most sensitive stage of water deficit in fruiting vegetables are transplanting, flowering, yield formation and also immediately after transplanting. At flowering, water deficit causes flower drop (Petrikova et al. 2006, 2012).

4 Effects of Climate Change on Plant Diseases

The change in climate affects the reproduction, spread and severity of many plant pathogens which ultimately affects the growth and cultivation of different crops (Das et al. 2016). The effect of climate change depends upon the system of plant pathogen. Annual vegetable crops are more flexible than perennials, when it comes to adopt new cultivars and cultural practices. The 3 major elements of disease triangle, *viz.*, host, pathogen and environment and there elements are highly affected by change in climate (Legreve and Duveiller 2010). The presence of any disease will altered under high temperature condition, high concentration of atmospheric CO_2 and changes in precipitation patterns (Ghini et al. 2008; Chakraborty and Newton 2011). Proliferation and germination of fungal spores of various pathogen as well as initial disease development increases under high moisture and temperature (Agrios 2005). Due to continuous change in climate, the population of insect vector and host plant are affected and spread the plant viruses (Jones 2009). Due to increasing rate of disease incidence, crop treated with higher application rate of fungicide, hence increasing costs for farmers, consumers and the likelihood for the development of fungicide resistance (Juroszek and Tiedemann 2011). The efficacy of various insecticide, pesticides and fungicides are weather dependent *i.e.,* affected by the change in temperature, intensity of precipitation. Cole crops are insect pollinated crops *i.e*, entomophilus in nature thus change in climate affects the distribution of insect species which ultimately affects the pollination process. The biology and ecology of insect-pest influenced by climate change (Jat and Tetarwal 2012). Insects with shorter life cycle such as aphids and diamond back moth increases their fecundity rate and complete their life cycle earlier under high temperature. While, disease cycles of air borne pathogens become faster due to high temperature and due to reduction in frost they increase their survival (Termorshuizen 2008; Boonekamp 2012) (Tables 3 and 4).

5 Impact of Climate Change on Seed Production

The availability of quality seed is of utmost importance for increasing the vegetable production. Use of quality seeds of improved varieties is the most strategic resource for higher and better vegetable yields. The change in climate has adverse effect on seed production of vegetable. Kumar et al. (2009) reported that due to high temperature and rainfall seed production of cabbage var. Golden acre reduced around 40% per unit area. In case of cauliflower, fluctuation in temperature in February month at the time of curd formation reduced the seed yield by 49.17–100% (Gill and Singh 1973). Priyanka et al. (2018) in cauliflower reported that due to different weather parameters in Himachal Pradesh from 1991 to 2016 the seed yield reduced from 380.2 kg/ha to 216.0 kh/ha, respectively. The maximum temperature exhibited negative correlation and forenoon humidity showed positive correlation with cauliflower seed yield.

6 Management Strategies for Adopting Climate Change

Adoption of effective and efficient management strategies is the best way to overcome with the adverse effect of climate change on both quality and production of cole crops. Various crop management practices such as conservation tillage, organic farming, drip irrigation, mulching with crop residues and plastic mulches help in conserving soil moisture. Growing crops on raised beds is one of the important practices for conserving excess soil moisture due to heavy rains (Abewoy 2018).

Adoption of grafting technique to improved environmental-stress tolerance of horticultural crops is considered as a common practice in vegetable production in Asian countries which is an advanced tool as compare to slow breeding methods (Martinez et al. 2010).

Genetic improvement of vegetable crops *i.e*, development of varieties which has ability to make them able to withstand the adverse effects of climatic factors such as heat stress, drought, cold, flood, moisture stress are appropriate adoption strategy

Table 3 Major diseases and physiological disorders of cole crops

Sl. no.	Disease/ disorder	Symptoms	Reason	Control
1	Alternaria leaf spot *(Alternaria brassicae)*	Small dark spots appear on leaves during initial stage, but later it develops into tan spots with target-like concentric rings.	High humidity for long periods, cool temperature, and rain favours the development of this disease.	Always used clean certified seed, crop rotation with non-host crops, avoid overhead irrigation.
2	Tipburn	Tips and internal leaf edges of the crop turn brown or necrosis (Guerena 2006) and these brown spots tend to break down during storage	Drought, especially combined with high temperature, application of excess nitrogen, calcium and water stress.	Avoid high temperature, water management and supply adequate amount of nitrogen.
3	Riceyness	Curds become uneven, fuzzy that results in a granular apperance hence, reduces market ability (Norman 1992).	Temperature more than 68° F during curd development favour its development.	Use of high resistant cultivars. Few hybrids can develop heads at 68 to 80° F (Dainello 2003).
4	Hollow stem	Stem of the crop become hollow, core and pith cracks.	High temperature and excessive nitrogen	Sowing at proper time and apply adequate nitrogen.
5.	Buttoning	The curd and flower buds become small and loose heads are formed.	Exposure of immature plants to low temperature for prolonged period.	Avoid low temperature during early stage.
6.	Bolting	Pre-emergence of flower buds	High temperature during early stages of development	Avoid high temperature during early stage.
7.	Blindness	No head formation or multiple small heads form	Low temperature (Fritz et al. 2009; Verma 2009).	Sowing at proper time
8.	Bursting/ splitting of heads	Head split or burst	Heavy rain or delayed harvest	Avoid over watering and timely harvesting
9.	Leafy cuds or leafiness	Green leaves appear between sections of the curd (Loughton 2009)	High temperature	Proper sowing time

Source: Wein (1997)

(Koundinya et al. 2014). Adaption of improved advanced biotechnology method such as tissue culture, embryo rescue techniques and genetic engineering are useful methods to develop resistant varieties that can cope with available climatic change. So many varieties of cole vegetables has already been identified (Table 5) and reported by various vegetable breeders (Selvakumar 2014).

Table 4 Important insect-pest and effect of climate change on cole crops

S.No.	Pest	Effect of climate change
1	Diamond Back moth (*Plutella xylostella*)	Increase in population of pest and more frequent overwintering.
2	Cabbage butterfly (*Pieris brassicae*)	Increase in range, abundance and diversity.
3	Cabbage root fly (*Delia radicum*)	By increase in mean temperature of 3 °C, population become active a month earlier in UK.

Source: Yukawa (2008), Cannon (2008)

Table 5 Identified cole crops varieties to tolerate against the harmful effect of climate change

S. no.	Crop	Varieties	Tolerance
1.	Early cauliflower	Pusa Meghna	Form curd at high temperature
		Arka Kanti	Tropical variety form curd at high temperature
2.	Late cauliflower	Pusa Snowball-1, Pusa Himjyoti	Form curd at low temperature
3.	Cauliflower	Dania	Tolerance to stress condition
3.	Cabbage	Pusa Ageti	1st tropical variety

Source: Selvakumar (2014)

7 Conclusion

Cole crops are major group of vegetable which belongs to species *Brassica* and play important role in human diet. Now-a-days, due to evaluation of different cultivars for different season, cole crops are being grown round the year but due to change in climatic situation, its hamper both production and quality of cole crops. Temperature play major affect on crop plants among all climatic factors. Various practices like mulching, protected cultivation, organic farming, post harvest technology, irrigation management and improvement of cultivars with different breeding techniques such as tissue culture, genetic engineering, molecular marker etc. are very primitive and fruitful method for increasing both quality and yield under combat climate change.

References

Abewoy D (2018) Review on impacts of climate change on vegetable production and management practices. Adv Crop Sci Technol 6(1):330. https://doi.org/10.4172/2329-8863.1000330

Abou-husscin SD (2012) Climate change and its impact on the productivity and quality of vegetable crops (review article). J Appl Sci 8(8):4359–4383

Agrawal M, Krizek DT, Agrawal SB, Krarner GF, Lee EH, Mirecki RM, Rowland RA (1993) Influence of inverse day/night temperature on ozone sensitivity and selected morphological and physiological responses of cucumber. J Am Soc Hortic Sci 118:649–654

Agrios GN (2005) Plant pathology, 5th edn. Elsevier, London, pp 249–263

Bisbis MB, Gruda N, Blanke M (2018) Potential impacts of climate change on vegetable production and product quality – a review. J Clean Prod 170(1):1602–1620

Boonekamp PM (2012) Are plant diseases too much ignored in the climate change debate? Eur J Plant Pathol 133:291–294

Cannon RJC (2008) Annexure-1. In: Climate-related transboundary pests and diseases, technical background document from the expert consultation held on 25 to 27 February 2008. Rome, FAO

Caplan LA (1988) Effects of cold weather in horticultural plants in Indiana. Purdue University Cooperative Extension Publication, West Lafayette

Chakraborty S, Newton AC (2011) Climate change, plant diseases and food security: an overview. Plant Pathol 60:2–14

Cheeseman JM (2008) Mechanisms of salinity tolerance in plants. Plant Physiol 87:547–550

Dainello FJ (2003) Cauliflower. Department of Horticultural Sciences, Texas A&M University Web Page, College Station. Downloaded April 2005. http://aggie-horticulture.tamu.edu/extension/vegetable/cropguides/cauliflower.html

Das T, Majumdar MHD, Devi RKT, Rajesh T (2016) Climate change impacts on plant diseases. SAARC J Agric 14(2):200–209

Data S (2013) Impact of climate change in Indian horticulture – a review. Int J Environ Sci Technol 2(40):661–671

Fritz VA, Rosen CJ, Grabowski MA, Hutchison WD, Becker RL, Tong CBS, Wright JA, Nennich TT (2009) Growing broccoli, cabbage and cauliflower in Minnesota. University of Minnesota Extension Bulletin, Minneapolis, pp 1–15

Ghini R, Hamada E, Bettiol W (2008) Climate change and plant disease. Scientia Agricola (Piracicaba, Brazil) 65:98–107

Gill HS, Singh JP (1973) Effect of environmental factors on seed yield of late cauliflower in Kullu Valley. Indian J Agric Sci 43(3):234–236

Guerena M (2006) Cole crops and other brassicas: organic production. A Publication of ATTRA, Butte, pp 1–19

Hatfield JL, Prueger JH (2015) Temperature extremes: effect on plant growth and development. Weather Clim Extremes 10:4–10. https://doi.org/10.1016/j.wace.2015.08.001

Hochmuth GJ (2003) Progress in mineral nutrition and nutrient management for vegetable crops in the last 25 years. HortScience 38:999–1003

Jahangir M, Kim HK, Choi YH, Verpoorte R (2009) Health-affecting compounds in *Brassicaceae*. Compr Rev Food Sci Food Saf 8:31–43

Jamil M, Rha ES (2004) The effect of salinity (NaCl) on the germination and seedling of sugar beet (*Beta vulgaris* L.) and cabbage (*Brassica oleracea capitata* L.). Korean J Plant Res 7:226–232

Jat MK, Tetarwal AS (2012) Effect of changing climate on the insect pest population national seminar on sustainable agriculture and food security: challenges in changing climate, March 27–28, Hisar, India, pp 200–201

Jones RAC (2009) Plant virus emergence and evolution: origins, new encounter scenarios, factors driving emergence, effects of changing world conditions, and prospects for control. Virus Res 141:113–130

Juroszek P, Tiedemann AV (2011) Potential strategies and future requirements for plant disease management under a changing climate. Plant Pathol 60:100–112

Koundinya A, Sidhya P, Pandit MK (2014) Impact of climate change on vegetable cultivation – a review. Int J Environ, Agric Biotechnol 7(1):145–155

Kumar PR, Yadav S k, Sharma SR, Lal SK, Jha DN (2009) Impact of climate change on seed production of Cabbage in North Western Himalayas. World J Agric Sci 5(1):18–26

Legreve A, Duveiller E (2010) Preventing potential diseases and pest epidemics under a changing climate. In: Reynolds MP (ed) Climate change and crop production. CABI, Wallingford, pp 50–70

Lingyun W, Ashraf MHN, Facci M, Wang R, Paterson PG, Ferrie A, Bernhard HJJ (2004) Dietary approach to attenuate oxidative stress, hypertension, and inflammation in the cardiovascular system. Proc Natl Acad Sci U S A 101(18):7094–7099. https://doi.org/10.1073/pnas.0402004101

Loughton A (2009) Production and handling of broccoli. Factsheet ISSN 1198-712X. Queen's Printer, Ontario

Martinez RMM, Estan MT, Moyano E, Garcia AJO, Flores FB (2010) The effectiveness of grafting to improve salt tolerance in tomato when an excluder genotype is used as scion. Environ Exp Bot 63:392–401

Massonnet C, Costes E, Rambal S, Dreyer E, Regnard JL (2007) Stomatal regulation of photosynthesis in apple leaves: evidence for different water-use strategies between two cultivars Catherine. Ann Bot 100:1347–1356

Maynard DN, Hochmuth GJ (1997) Knott's handbook for vegetable growers, 2nd edn. John Wiley & Sons Inc, New York

Nieuwhof M (1969) Cole crops, World crops series. Leonard Hill, London, p 353

Nilsson T (1988) Growth and carbohydrate composition of winter white cabbage intended for long-term storage. II. Effects of solar radiation, temperature and degree-days. J Horticult Sci 63:431–441

Norman JC (1992) Tropical vegetable crops. Arthur H. Stockwell LTD, Elms Court, p 252

Peirce LC (1987) Vegetables: characteristics, production and marketing. Wiley, New York

Pena R, Hughes J (2007) Improving vegetable productivity in a variable and changing climate. SAT e J 4(1):1–22

Petrikova K, Jansky J, Maly I, Peza Z, Polackova J, Rod J (2006) Zelenina- pestovani ekonomika, prodej. ProfiPress, Praha, p 240

Petrikova K, Hlusek J, Jansky J, Koudela M, Losak T, Maly I, Pokluda R, Polackova J, Rod J, Ryant P, Skarpa P (2012) Zelenina. Profi press, Praha, p 191

Potopova V, Stepanek P, Mozny M, Turkott L, Soukup J (2015) Performance of the standardized precipitation evapotranspiration index at various lags for agricultural drought rsk assessment in the Czech Republic. Agric For Meteorol 202:26–38

Priyanka S, Mohinder S, Bhardwaj SK, Rasna G (2018) Impact of long term weather parameters on seed production of cauliflower. Pharma Innov J 7(7):521–523

Radovich T, Kleinhenz M, Honeck NJ (2004) Important cabbage head traits and their relationships at five points in development. J Veg Crop Prod 10:19–32

Ray M, Mishra N (2017) Effect of weather parameters on the growth and yield of Cauliflower. Environ Conserv J 18(3):9–19

Reblova Z (2012) Effect of temperature on the antioxidant activity of phenolic acids. Czech J Food Sci 30:171–177

Rozema J, Flowers T (2008) Crops for a salinized world. Science 322:1478–1480

Saure MC (1998) Causes of the tip burn disorder in the leaves of vegetables. Sci Hortic 76:131–147

Selvakumar R (2014) A textbook of Glaustas Olericulture. New Vishal Publications, New Delhi, pp 371–374

Termorshuizen AJ (2008) Climate change and bioinvasiveness of plant pathogens: comparing pathogens from wild and cultivated hosts in the past and the present pests and climate change' December 3, pp 6–9

United States Global Change Program Research (2000) Global climate change impacts in the United States: agriculture. United States Global Change Program Research, Washington, DC, pp 77–78

Verma P (2009) Physiological disorders of vegetable crops. Alfa Beta Technical Solutions, Jaipur, p 170

Warland J, Mckeown AW, Mcdonald MR (2006) Impact of high air temperatures on Brassicacae crops in Southern Ontario. Can J Plant Sci 86(4):1209–1215

Wein HC (1997) The physiology of vegetable crops. CAB International, Wallingford, p 662

Wurr DCE, Fellows JR, Phelps K (1996) Investigating trends in vegetable crop response to increasing temperature associated with climate change. Sci Hortic 66:255–263

Xiangqun G, Talalay P (2004) Induction of phase 2 genes by sulforaphane protects retinal pigment epithelial cells against photooxidative damage. Proc Natl Acad Sci U S A 101(28):10446–10451

Yukawa J (2008) Annexure-3: northward distribution range extensions of plant pests, possibly due to climate change: examples in Japan. In: Climate-related trans boundary pests and diseases, technical background document from the expert consultation held on 25 to 27 February 2008. Rome, FAO

Impact of Climate Change on Root Crops Production

Menka Pathak, Satyaprakash Barik, and Sunil Kumar Das

1 Introduction

Climate change represent a warmer and unpredictable climate. Changing climate is a long-term constraint. Anthropogenic activities effect climate in various forms such as heatwaves, altered transmission of disease, and extreme weather events take 200,000 human lives annually (Patz et al. 2005). But it also affects humans indirectly. Despite, variation in climate after the decade of the 2030s are seriously vulnerable upon agreements to reduce universal radiation of greenhouse gases as soon as feasible. In our atmosphere, the greenhouse gases benefit produce the earth a warm and habitable place. They develop like the outdoor cover of a greenhouse, trapping heat from the sun. Greenhouse gases like carbon dioxide, methane and nitrous oxide are split generally human activity. Transmission of carbon dioxide from the scorching of fossil fuels have introduced in a new era where human activities will generally regulate the transformation of sphere as carbon dioxide in the atmosphere is prolonged alive, it can forcefully lock Earth and future generation into a range of impacts, some of which could become very serious.

Climate crisis have a serious shock on the increase of tropical root crops (TRC). Element like fluctuation precipitation, rise intensity of wet and dry periods, rising temperatures, sea level increase affect soil salinity will impact negatively and also occasionally positively on root crops production. Roots acquire tremendous scope to the physicochemical framework of the soil and to adapt their improvement properly, thus even for plants under biotic stress, play an imperative function in protect the nutritional and improvement action of the plant (Maurel et al. 2010). Acting as a bridge between the soil and plant, root system is exposed to many environmental stresses such as drought/moisture stress, waterlogging as well as salt stress (Wells

M. Pathak (✉) · S. Barik · S. K. Das
Department of Vegetable Science, College of Agriculture, Odisha University of Agriculture and Technology, Bhubaneswar, Orissa, India

S. S. Solankey et al. (eds.), *Advances in Research on Vegetable Production Under a Changing Climate Vol. 1*, Advances in Olericulture,
https://doi.org/10.1007/978-3-030-63497-1_6

and Eissenstat 2003). Root development has been sophisticated due to the different possible bias combine with these different methods and action of roots. For example, high temperature has a vital impact on growth and development of root system. The relationship between high temperatures and plant physiological activity ranges from photosynthesis to carbon sequestration as well as nutrient uptake potential of plant (Mcmichael and Burke 2002).

2 Why Is Climate Change Crucial?

Climate change constitute both endanger and opportunities at global level. Societies and Individuals can take profit of opportunities and decrease endanger by knowledge, plan for and acclimate to a changing climate. The importance of climate fluctuation or climate diversity are probably greater compelling for the poor in developing countries than for those living in more rich nations. Generally, the underprivileged are vulnerable on economic activities that are keen to the climate. For example, Local weather or climatic conditions influence in agriculture or forestry activities. A variation in climatic conditions could exactly impact productivity levels and reduce employment. High temperatures induced by climate change alter developmental and physiological plant processes that, ultimately, impact on crop earnings and nature. Plant roots are important for water and nutrients uptake, but variation in soil temperature modify this process limiting crop growth.

3 Importance of Root Vegetables

Worldwide, Vegetables exhibit an importance to the nutritional demand of the world's individual population, especially because vegetables are cheapest and best sources of vitamins, minerals as well as proteins. Besides nutritional values, vegetables are protective food against heart disease and cancer. Fresh vegetables are main component of a lively and well-balanced diet, because they surplus distinct biologically relevant components to human creature. Elevated mean temperatures and low soil moisture are the major sources of low productivity in the tropical zone. Root vegetables are those crops have prominent and fleshy underground structures and are direct sown winter season crops. Radish, carrot, turnip, beetroot, parsnip, rutabaga, horseradish are some of the most nutrient-dense root vegetables in the world include etc. While each root contains its own set of health benefits because root vegetables develop belowground, they ingest a big quantity of supplements from the ground. They are filled with a large adsorption of antioxidants, Vitamins A, B, C, and iron, helping to purify our system. They are also with slow-burning carbohydrates and fiber, help control our blood sugar and digestive system. Radish is popular as garden crop and cash crop. Red radish is the most common type of radish. Other radishes available are white radish winter (black or red) and Oriental radish (also called daikon), radish. White radishes and red are used popularly as salad items. Red radishes available in various shapes, from oval to round. During Indian summer climate the roots bear to grow deeper, produce round roots to take oval shape. Since round roots have set accepted among available varieties, oval and oblong cultivars are becoming less popular. Some cultivars are known as "short top," showing that they produce short sized leaflet during hot seasons. Short top cultivars are usually very less sensitive to bolting than regular cultivars. On the alternative, short top cultivars are not able to produce required leaves for regularly harvest during low temperature climatic condition. Red roots have high amount of ascorbic acid. Roots are used in treating piles and urinary complaints. Radish is a very good source of vitamin C- (140 mg/160 g). Major sugar content in radish is Glucose. Red fleshed winter cultivars of radish have anthocyanin ranges from 12.2 to 52 mg/100 g. Pink-skinned radish has good amount of ascorbic acid.

Carrot is commonly as a vegetable after heating but is also consumed fresh in salads and provided as an integral part in soups and sauces. Smell of carrot is due to the presence of glutamic acid. Carrot juice is a rich source of carotene. A special type of beverage known as *Kanji* is processed from black carrot (Asiatic carrot root) which is used as a taste. Tropical red carrot used for dessert preparation called *Gajar Halwa* is very famous dish in North India. Orange colored (Temperate) carrot contains more extra carotene (a precursor of Vitamin -A) than Asiatic type and is rich in thiamine and riboflavin. Asiatic Consist of more anthocyanin pigment but poor in balanced values. Carrot green leaves contain appreciable bulk of protein, minerals, and vitamins. Carrot juice is used for coloring butter and other foods. Carrot roots rich in sucrose i.e., 10 times more than glucose and fructose. The free sugar, glucose, fructose and sucrose give to sweetness whereas volatile mono and sesquiterpenoids contributes to harshness. Presence of polyacetylenes, give to the unpleasant odour in carrots. Carrot seed germination inhibitors called as Carrotal. Dietary supplementations of a mixture of carrot and orange juice have been found to control the oxidation of low density lipoproteins in addicted cigarette smokers. Purple or red Asiatic carrots having anthocyanin and lycopene are helpful antioxidants. Carotene lower the risk of cataract formation.

Beet root after sugarcane is the instant most important crop for the preparation of sugar. Beet root is the rich source of protein, carbohydrate, Ca, P and vitamin C. The color of root is due to the presence of red- violet pigment Beta – Cyanins and yellow pigment Beta- Xanthins.

Rutabagas acquire broad (up to 6 inches), round white or tan roots, with white or yellow flesh. Yellow- fleshed verities are the most attractive. Rutabagas have smooth, fleshy leaves identical to collard leaves. Rutabagas roots are commonly ingest as a cooked vegetable. They can be saved for different months in a root cellar and were a suitable winter vegetable in the days before refrigeration and long-area sailing of fresh goods. Rutabaga utilization has reject as natural produce has become more available throughout the year.

Turnips are adequate lesser than rutabagas (2–3 inches in diameter) with white flesh. Most cultivars have purple shoulders and white external color. Some varieties are fully white with no purple shoulders. Turnip leaves are hairy, and are generally eaten as a cooked vegetable. Some cultivars that have been matured for leaf production outcome very less roots. Turnip Young leaves are rich source of vitamin A (1500 IU/100 g), vitamin-C and Iron.

Parsnip is a root vegetable jointly related to carrot and parsley. It is a biennial plant developed as an annual. Parsnip is a great source of important nutrients, packing a generous quantity of Vitamins, minerals and fiber into individual serving. In particular, parsnips are an enormous source of vitamin C, Vitamin K and Folate, as well as several other important micronutrients.

4 Effect of Climate Change on Root Crops

The normal physiological functions of plant are vulnerable to different biotic and abiotic factors. Every plant has its own climatic requirements, in the availability of which plant reaches its true genetic potential of yield and other economical components. The absence of optimum environmental condition leads to decline in vigor of the plant. However in unfavorable conditions, crops can also be grown thanks to the modern protected structures with temperature, humidity and CO_2 supplementation.

Radish: Radish is a cool season crop but cultivars been grown up which can be developed in summer and spring. The main season crop does not need any special consideration but for summer and spring crops, choice of the variety is crucial. Radish varieties are very distinct to climatic requirements mainly the temperature. Main season cultivars grow good aroma, taste and root size when the temperature is between 10 and 15 °C. In temperate types, best root growth occurs at the beginning at 20–30 °C and later at 10–14 °C. Higher vegetation growth is favored at temperature above 25 °C. In hot weather roots become bitter and hard before obtaining marketable maturity. Long photoperiod and warm temperature favours early bolting even before proper root development. Ordinarily plants bolt when days are 8–10 h long. At less temperature, pungency is decreased. Radish is tolerant to frost.

Radish is grown up on all forms of soil but a light breakable soil is considered best. Radish can be grown in moderately acidic soils having a pH of 5.5–6.8. Sandy loam soil is excellent for rapid growth of roots. Heavy soils are not fascinating as they turn to produce misshapen roots with large, small laterals. The soil is thoroughly able to grant proper development of the roots.

Carrot: The climatic requirements of carrot are more or less identical as that of radish but it requires a relatively long growing season than radish. Though carrot is a cool season crop but a little part of the varieties can tolerant relatively high temperature. The color development and growth of roots is damaged by temperature. Optimum temperature for growth is 18–24 °C and for germination is 7–24 °C. The best root color develops at 15–20 °C. Temperature higher than 30 °C, especially in next stages of development, activate unsatisfactory solid flavor and roughness in the roots. Shape of the roots is not affected by the increase in temperature. Temperate types demand low temperature of 5–8 °C for 40–60 days before flowering to break inactivity. Carrot use deep and sloppy-loamy soil for best root development. Bulky soils may research root development and develop forking of roots. Highly acidic soils do not contribute good quality carrots. The maximum yield is expected when the soil pH is about 6.5.

Beetroot-It is a winter season crop and can tolerate freezing temperature. Unlike other root crops, beetroot does not have tropical type varieties. All varieties are temperate types and biennial in nature. Beetroots develop best color, sugar content, root quality and shape when temperature remains between 15 and 19 °C. The germination is slower and erratic if temperature is below 7 °C. Exposure of crop to temperature between 4 and 10 °C for 15 days or more induces pre-mature bolting. At temperature above 25 °C, the root color, texture and quality are adversely affected. Roots become stingy and tough in response to high temperature. Deep well drained loams or sandy loam soils are suitable for beetroot cultivation. In heavy soils, the roots become misshapen. Germination is also poor in heavy soils. The desirable soil pH range is 6.0–7.0.

Rutabaga – Rutabaga is a cool-climate crop, cold hardy or drought tolerant. Their moist dynamic root growth take place during season of low temperature. The leaves protect their feeding quality even after repeated vulnerability to frost. Rutabaga increase best in a slightly wide, productive and kind of acid soil. Rutabaga will not do accurately in soils that are excessive, rainy or poorly depleted.

Parsnip- The excellent temperatures for spreading parsnips are 16–20 °C. Roots are not injured by freeze, but chill temperature may result in failure of roots mature to flowering or "bolting". Well- drained, underground soils are required for parsnips with a sloppy, granular texture without stones to facilitate excellent development of the root. Compared to carrots, parsnips are more susceptible to acidic soils and superbly need a pH between 6.0 and 6.8. Parsnips do not require exposure to full sun and can be develop in moderately cloudy plots. Be informed that parsnips have a prolonged growing season and may immerse the soil for around a year. Beetroot is one of the few vegetables that can successfully be grown on saline soils.

Horseradish-Horseradish is an annual crop, generally cultivated in temperate regions of the globe, although up to some extent perennial cropping system is also adopted. Horseradish develop best at cool temperatures (60–65 °C). Horseradish require a moderately cold fall, when the root size and aroma develop. During the growing season, any of the ground leaves that shift brown. It needs high temperatures during the summer period and low temperatures during the late summer period as well as fall for the enlargement of the root. Horseradish grow up in all soil forms provide that they are well depleted, high in living matter, and productive. The suggested pH is 6.0–7.5.

Turnip-Turnip is a winter season, hardy crop and tolerates frost. Moderate and rainy climate is favorable conditions for growing turnip. Despite, it can also be developed where summers are sunny. The roots flourish good aroma, taste and size at 10–15 °C. For appropriate development roots favor Short day length and cool weather. High temperature and Long days promote early bolting without development of roots. In hot weather, roots become obscene, stiff and pungent. Turnip can be grown up in all forms of soils but well drained sandy loam soils, having ample humus, are best suited. Excessively light sandy soils or clay soils should be avoided. In such soils, plant growth is obstruct; roots are split and unsuitable for marketing. The optimum soil pH is 5.5–6.8.

5 Factors Responsible for Climate Change on Vegetable Production

The severity of environmental stresses i.e., extreme temperatures, reduced water accessibility, water logging and saltiness will be main hinder factors in enhancing vegetable production. These factors may also negatively affect the soil productivity and raised soil destruction (De la Pena and Hughes 2007). The responsible factors for climate change and its effect on vegetable crops are illustrated below:

5.1 Temperature

Optimum temperature plays an important role in the growth and development of root crops that ultimately increases the productivity of root vegetables. During their improvement, elevated temperature can influence photosynthesis, respiration, membrane stability and plant hormone levels as well as primary and secondary metabolites. Vegetables can be classified into different groups based on their temperature requirement especially temperature during night time for optimum growth and yield of the plant. Crops originated in tropical climate require relatively high temperature than the temperate crops which needs low temperature for their growth. However, when temperature exceeds too much even in tropical crops leads to destruction of protoplast and ultimately leads to death of cells.

Seedling Germination and Emergence Temperature of soil has a vital impact on growth and development of microorganisms, germination rate of seed, and elongation of root and potential of water as well as nutrient absorption. In general, the rate of these physiological processes positively correlates with increasing temperature. Most root crops being cool season vegetable crops, show good germination in 3–17 °C temperature range. Again as these crops are direct sown crops, soil temperature further plays an important role in faster germination of seeds, vegetative growth as well as root traits viz., length and diameter of root, number of root tips, and number of lateral roots as well as density of roots. The extent of theses correlation varies according to the environmental conditions. A little high winter temperatures could facilitate the autumn crop to be moved farther into the winter, but high summer temperatures could inhibit the production and yield of spring crops. Radish, rutabagas, and turnips are cool season crops. Radishes mature during long days and high temperatures prolonged elongated leaves and misshapen roots and become pungent and pithy. However, cultivars have been grown up that produce high quality roots during warm, long days.

Growth and Development Plant growth is enhanced upon the availability of optimum moisture/humidity, temperature, light, soil nutrient and carbon dioxide. Temperature is of vital consequence in the formation of an elevated harvest index. The rate of maturity of the crop is directly correlated with the prevailed environmental condition especially temperature. A fluctuation in the temperature particularly escalating temperature tends to increase in formation of tropospheric ozone that causes oxidative stress for the crops. As a result, it reduces photosynthetic that check the plant growth and development (Adebayo 2010; Ainsworth et al. 2012). Modest day and relatively low night temperatures during root formation stage helps in better accumulation of carbohydrate in root crops. According to Rosenfeld et al. 2002, under high temperatures, when the carrot grown up then the rate of synthesis of terpene increases, resulting bitter taste of carrots. Root growth is also affected by the degree of tillage, formation of hardpans below the tilled layer as well as by controlled growing media such as small size of container, etc. Asymmetrical root system occurs due to various forces like compression and tensile forced due to which leeward and windward roots enter into stress, respectively (Nicoll et al. 2008). So, the development of cultivars suitable for planting in summer as well as winter season in all agro climatic areas has made it possible for all round year production of carrot. (Simon et al. 2008).

Yield and Quality According to Bloksma et al. (2003), temperature played a key role in enhancing quality and yield of carrot roots. Simon et al. (1982) observed that the growth temperature of carrots affects the sugar, level of volatiles, carotenoids and texture of the roots. According to Rubatzky et al. (1999), during storage root development at temperatures above 25 °C, the respiration rate of the plant rises, which results in lower productivity. Rutabagas developed 90–100 days after seeding and produce the best character roots under cool conditions. According to Brewester (1994) income of a crop is determined by the following: the amount of light absorbed by its leaves while harvestable dry matter is produced; absorbed light is transformed into sucrose by photosynthesis; photosynthetic product portion transferred to the harvested plant part; the combined between the photosynthetic sucrose and biochemical component of the harvested material and; the weight losses due to decomposition and respiration after the above photosynthetic and biosynthetic processes have taken place. However, crops like radish utilize a cold period for seed production. The yield and quality of a number of vegetable crops are adversely affected by high temperatures, which decrease the marketability of produce. According to Mazza 1989, Carrot quality can be classified by external quality parameters as per root length and diameter, pigment on the skin, and absence of deformities as well as internal quality parameters viz., fruit firmness, sensory quality, carotene content, and terpene content. Thus root vegetables producer, packing personnel and transporter must pay attention to temperature fluctuations during growing period to harvest at the apportion time. Mineral accumulation also damaged by high temperature and/or direct sunlight.

External Quality Shape of roots can vary from conical, cylindrical to spherical. According to Rosenfeld et al. (1998), shape of carrot root depends upon genotype and climate conditions during root growth. High temperatures (>25 °C) leads to expansion of sloping of shoulders, giving the appearance of a neck, while availability of optimum temperatures gives rise to lightly rounded or square to the root. When temperature exceeds 25 °C, storage material in the roots is also inhibited, but low temperature range of 10–15 °C enhances length of root (Joubert et al. 1994). Bacteria viz., *Agrobacterium rhizogenes* and *Erwinia chrysathemi* lead to misshapen roots and ultimately loss of yield. Nematodes like root knot nematode (Meloidogyne hapla) in a temperature range between 12 and 18 °C during the crop growth stage leads to crop losses.

Internal Quality Important character inherited characteristic like sweet taste, harshness and pungency, taste and carotene content. In any part of the world, Carrots are not a large essential food, but they are highly valued in many countries due to their nutritional quality. The production characteristic of carrots is governed by place of sowing and also influenced by combined factor of rainfall, growth system and diurnal temperature of the place and also influenced by duration of cropping (Hogstad et al. 1997). Terpenes in the presence of high temperature are the chief component to impart bitter taste in carrot roots (Simon et al. 1980). Climatic conditions have an ample effect on the taste of carrots (Baardesth et al., 1995). Bhale (2004) demonstrated that high rate of temperatures may cause damage to root cells and can subdue their desirable characteristic. Carrot taste manly depend on the genetic behaviour, location of sowing and climatic conditions over the cropping period. (Hogstad et al. 1997; Kjellenberg 2007). According to Rosenfeld et al. (1998) temperature was the most vital component defining the sensory characteristic and chemical factor related to growth and productivity. The sensory outline of carrots cultivated in controlled climate compartments correlate well with those of naturally cultivated environment without altering any naturally grown behaviour (Rosenfeld 1998). Low temperatures of various ranges such as (9, 12 as well as 15 °C) promoted the sweet and acidic taste, juiciness and crispiness of carrots vis a vis when subjected to high temperatures such as (18 and 21 °C) it led to a astringent flavour and resulted in hardness of roots grown in phytotrons (Rosenfeld 1998). Thus, it can be well articulated that low temperatures in a range of 12–15 °C accelerated carrot yield and some other characteristic like internal and external behaviour significantly.

Flower Induction and Dormancy Bolting or flowering can be a constrained. Normally rutabagas are biennials, i.e., after a cold period they form a swollen root during the first year of growth and flowering stems in the second year of growth. If transplants are exposed to low temperatures (below 50 °C) when they are less than 10 weeks old, this will cause the development of flowering stems. The extent of the low temperature period needed to cause flowering diversify with the cultivar being developed. Field plantings or seedlings can be damaged by low temperatures but uproot more than 10 weeks old would depend upon several nights of freezing temperatures to promote flowering.

In radish and carrot, high temperature coupled with long day photoperiod system promote flowering. Biennial vegetables like root crops, require optimum chilling period during winter for the formation of seed. However, upon the initiation of the seed stalk (bolting), crop quality starts degrading significantly.

In case of Radish, European or Temperate type (Biennial) – Short duration (25–30 days), produce good quality roots with less pungent. Small size, yield low (75–100q/ha), requires chilling temperature for bolting, seeds are not produced under North Indian plains like White Icicle, Pusa Himani, Rapid Red White Tipped, Scarlet long and Scarlet Globe and Asiatic or Tropical Type (Annual)-More pungent, long duration (45–55 days), large size roots, High yield (250–300q/ha), do not require chilling temperature for bolting, sets seed freely under tropical condition, can be produced under North Indian plains also.

In case of Carrot- Annual - For root production and Biennial - For flowering and seed set.

Asiatic type are high yielding and produce seed under tropical condition, but they are poor in carotene content and other quality attributes.

Japanese White- Seed produced only in hills, less bolting.

Pusa Kesar- Roots stay for about a month longer in the field than local Red without bolting, sets seeds freely in plains.

Pusa Meghali- Suitable for early sowing (mid-August- early October).

Nantes Half Long- suitable for sowing from mid-October to early December.

Pusa Yamdagni, Early Nantes, Chantaney, Imperator- Suitable for sowing from mid-October to early December.

Reproductive Development High temperature impart dominance upon reproductive development mainly in two ways. Firstly, the reproductive development rate is enhanced, minimizing the period required for seed-filling as well as for maturation of fruits. Thus, this results in reduction in singular seed production and root density and up to some extent reduction of soluble solids in the roots. Moreover, reproductive development is inhibited in most of the cops on slight raising of temperature more than the optimum temperature range. Thus, cool-season vegetable crops, being extensively grown in the tropical climates, are more susceptible to global warming.

Root Ripening High soil temperature near the root surface occurred due to prolonged exposure to sunlight fasten the ripening as well as other physiological processes. Typical climacteric patterns happened at 20, 25, 30 and 35 °C along with the climacteric maximum rapidly increasing with temperature, but only an inhibited respiratory rate with time was recorded at 40 °C. The exposure to exogenous ethylene or propylene increases the ripening response up to 35 °C. However, at 40 °C the respiratory rate was increased, but ethylene production and normal ripening did not occur.

Rapid Cooling techniques have been used in vegetable crops since the 1920s to remove field heat from fresh produce, based on the thumb rule that for every 10 °C decrease in ambient temperature, shelf of the produce is extended two to three fold. Generally, vegetable crops are cooled after harvest and prior to packing of the produce. Rapid cooling reduces the respiration rate significantly. Reduction of rate of respiration and enzymatic activity leads to slowing of senescence rate of produce, continuance of firmness, reduction of water loss as well as pathogenic microbial activity (Talbot and Chau 2002). Hydro-cooling, air cooling and vacuum cooling types of rapid cooling methods require significant energy for their operations (Thompson 2002). Hence, it is estimated that upon high temperature climatic conditions, horticultural produces are harvested with higher pulp temperatures, thus ultimately needing much more energy for optimum cooling and price of the produce goes up.

Antioxidant Activity Antioxidant in root vegetables crops fluctuate according to the temperature prevailed in the rowing area. High temperature lead to increase in the levels of flavonoids as well as antioxidant capacity of the roots. Warland et al. (2006) also documented the impact of higher temperatures towards reduction in vitamin content in the vegetable crops.

Physiological Disorders and Tolerance to High Temperatures Any type of irregularity in economically crucial component of vegetables or other parts that sharing to yield and quality of root vegetable is physiological disorder. Physiological disorder provoke as deficiency of micronutrient, unusual fluctuation in temperature, Poor soil conditions and improper moisture availability during farming. This condition is big crucial in physiological disorder in root vegetable crop as character and marketable yield is very a major factor in vegetable market. If a farmer has a homogenous, healthy and good- looking commodity, he will get higher prices. Susceptibility of root vegetables crops towards high temperatures leads to appearance of a number of physiological disorders and other internal and external symptoms. Absence of moisture causes erratic maturity and poor quality roots (Table 1). These root crops are also affected to manganese (Mn) deficiency, which create a general yellowing of plant leaves, appearing about midway in the season. It is mainly a constraint in soils with a pH above 6.5. Some Other disorders related to high or low temperatures are shown in Table 1.

Table 1 Physiological disorders of vegetables caused by high or low temperature (Wien 1997)

Crop	Disorder	Aggravating factor
Carrot	Low carotene content	Temperatures <10 °C or > 20 °C
	Forking	When hard soil does not allow straight growth.
	Poor extent, poor development or pithiness	When harvesting is delayed due to improper climatic conditions
	Cavity spot	Associated with an increase accumulation of K and decreased accumulation of Ca.
	Splitting/cracking	Sudden change in soil moisture status
Radish	Akashin	High day and high night temperature 30 °C and 20 °C, as well as by soil moisture.
	Bolting	Seed vernalizing crop in response to low temperature.
	Elongated root or forking	Excess moisture during root development, occur in heavy soil due to soil compactness.
	Pungency	High temperature and water stress condition.
	Root splitting	Cultivation in soil containing high N
	Pithiness/hollow root wart	Caused due to high temperature three weeks before harvest, soil moisture stress. High temperature during 16–30 days of sowing.
	Hollow root/wart disorder	High temperature during 16–30 days of sowing. Soil moisture stress
	Wart disorder/scab	Soil moisture stress
Turnip	Cracking /Brown heart	Boron deficiency. More common in acidic soils.
Beetroot	Poor root development, zoning	High warm weather high temperature, boron deficiency
Rutabaga	Brown heart or water core	Boron deficiency

5.2 *Carbon Dioxide Exposure*

The Earth' atmosphere consists mostly of nitrogen (78.1%) and oxygen (20.9%) with argon (0.93%) and carbon di oxide (0.031%) constitute next most abundant gases (Lide 2009). The greenhouse effect is a complex mixture of the effects of CO_2, water vapour and minute amounts of gases including methane, nitrous oxide, and ozone, which consume the radiation leaving the surface of earth (IPCC 2001). Plants have been precisely damaged by increasing atmospheric CO_2 concentration they are the first molecular link between the atmosphere and the biosphere. Carbon dioxide levels have the ability to affect all environment, from microscopic cellular organisms to the macroscopic agro-ecosystems.CO_2 has the capacity to influence major physiological processes namely photosynthesis, respiration and transpiration. In most crops including vegetable crops, elevated [CO_2] level improves the water use efficiency of the crop (WUE) due to the declines in stomatal conductance (Rogers and Dahlman 1993), thus ultimately reducing susceptibility to drought and irrigation requirements. However, the impact of reduced transpiration on root vegetable crops is unlikely to be large as vegetables are cultivated only in the regions with irrigation facility. Wolfe et al. (1998) concluded that the greatest increases in productivity as a result of elevated [CO_2] exposure will appear when soil N and P availability are high. Changes in CO_2 concentrations in the atmosphere can alter plant tissues in terms of growth and physiological behaviour. In summary, that increased atmospheric CO_2 alters net photosynthesis, biomass production, sugars and organic acids contents, stomatal conductance, firmness, seed yield, light, water, and nutrient use efficiency and plant water potential. It is also observed that the elevated behaviour of moisture mixed with CO_2 levels lead to high shoot: root ratios, and it continues to increase in damp soil condition (Morgan et al. 2004). Another hypothesis recommend that the shoot: root ratio may decrease with increased CO_2, Since the wet condition may led to reduced N concentration and the increase in N demand may led to further increase in demand for more large-scale root exploration of the soil (Milchunas et al. 2005). CO_2 enriched climatic atmosphere during controlled production were utilized to attain seedlings with lower shoot: root ratios, mainly resulted in rich biomass of root (Wullschleger et al. 2002; Cortes et al. 2004). Some authors have also mentioned that storage of water along with increased CO_2 levels leads to elevated shoot: root ratios, which may be larger in year with more moisture in the soil. (Morgan et al. 2004).

5.3 Interactions of High Temperature and Carbon Dioxide

Generally, temperature of the ambient environment increases with increase in CO_2 concentration of the atmosphere (Boote et al. 1997), which may be due to ability of the plants to promote added carbohydrate when growth rate is fast and this decreases downward adjustment to [CO_2] (Wolfe et al. 1998). However, carrot, being an indeterminate crop, produce high yield with increase in temperature (Wheeler et al. 1996). This reaction may be due to the fact that temperature-sensitive cultivars (requiring only a 4.0–5.8 °C temperature increase to offset [CO_2] increases) being a short- season type. Thus, all-around climatic change may impact the potential of the related cultivars. Stomatal closure due to high [CO_2] inhibit water losses and thereby inhibiting the negative impact of increased transpiration at higher temperatures.

5.4 Drought

A drought is the most important factor of climate change, responsible for the great famines in past a century. Around 45% of world agricultural lands own extreme drought condition (Bot et al. 2000) that badly affects the agriculture ecosystem and consequently food security of the world. It also influences the function, structure and productivity of soil ecosystem (Lal et al. 2013), which reduces microbial activities and available nutrient uptake by crops. Severe water stress along with increased temperature will reduce productivity and quality of vegetables crops due to increase in evapotranspiration, abridged precipitation and reduced soil microbial activities leads to escalation of solute concentration and eventually sinking the water potential, disrupting membranes and photosynthesis processes, leading to cell death. Due to rapid rates of transpiration cause desiccating effects and wilting of plants (Yusuf 2012). Heat stress decreases the anthesis duration, total soluble protein and nitrogen in leaf as it suppresses the photosynthetic processes, rubisco protein and sucrose phosphate synthase activity (Xu and Zhou 2006). Vegetable crops requires irrigation under certain critical stages, if not so, the productivity and quality of crops drastically reduced or even cause death of plants. Critical stages of irrigation of some vegetables are illustrated in Table 1. In root vegetable crops, the yield and quality are significantly reduced under water deficit condition as it suppresses the translocation of carbohydrates from the leaves to the storage organs.

All the root vegetable crops require a regular amount of water during the growing season. Water recycled to cool and slow the rise of the crop as well as to speed growth and character. Irrigation stimulate soil insecticides or bring fertilizer to the plant roots. Development of root is heavily determined by thriving situations such as moisture deficiency (Sangakkara et al. 2010). The root growth under drought condition over a sustained period may led to high tolerance to tough climate. This negative impact of drought stress cultivation is usually more on Shoot growth in comparison to the root. This decrement in susceptibility of roots appears due to reaction of the increased osmotic change in behaviour of roots in water deficient soil condition which grant the increased ability of water uptake, and is also due to increasing root walls (Sharp et al. 2004). Thus, decrease in the shoot: root ratio under water stress is a mainly an aspect of investigation and research, It is developed either by an further reduction in root growth or it may be due to larger decline in shoot growing capabilities in comparison to root growth (Franco et al. 2006). Besides the shoot: root ratio, dry biomass, root length, root diameter, fresh biomass, and surface area, and thickness of cortex and behaviour of root viz., root turnover, hardening, as well as hydraulic conductivity may be steadily affected by water stress conditions. Plants subjected to scarcity of water develop huge root system for efficient utilization of the available water. Deep root system is also one of the most important aspect in deciding the potential of plant to absorb as well as to utilize water from soil (Koike et al. 2003). Such behaviour causes shorter, thicker and denser roots and reduction in the shoot: root ratio up to 60% observed in the preconditioned plants. This action was also connected to shifting of the assimilates towards the root. On the other hand, drought induced inside the root passage of plant may have impact on carbon sequestration. However, root movement can decline due to drought stress. The hardening of roots, as explained by enhanced ratio of brown roots, is continual in drought stressed plants (Franco et al. 2006). The combined effects of moisture stress and low temperature led plants with reduced shoot: root ratios, higher percentages of brown roots as well as xylem vessels in their stems and roots. *Root elongation and enlargement stage are the most sensitive stage of these vegetables against moisture stress. Root vegetables like radish and carrots require an even and abundant supply of water throughout the season. Drought stress causes small, woody, pithy, forked and badly flavored roots. Depending on soil moisture-holding capacity in horse radish, irrigation during dry periods, particularly in late*

Table 2 Critical stages of drought stress and its impact on vegetable crops

Vegetable crops	Critical period for watering	Impact of water stress
Carrot, radish, and turnip	Root enlargement	Distorted, rough and poor growth of roots, strong and pungent odour in carrot, accumulation of harmful nitrates in roots

summer to fall, can improve marketable yield. Horse radish is quite drought tolerant, but the roots become woody and has a weak flavour if stressed too much *Studies showed that most primary-root development occurs during the mid- to late fall period rather than during the summer.* Sugar beet is adapted to a wide range of climatic conditions. It is tolerant to moderate soil water stress (Hills et al. 1986). When sugar beet plants lose water from their leaves faster than their roots can absorb it from the soil, internal water deficit develops, growth is slowed, and the plants may wilt. Even with plenty of water, sugar beet may wilt slightly during the afternoon on hot, dry days. Such wilting does not indicate a need for irrigation. However, if the plants wilt early in the day, or if recovery is slow in the late afternoon as temperatures and light intensity decline, irrigation is needed. In arid and semiarid areas, irrigation may supply all or most of the water that crops need. In more humid production areas, irrigation is used primarily to supplement infrequent or irregular precipitation during short-term droughts. (Table 2)

Impaired stomatal conductance due to ozone exposure can reduce root growth, affecting crops such as carrots, and beet roots.

5.5 *Flooding*

Flooding may occur as an overflow of water from water form, such as a river, lake, or ocean, emerge in some of that water avoiding its regular boundaries or it may occur due to an accumulation of rainwater on waterlogged ground in an sectional. Micronutrients (boron, chlorine, copper, iron, manganese, molybdenum and zinc) are vital for plant growth and play a significant role in quantity and quality production of vegetable crops Flooding, nutrients extract and float out from upper abundant soils together with corrosion of soils May results in nutrient lack soils that precisely disturb the productivity of vegetables. (Tuomisto et al. 2017). The decomposition rates of organic sources may also hampers due to rapid loss of chemical nutrients by soil biology (Pandey et al. 2007). Under the flooded condition, an aerobic process repressed as a reduction of oxygen around the root zone of vegetable plants. Under flooded condition, roots respiration hampers because soil air replaced by water resulting in improper nutrient uptake and also leads prone to soil borne diseases. Severe flooding causes rotting of roots, shrinkage of roots in storage and necrotic spots in leaves. The Horse radish roots convert very soft and have a strong spice if over watered. Water horseradish once a week (1–2 inches of water) so it enter to a measure of 18–24 inches. Waterlogging leads to decrease in the increase in oxygen content towards the waterlogged organs, which may negatively change development of plant growth. Thus, root function, is a result of lacking oxygenation, which can revise plant growth and disturb the crop yields through conflict with water relations, mineral nutrition, and hormone balance (Horchani et al. 2009);. Over-irrigation, excessive rainfall, or flooding, Oxygen deficient can appear in soils leading to root anoxia. Low oxygen diffusion and reduced aeration can severely affect rhizogenesis as well as morphological, anatomical, and physiological properties of the root system (Ochoa et al. 2003). Deficient or excessive moisture content in soil can disturb the growth and development of roots. Low soil moisture leads to production of thin and long roots, elevated high soil moisture results in production of thick and pale coloured root growth (Rubatzky et al. 1999).

5.6 Photoperiod

Sunshine and photoperiod have significant influence on the productive potential of crop. Photoperiod ranges from 12 hour day at the equator to repeated light or darkness throughout the 24 hours for a section of the year at the north and south poles (Decoteau 1998). For the development of storage roots all root crops, i.e., carrot, radish, turnip, beetroot, etc. require abundant sunshine. Carrots grow at constant temperatures 9–21 °C under controlled environmental conditions with 16 hours of artificial light and found a significant increase resulted due to combination of temperature and photoperiod. Carrots can produce yield in long days but requires relatively low temperatures for production of flower. Hence, Suojala (2000) noted that long photoperiod (16 h) has a far reaching impact on carrot leaves in comparison to short photoperiod (8 h). The inherited quality of carrot roots, mainly the content of sucrose recovered with long photoperiods. Samuqliene et al. (2008) shown that the rise in sugar content in the apical meristems of carrots over initiation of flower growth specially under long day condition (16 h at 21/17 °C) treatments and reduced quantity of sucrose was recorded under short day conditions (8 h at 4 °C). The carotene pigment in carrot was also inhibited when light hours were decreased and temperatures are altered below and above the temperature required for optimal growth (10 to 25 °C).

5.7 Relative Humidity

Vegetable plants are very sensitive to the fluctuation in atmospheric moisture. Under elevated relative humidity (85–90%) conditions, the reproductive processes of vegetables are affected in respect to anthesis, anther dehiscence, pollination and fertilization. If plants exposed to high temperature, it requires high humidity because of transpiration. High relative humidity responsible for low transpiration rates, decrease in nutrients (nitrogen, phosphors, potassium, calcium and magnesium), rapid spread of diseases like black spot, powdery mildew, blossom-end rot (along with Ca deficiency) etc., reduced fruit quality and leaf area (Triguii et al. 1999). High RH reduces the quality and yield of tomato as it reduces the leaf area index, decreases the translocation of Calcium towards leaves, death of apical parts. (Holder and Cockshull 1990; Barker 1990). Some of the advantages are also available under mild RH stress conditions like hardening (under low RH), osmotic adjustment, leaf turgor regulation etc. Rainfall is crucial circumstances, especially when vegetables are mature under dryland conditions. Suitable soil precipitation is needed for acceptable crop establishment, good yields and good quality. This moisture may have from rainfall or irrigation. High rainfall incident may cause flood damage, fractional drench on certain soil types, and will regularly favour diseases development. Humidity, or air moisture content, may also play a character.

5.8 Nutrient Deficiency or Excess

For the raising of a successful crop, proper fertilization management is of utmost need. Under drastic climatic condition, there arise a significant competition among the plants for minerals and water. Similarly, presence of a specific nutrient in the soil in higher amount has negative impact on plant growth. Deficiency and excess amount of nutrient generally affect development of root, however the extent of impact vary according to family or species of plant (Francini and Sebastiani 2010). Mg-deficiency did not affect root Dry weight as well as shoot: root ratios. However, root growth is generally negatively damaged by nutrient toxicity.

5.9 Heavy Metal Stress

Heavy metals are a group of pollutants recognized as unsafe to human health and human hazard occurs through all environmental media. Considering metals are naturally occurring chemicals that do not breakdown in the environment and can accumulate in soils, water and the sediments of lakes and rivers, it is very important to evaluate the contribution and pattern of their natural emissions. Heavy metals play a vital role in plant bionomics and presence in excess amount restrict the growth of roots disturb the growth of roots, alike in limited bulk as fragment material (Hagemeyer and Breckle 2002). Equivalent Al is the large toxic component in largest acid soils, and also one of the most deleterious factors for plant growth in acid soils. Al cations restrict growth and development of root system, therefore, hinder normal physiological function of (Matsumoto 2002). Modifying diversity in root planning may be crucial for plants to tolerate soils with tremendous content of heavy metals (Hagemeyer and Breckle 2002).

5.10 Salinity

In hot and dry environments, high evapotranspiration reaction in substantial water loss, thus residue salt over the plant roots which hinder with the plant's capacity to uptake water. Physiologically, salinity require an initial water loss that results from the approximately high solute concentrations in the soil, causes ion distinct stresses derived from modified K+/Na + proportion, and leads to a strengthen in Na^+ and Cl^- concentrations that are harmful to plants (Yamaguchi and Blumwald 2005). During dry conditions, considerable water loss occurs due to high evapo-transpiration rate and accumulation of salt around roots that restricts the uptake of water and nutrient by the plants (Taiz and Zeiger 2006). High salinity in soil alters the K^+/Na^+ ratios results in ion-specific stresses that make Na^+ and Cl^- concentrations which is harmful to plants. It affects the plants in different mechanisms like decrease photosynthesis, loss of turgidity, tissue necrosis, leaf curling, leaf abscission, wilting and even death of plant under severe conditions (Cheeseman 1988; Yamaguchi and Blumwald 2005). The excess salinity in soil affects the ontogeny of plants and decreases the productivity of vegetables. The effects of accumulated salts may encounter by the plants through various mechanisms such as osmotic stress tolerance and salt elimination (Munns and Tester 2008). Vegetable productivity may also affects through irrigation as the ground water is contaminated with deposition of salts in bedrocks or by various natural or manmade actions. Salinity causes osmotic stress leading to inhibited water potential as well as ionic stress due to the changes in the amount of particular ions inside the medium as well as the root tissue system (Bernstein and Kafkafi 2002).

5.11 Air Pollutants

Ozone in the atmosphere is the phytochemical reaction among carbon monoxide, methane and other hydrocarbons in the presence of nitrogen category. Presently, tropospheric ozone concentrations found above preindustrial levels in most agricultural areas of the world. Air pollutants (NO_2, CO_2, CH_4 etc.) perform with hydroxyl radicals in the presence of solar radiation, results in formation of ozone at ground level, which may cause oxidative damage to photosynthetic activity in most of the vegetables (Wilkinson et al. 2012). Under drought conditions, higher atmospheric CO_2 level helps in maintaining biomass production. It was concluded from the previous studies that 1000–3000 μmol/mol CO_2concentrations increases both antioxidants and yield but at the same time very high concentration of CO_2 (3000–5000 μmol/mol) results in decreased protein content, growth, yield and quality of crops (Kimball and Idso 1983; Tuab et al. 2008). Under high CO_2 concentration will lower O_3 level uptake by crop plants due to reduced stomatal conductance, to that reduce injury to the plant and sustaining crop production (McKee et al. 2000).

The significant impact of ozonosphere on flora have been evaluated under lab as well as ground conditions. Ozone enters plant tissues through the stomata and causes primitive biological damage in the palisade cells (Mauzerall and Wang 2001). The negative impact may be due to difference in in membrane permeability and may or may not be manifested as visible injury or reduced yield (Krupa and Manning 1988).

Visible injury syndrome is manifested in the form of bronzing, leaf chlorosis as well as premature senescence (Felzer et al. 2007). Leaf tissue stressed could influence the rate of photosynthesis, biomass production and enhancement of overall appearance, colour as well as aromatic compounds. Today, report of the related information linked to plant responses to ozone exposure explain that there is huge variation in category response. Greatest impact in vegetable crops may develop from changes in carbon transport. Underground storage organs like roots obtain carbon in the form of starch and sugars, which are considered as quality measures for fresh as well as processed crops. Upon irregularities in carbon transportation, quality of the produce is drastically reduced. Exposure of other crops to high concentrations of atmospheric ozone level can give rise to external as well as internal disorders, which can either appear together or independently. These physiological disorders reduce the postharvest quality of fruit and vegetable crops desired for both fresh market and processing by causing symptoms like yellowing, variations in starch and sugars content of underground organs, etc. Reduced biomass production directly affect the size, other important visual quality parameters. Moreover, faulty stomatal conductance due to ozone risk can inhibit root growth in root crops (Felzer et al. 2007).

6 Mitigation of Climate Change Impacts

Adverse effects of climate crisis can be mitigating through adaptation of energetic and potent measures precisely build up the yield and quality of the vegetables. There are different methods to mitigate the effects of climate change are enlisted below:

6.1 Strengthen Vegetable Management Systems

The production yields of the vegetable grown in lowland topography subjected to hot and moist climate can increase by various management methods. AVRDC – The World Vegetable Centre has developed various mitigation strategies to combat areas subjected to drought and flooding, to control salinity of soil, and also to production of plant well enriched with nutrients.

6.2 *Water-Saving Irrigation Management*

Water management plays an important role in productivity of the crops. The optimum water is also dependent on the weather and climate of the places, stage of crop production, crop diversity, soil water absorption capabilities and texture of the soil, also influenced by irrigation method employed for production of crops. (Phene 1989).

6.3 *Agronomical Practices That Conserve Water*

Various crop growing methods such as use of elevated beds, use of desired shelters and mulching help in retention of moisture capabilities of soil and also prevent the soil degradation and prevent the damage of crops from excessive rains, elevated temperature of atmosphere and flooding. Protected horticulture is a process of growing crops in a controlled environment. These practices can be used independently or in combination, to provide favourable environment to save plants from harsh climate and extend the duration of cultivation or off-season crop production. Pusa Sweti is an offseason variety of Turnip. Variegata - a cultivar that is less invasive, has cream- variegated leaves, and tolerates partial shade of Horseradish.

6.4 *Enriched Stress Resistance Through Grafting*

Grafting is a technique to combine two living parts of plant (rootstock and scion) to produce an individual plant.it mainly result in prevention of soil borne disease which effect the production capacity as well as quality. But it may lead to development of soil stresses in form of salinity of soil, low soil temperature, drought and flooding if suitable tolerant rootstocks are used.

6.5 *Promoting Climate-Resilient Vegetables*

Enriched, vegetable germplasm is ergonomically well suited opportunity for grower to conflict the challenges of a dynamic climate. Maximum current producer show very less genetic vulnerability including environmental stresses. Breeding new cultivars, basically for intensive, high profit production systems in developed countries, under ideal growth circumstances may have counter-preferred for characteristic which would provide the adaptation or tolerance to low profit and not favourable cultivating conditions.

6.6 High Temperatures Tolerance

The main factor to achieve high yields with heat tolerant varieties is the expanding of their genome base through crossing among heat tolerant tropical climate and disease resistant temperature or low temperature climate. Also, heat tolerant cultivars are compulsory to meet the changing climatic condition and this must be in line with the non-heat tolerant cultivars subjected to minimal stress condition. Various crops are altered and modified for ability to cope with heat tolerance. Pusa Chetki, Pusa Reshmi and Punjab Safed are two heat tolerant varieties in radish. Pusa Vrishti, Pusa Kesar are two heat tolerant varieties in carrot.

6.7 Drought Tolerance and Water-Use Efficiency

Plants combat drought stress in several ways. Plants may overcome drought stress by reducing their life cycle in water deficient condition (Chaves and Oliveira 2004). Transmission of this genome will led to increase in tolerance of vegetable crops to dry surroundings. OOTY-1 Variety is a heat tolerant variety of Carrot.

6.8 Saline Soils Tolerance

Various research are directed to develop the salt tolerance crops through conventional breeding programs but they have little success due to the genetic and physiologic multiplicity of this trait (Flowers 2004). Development in breeding for salt tolerance requires qualified screening methods, best availability of genetic variability, and capability to transfer the gene to the targeted species of research. Screening for salt tolerance in the natural working condition is not recommended because of the uneven level of salinity present in the soil.

7 Recommendations

It could be recommended that the scientists in developing countries should be working on how they could mitigate and adapted the climate change in agriculture because the agriculture is a major source for income. Thus, the researchers should carrying out their research on development varieties of crop plants which can successively develop under conditions of drought stress, water scarcity, heat stress and higher levels of water and soil salinity, as well as being naturally resistant to certain diseases and pests. Also, to change planting dates of vegetable crops to suit the new weather conditions and the planting of varieties in the proper climatic region to raise the yield from the water unit.

8 Conclusion

Climate change and Agriculture are linked with each other in multiple forms, as biotic and abiotic stress which a plant is subjected is mainly due to changing climatic pattern, which have detrimental effects on the natural growing condition of any crop. The soil and cropping pattern is influenced by climatic change by various ways, such as fluctuation in annual rainfall required for sowing the crops, average temperature of the place, heat waves, adaptation of pests and microbes, global warming and ozone layer depletion and variations in sea level. Last few decades, it has been recorded that there is tremendous change in the climate as it is regular process, which indirectly influence the production, productivity and quality of root vegetable crops. Global warming and pollutions (mostly soil, air and water) are the two main causes of the climate change. According to reports, it seems that it may mostly impacts on the world food security in near destiny. Climate-smart agriculture is the only panacea to control the negative impact of climate diversity on crop production, and it need to be taken care before it could led to huge losses in production capabilities. So, for preserve the quality and quantity of vegetables, the impacts of climate change can be alleviate by maintain the proper approaches like develop cultivars tolerance to biotic and abiotic stresses; forest, soil and water conservation; utilisation of renewable energy, protected cultivation, Hi-tech and judicious management land resource, cropping sequence etc. Distinct extension programmes need to be made for confirming these approaches with the farmers, so that they can quickly use these technologies in vegetable cultivation practices.

References

Adebayo AA (2010) Climate: resource and resistance to agriculture. Eight Inaugural Lecture. Federal University of Technology, Yola

Ainsworth EA, Yendrek CR, Sitch S, Collins WJ, Emberson LD (2012) The effects of tropospheric ozone on net primary productivity and implications for climate change. Annu Rev Plant Biol 63:637–661

Baardesth P, Rosenfeld HJ, Sundt G, Skred P, Lea P, Slinde E (1995) Evaluation of carrot varieties for production of deep-fried carrot chips. I. Chemical aspects. *Food Res Int* 28(3):195–200

Barker JC (1990) Effects of day and night humidity on yield and fruit quality of glasshouse tomatoes (*Lycopersicon esculentum* Mill.). J Horticul Sci 65(3):323–331

Bernstein N, Kafkafi U (2002) Root growth under salinity stress. In: Waisel Y, Eshel A, Kafkafi U (eds) Plant roots: the hidden half, 3rd edn. Marcel Dekker, New York, pp 787–805

Bhale SD (2004) Effects of ohmic heating on color, rehydration and textural characteristics of fresh carrot cubes. Agricultural University, India

Bloksma J, Huber M, Northolt M, Van Der Burgt GJ, Adriaansentennekens R (2003) The inner quality concept for food, based on life processes. Louis Bolk Instituut, Berlin

Boote KJ, Pickering NB, Allen LH Jr (1997) Plant modeling: advances and gaps in our capability to predict future crop growth and yield in response to global climate change. In: Allen LH Jr, Kirkham MH, Olszyck DM, Whitman CE (eds) Advances in Carbon Dioxide Effects Research, ASA Special Publication No. 61. ASA/CSSA/SSSA, Madison, WI, pp 179–228

Bot AJ, Nachtergaele FO, Young A (2000) Land resource potential and constraints at regional and country levels; world soil resources reports 90; land and water development division. FAO, Rome

Brewester JL (1994) Onions and other vegetable alliums. Horticulture Research International, Wellesbourne

Chaves MM, Oliveira (2004) Mechanisms underlying plant resilience to water deficits: prospects for watersaving agriculture. J Exp Bot 55:2365–2384

Cheeseman J (1988) Mechanisms of salinity tolerance in plants. Plant Physiol 87:104–108

Cortes P, Espelta JM, Save R, Biel C (2004) Effects of a nursery CO_2 enriched atmosphere on the germination and seedling morphology of two Mediterranean oaks with contrasting leaf habit. New For 28:79–88

De la Pena R, Hughes J (2007) Improving vegetable productivity in a variable and changing climate. J SAT Agric Res 4:1–2

Decoteau D (1998) Plant physiology: environmental factors and photosynthesis. Department of Horticulture, Pennsylvania State University, USA

Felzer BS, Cronin T, Reilly JM, Melillo JM, Wang (2007) Impacts of ozone on trees and crops. C R Geosci 339:784–798

Flowers TJ (2004) Improving crop salt tolerance. J Exp Bot 55:307–319

Francini A, Sebastiani L (2010) Copper effects on Prunus persica in two different grafting combinations (*P.persica* X *P.amygdalus* and *P.cerasiferae*). J Plant Nutr 33:1338–1352

Franco JA, Martinez-Sanchez JJ, Fernandez JA, Banon S (2006) Selection and nursery production of ornamental plants for landscaping and xerogardening in semiarid environments. J Hortic Sci Biotech 81:3–17

Hagemeyer J, Breckle SW (2002) Trace element stress in roots. In: Waisel Y, Eshel A, Kafkafi U (eds) Plant roots: the hidden half, 3rd edn. Marcel Dekker, New York, pp 763–785

Hills FJ, Johnson SS, Goodwin BA (1986) The sugarbeet industry in California. University. California Agriculture Experiment Station. Bull.: 1916

Hogstad S, Risvik E, Steinsholt K (1997) Sensory quality and composition in carrots: a multivariate study. Acta Agric Scand 47:253–364

Holder R, Cockshull KE (1990) Effects of humidity on the growth and yield of glasshouse tomatoes. J Horticul Sci 65(1):31–39

Horchani F, Khayati H, Raymond P, Brouquisse R, Aschi-Smiti S (2009) Contrasted effects of prolonged root hypoxia on tomato root and fruit (*Solanum lycopersicum*) metabolism. J Agron Crop Sci 195:313–318

IPCC Climate Change (2001) Working group II; impacts, adaptions and vulnerabilityhttps://www.grida.no/climate/ipcc_tar/wg2/005.html. Accessed 13 March 2009

Joubert TG, La G, Boelema BH, Daiber KC (1994) The production of carrots. Vegetable and Ornamental Plant Institute, Agricultural Research Council, Roodeplaat

Kimball BA, Idso SB (1983) Increasing atmospheric CO_2: effects on crop yield, water use and climate. Agric Water Manag 7:55–72

Kjellenberg L (2007) Sweet and bitter taste in organic carrot. Swedish University of Agricultural Sciences, Alnarp

Koike ST, Subbarad KV, Davis RM, Turini TA (2003) Vegetable disease caused by soil borne pathogens. Division of Agriculture and Natural Resources, Publication 8099, University of California, USA

Krupa SV, Manning WJ (1988) Atmospheric ozone: formation and effects on vegetation. Environ Pollut 50:101–137

Lal S, Bagdi DL, Kakralya BL, Jat ML, Sharma PC (2013) Role of brassinolide in alleviating the adverse effect of drought stress on physiology, growth and yield of green gram (*Vigna radiata* L.) genotypes. Legum Res 36:359–363
Lide DR (2009) CRC handbook of chemistry and physics, 90th edn. CRC Press, Boca Raton, p 2804
Matsumoto H (2002) Plant roots under aluminium stress: toxicity and tolerance. In: Waisel Y, Eshel A, Kafkafi U (eds) Plant roots: the hidden half, 3rd edn. Marcel Dekker, New York, pp 821–838
Maurel C, Simonneayu T, Sutka M (2010) The significance of roots as hydraulic rheostats. J Exp Bot 61:3191–3198
Mauzerall DL, Wang X (2001) Protecting agricultural crops from the effects of tropospheric ozone exposure: reconciling science and standard setting in the United States, Europe, and Asia. Annu Rev Energy Environ 26:237–268
Mazza G (1989) Carrots. In: Eskin NA (ed) Quality and preservation of vegetables. CSC Press, Boca Raton, pp 75–119
McKee IF, Mulholland BJ, Craigon J, Black CR, Long SP (2000) Elevated concentrations of atmospheric CO_2 protect against and compensate for O_3 damage to photosynthetic tissues of field-grown wheat. New Phytol 146:427–435
Mcmichael BL, Burke JJ (2002) Temperature effects on root growth. In: Waisel Y, Eshel A, Kafkafi U (eds) Plant roots: the hidden half, 3rd edn. Marcel Dekker, New York, pp 717–728
Milchunas DG, Mosier AR, Morgan JA, Lecain DR, King JY, Nelson JA (2005) Root production and tissue quality in a shortgrass steppe exposed to elevated CO2: using a new ingrowth method. Plant and Soil 268:111–122
Morgan JA, Mosier AR, Milchunas DG, Lecain DR, Nelson JA, Parton WJ (2004) CO_2 enhances productivity but alters species composition and reduces forage quality in the Colorado shortgrass steppe. Ecol Appl 14:208–219
Munns R, Tester M (2008) Mechanisms of salinity tolerance. Annu Rev Plant Biol 59:651–681
Nicoll BC, Gardiner BA, Peace AJ (2008) Improvements in anchorage provided by the acclimation of forest trees to wind stress. Forestry 81:389–398
Ochoa J, Banon S, Fernandez JA, Franco JA, Gonzalez A (2003) Influence of cutting position and rooting media on rhizogenesis in oleander cuttings. Acta Hortic 608:101–106
Pandey RR, Sharma G, Tripathi SK, Singh AK (2007) Litterfall, litter decomposition and nutrient dynamics in a subtropical natural oak forest and managed plantation in Northeastern India. For Ecol Manage 240(1–3):96–104
Patz JA, Campbell-Lendrum D, Holloway T, Foley JA (2005) Impact of regional climate change on human health. Nature 438:310–317
Phene CJ (1989) Water management of tomatoes in the tropics. In: Green SK (ed) Tomato and pepper production in the tropics. AVRDC, Shanhua, pp 308–322
Rogers HH, Dahlman RC (1993) Crop responses to CO_2 enrichment. Vegetatio 104/105:117–131
Rosenfeld HJ (1998) The influence of climate on sensory quality and chemical composition of carrots for fresh consumer and industrial use. Acta Horticulture 476:6976
Rosenfeld HJ, Aaby K, Lea P (2002) Influence of temperature and plant density on sensory quality and volatile terpenoids of carrot (*Daucus carota* L.) root. J Sci Food Agric 82:1384–1390
Rubatzky VE, Quiros CF, Simon PW (1999) Carrots and related vegetable Umblelliferae. CABI Publishing, New York
Samuqliene G, Urbonaviciute A, Sabajeviene G, Duchovskis P (2008) Flowering initiation in carrot and caraway. Scientific Works of the Lithuanian Institute of Horticulture and Lithuanian University of Agriculture, USA
Sangakkara UR, Amarasekera P, Stamp P (2010) Irrigation regimes affect early root development, shoot growth and yields of maize (*Zea mays* L.) in tropical minor seasons. Plant, Soil & Environ 56:228–234
Sharp RE, Porokyo V, Hejlek JG, Spollen WG, Springer GK, Bohnert HJ, Nguyen HT (2004) Root growth maintenance during water deficits: physiology to functional genomics. J Exp Bot 55:2343–2351

Simon PW, Peterson CE, Lindsay RC (1980) Genetic and environmental influences on carrot flavour. J Am Soc Hort Sci 105:416–420
Simon PW, Peterson CE, Gaye MM (1982) The genotype, soil, and climate effects on sensory and objective components of carrot flavour. J Am Soc Hort Sci 107(4):644–648
Simon PW, Freeman RE, Vieira JV, Boiteux LS, Briard M, Nothnagel T, Michalik B, Kwon Y (2008) Carrot. In: Prohens J, Nuez F (eds) Handbook of plant breeding, vol 2. Springer, New York, pp 327–357
Suojala T (2000) Pre and postharvest development of carrot yield and quality. Publ. 37, Department of Plant Production, University of Helsinki, Finland
Taiz L, Zeiger E (2006) Plant physiology, 4th edn. Sinauer Associates, Sunderland, p 690
Talbot MT, Chau KV (2002) Precooling strawberries agricultural and biological engineering department, Florida cooperative extension service. Gainesville: Institute of food and Agricultural Sciences, University of Florida (11p, Bulletin 942)
Thompson JE (2002) Cooling horticultural commodities. In: Kader AA (ed) Postharvest technology of horticultural crops, 3rd edn. University of California's Division of Agriculture and Natural Resources (Pub. 3311), Oakland, pp 97–112
Triguii M, Barringtoni SF, Gauthier L (1999) Effects of humidity on tomato. Can Agric Eng 41(3):135–140
Tuab DR, Miller B, Allen H (2008) Effects of elevated CO2 on the protein concentration of food crops: a meta-analysis. Glob Chang Biol 14:564–575
Tuomisto HL, Scheelbeek PFD, Chalabi Z, Green R, Smith RD, Haines, Andy Dangour AD. (2017) Effects of environmental change on agriculture, nutrition and health: a framework with a focus on fruits and vegetables. Wellcom Open Res 2:21
Warland J, McKeown A, McDonald MR (2006) Impact of high air temperature on Brassicaceae crops in southern Ontario. Can J Plant Sci 86:1209–1215
Wells CE, Eissenstat DM (2003) Beyond the roots of young seedlings: the influence of age and order on fine root physiology. J Plant Growth Regul 21:324–334
Wheeler TR, Ellis RH, Hadley P, Morison JIL, Batts GR, Daymond AJ (1996) Assessing the effects of climate change on field crop production. Asp Appl Biol 45:49–54
Wien HC (1997) The physiology of vegetable crops. CAB International, Wallingford
Wilkinson S, Mills G, Illidge R, Davies WJ (2012) How is ozone pollution reducing our food supply? J Am Soc Hort Sci 63:527–536
Wolfe DW, Gifford RM, Hilbert D, Luo Y (1998) Integration of photosynthetic acclimation to CO_2 at the whole-plant level. Glob Chang Biol 4:879–893
Wullschleger SD, Tschapliniski TJ, Norby JR (2002) Plant water relations at elevated CO2. Implications for waterlimited environments. Plant Cell Environ 25:319–333
Xu ZZ, Zhou GS (2006) Combined effects of water stress and high temperature on photosynthesis, nitrogen metabolism and lipid peroxidation of perennial grass Leymus chinensis. Planta 224:1080–1090
Yamaguchi T, Blumwald E (2005) Developing salt-tolerant crop plants: challenges and opportunities. Trends Plant Sci 10:615–620
Yusuf RO (2012) Coping with environmentally induced change in tomato production in rural settlement of Zuru local government area of Kebbi state. Environmental Issues 5(1):47–54

Impact of Climate Change on Leguminous Vegetables Productivity and Mitigation Strategies

Hemant Kumar Singh, Pankaj Kumar Ray, Shashank Shekhar Solankey, and R. N. Singh

1 Introduction

Crop yield is affected their direct and indirect due to climate change. Climatic changes and environmental variation are expected to exacerbate the problems of food security in the future by force on agriculture. "Climate change may have more effect on small and marginal farmers, particularly who are mainly dependent on vegetables (Anonymous 2009)". However, There are still several issues about the impact assessment, adaptation and mitigation of climate change in agriculture

Climate change by global warming, which leads to rising temperatures in the world, has become a megatrend that will lead to major climatic changes in the future. In its Fourth Climate Change Report (2007), the United Nations Intergovernmental Panel on Climate Change (IPCC) presented considerable scientific evidence and they have become clearly recognized throughout the world. Furthermore, people have become more aware that global warming can not be avoided because of the continuing increase in emissions of greenhouse gas and the changes in the atmospheric structure. However, the official definition of climate

H. K. Singh (✉)
Krishi Vigyan Kendra, Kishanganj, Bihar, India Bihar Agricultural University, Sabour, Bhagalpur, Bihar, India

P. K. Ray
Krishi Vigyan Kendra, Saharsa, Bihar, India Bihar Agricultural University, Sabour, Bhagalpur, Bihar, India

S. S. Solankey
Department of Horticulture (Vegetable and Floriculture), Bihar Agricultural University, Sabour, Bhagalpur, Bihar, India

R. N. Singh
Directorate of Extension Education, Bihar Agricultural University, Sabour, Bhagalpur, Bihar, India

S. S. Solankey et al. (eds.), *Advances in Research on Vegetable Production Under a Changing Climate Vol. 1*, Advances in Olericulture,
https://doi.org/10.1007/978-3-030-63497-1_7

change by the United Nations Framework Convention on Climate Change (UNFCCC) is that "climate change is the transition that can be directly or indirectly attributed to human activity that alters the composition of the global environment, as well as the natural climate variability observed over comparable periods".

When talking the pulses was rooted in the fact that pulses have health benefits and can help in providing healthy nutrition, maintenance of soil fertility, resilience of global warming, sustainable development and poverty eradication. Indian bean and French bean, Pea and Cowpea are the major leguminous vegetables. Other beans of less economic importance include broad bean, cluster bean, winged bean and lima bean etc. Pea and broad bean are cool crops during the season while other beans are crops during the warm season with all of sowing by direct seeded. This abnormal temperature increase will reduce total crop cycle duration, which will have adverse impact on pulse production. (Ali and Gupta 2012). Per decade increase in surface temperature of India is 0.3 °C. (Goswami et al. 2006). The production potential of pulses is likely to be affected by temperature and rainfall variations as well as differences of package and practices technique. Rise in winter and summer temperatures by 3.2 °C and 2.2 °C, respectively will be the most worrying event by year 2050.

Available carbon dioxide is increase in the composition of atmosphere will have an effect on crops in fertilization and fruit development as well as yield with the photosynthetic pathway C3 and thus promote their growth, development and productivity. India's average annual temperature increased by 0.46 °C over the last long period from (24.23 °C) year 1901 to (24.69 °C) year 2012 (Data Portal India 2013).

Based on current ambient temperature, temperature changes will reduce crop length, increase crop respiration rates at affect the survival, partitioning to economic products and distribution of populations of insect pests, decrease fertilizer-use efficiencies, increase the rate of evapo-transpiration and nutrient mineralization in soils,. Based on some of the past observations mentioned above, one of the main driving factors affecting the nutritional security of mankind on earth would be the effect of climate change on agriculture.

They also raised questions whether the climate change adoption strategies work properly or not in future. A major uncertainty is over the effect of increased concentration of CO_2 and its effect on crop growth (Parry et al. 2004). It is predicted that doubling of CO_2 concentration in atmosphere will enhance photosynthesis and nitrogen fixation and which ultimately will lead to increased yield of legume (Poorter and Nagel 2000). Hungate et al. 2004 predicted that increased CO_2 benefit will be available only when ample nitrogen will be available for to support growth. Zanetti et al. 1996 observed those non legumes which can fix atmospheric nitrogen are stimulated by increased CO_2 concentration under controlled conditions. Hence, cool grain legumes are likely to be less affected under adverse climate over non-legume crops (Andrews and Hodge 2010).

Climate changes are one of the main reasons for decline in vegetable production in the world; loss of economic yields for several legume vegetables. The several physiological processes and enzymatic activities are based on temperature; they are going to be mostly crop is not performed. Two major effects like salinity and drought are influencing by heat temperature and worsening the production of legume crops. Such effects of climate change also affect the occurrences of disease and pests, interactions of host-pathogen, pest distribution and ecology, migration to new places, time of arrival and their ability to overwinter, are therefore a major setback to the development of vegetables.

Physiology of the plant is enormously impacted by atmosphere fluctuation. Several stress on plats are showed due to environmental extremes and climate variability (Thornton et al. 2014). The Inter-Governmental Panel on Climate Change has anticipated the increment of temperature to be between 1.1 °C and 6.4 °C by the end of the twenty-first century (IPCC 2007) in Fig. 1. Boyer 1982 detailed that the changes of atmosphere composition have loss the economic yield up to 70%. As per FAO (Anonymous 2007) report, all the world's agricultural areas are threatened by climate change and just 3.5% of the areas are free from environmental confines. Although it is difficult to determine precisely the effect of abiotic stress on crop yield, it is known that abiotic stresses have a major impact on yield production based on the magnitude of complete damage of area under farm farming. In the upcoming year, the production of the several crops is expected to decline due to global warming, water scarcity and other effect of the atmosphere (Bonan and Doney 2018)

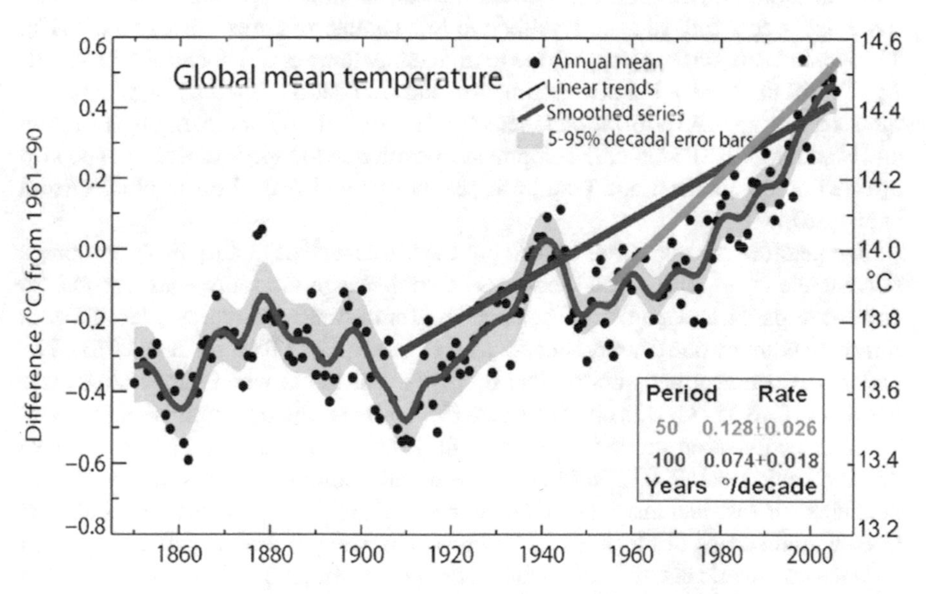

Fig. 1 Global temperature trained through the years (IPCC 2007)

2 Temperature

The earth's surface temperature is estimated to rise from 2.0 °C to 4.5 °C in the twenty-first century. As IPCC, 2014 suggests, "the period between the 19th and 21st centuries is considered to be the most warming time" (Pachauri et al. 2014). Extreme precipitation events may well cause flood-related destruction while the lack of rainfall over a longer period go to drought stresses (Khan et al. 2016). The optimal plant growth temperature is in the range of 10 °C to 35 °C. Due to environmental changes in many countries, wheat production is heavily affected by high temperature and may loss of yield by 6% with temperature rise (Asseng et al. 2015). Climate change which badly affects yield of legume crops by variation of temperature, as several, bio-chemical, plant physiological and metabolic activities rely on temperature. The found that high temperatures have affected the vegetable production in tropical and arid regions (Abdelmageed et al. 2014).

Due to high temperature the plant's physiological, morphological, molecular and biochemical response to change significantly affects of plant growth, development and economic yield of legume. Fruit failure due to high temperatures and loss of yield in tomatoes; this includes bud drop, irregular flowering, weak and lower pollen output, viability and dehiscence, abortion of ovules and less viability with low carbohydrate supply, and other reproductive abnormalities accorded by Hazra et al. 2007. In addition, Furthermore, "substantial loss in tomato production and productivity due to less fruit set, and smaller and low number of fruits" (Bhardwaj 2012). The pre-anthesis stage in pepper, due to high temperatures, did not affect the viability of pistil or stamen, but post-pollination and fruit set inhibited by high temperatures, indicating that fertilization is sensitive to stress of high temperature (Erickson and Markhart 2012). Colour development in chilli due to high temperature and also affected ovule abortion and poor fruit set, flower and fruit drop in chilli (Arora et al. 2010).

Temperature is a major roll of seed production as well as quality in legume crops (Christophe et al. 2011). Food legumes of cool-season are more sensitive to heat stress compared to legumes of hot-season. Temperature increases, also above a threshold point by one degree, is under heat stress in plant (Teixeira et al. 2013). The majority of tropical and sub-tropical crops, heat stress is when temperatures rise above 32 °C to 35 °C (Bita and Gerats 2013), however, the upper threshold for heat stress in cool-season crops is measured at a daily maximum temperature above 25 °C (Wahid et al. 2007). Vegetables are usually climate sensitive and thus temperatures increase and insufficient moisture in soil are the main reason of yields loss as they affect crops of biochemical and physiological processes such as decreased changed enzymatic activity and metabolism, photosynthetic activity, tissue thermal injury, low pollination and fruit setting etc.

Insects are cold-blooded organisms – their bodies' temperature is roughly the same as that of the environment. Temperature is most possibly the single most important environmental factor affecting insect behavior, distribution, growth, survival and reproduction. This gives a significantly growth of population in insect with temperature increase between 1 °C and 5 °C and less mortality rate due to winter, infected of plant in earlier stage and go to damage of crops by insect-pest under climate change (Harrington et al. 2010).

3 Drought

Drought is one of the most common atmospheric stresses which can affect crop production, growth and development. Drought continues to be a several difficulties for the farmers and researchers. Stress from drought is one of the major factors facing pulse producers in many countries. This can persistently lack productivity and extreme events can lead to losses in total yields and failure of crops. Due to heat stress and drought, yield may decrease by 50%, primarily in arid and semi-arid regions (Nam and Chauhan 2001). Essential sources of amino acids and quality of protein in French bean, pea, cowpea, Indian bean, broad bean, green gram, black gram, chickpea, pigeon pea, and lentil. Severe conditions of droughts occur from time to time in many food growing countries, with extensive effects on world food production and supply. The first seedling appearance and drought effect is low germination, and slow plant growth (Harris et al. 2002; Kaya et al. 2006). Plant exposure to stress of water like snap beans and vegetable peas varies with growth and developmental stages.

The physiological attributes that play a significant role in the resistance to drought can be assessed using remote sensing techniques in field environments, using non-destructive methods. The photosynthesis of the leaf plants can be regulated with "chlorophyll content measurement using portable chlorophyll meter and chlorophyll fluorescence while the stomatal conductance measurement indicates the severity of water stress" (Nemeskeri et al. 2015). It is a meteorological phenomenon that indicates a long-term lack of rainfall that caused a reduction in soil moisture and a lack of water with a potential water deficiency in plant tissues. It prevents the crop from achieving potential yield and reduces the development of legumes. Grain legumes generally dependent on rainfall, and are susceptible to erratic drought stress during their process of vegetative and reproductive growth. The availability of water is anticipated to be sensitive to climate change, and extreme stress level of water may affect crop productivity of legume vegetables. "Drought is a major problem in arid and semi-arid areas, which is the primary cause of yield loss worldwide, reducing average yields by more than 50% for most crops" (Sivakumar et al. 2016). Leaf is the most important organ in plant body as it is the kitchen of plant providing food i.e. photosynthesis rates to the whole plant body by the process of photosynthesis. Under drought, the leaf area may be reduced relative to un-stressed plants by adversely affecting leaf expansion during leaf growth by shrinking previously expanded leaves due to water loss (Scoffoni et al. 2014).

In older leaves the minimum number of new leaves to be developed under senescence or drought to start on previous. As we all are well known that water is the synonym to life. So, clearly the stress of drought has been found to significantly minimize the germination and seedling (Kaya et al. 2006). The primary impact of drought is impaired germination, weak establishment of crop stands (Harris et al. 2002). Drought stress assessed on peas impeded germination and seedling development on five cultivars (Okcu et al. 2005). Some crops are less susceptible to water scarcity in the early stages of vegetative growth (Nemeskeri et al. 2019; Kirda 2002), however during growing stage water scarcity leads to changes in many physiological attributes (Nemeskeri et al. 2018), resulting the less of fertility and yield loss.

Flowering ratio of drop in leguminous crops increased due to water stress (Fang et al. 2010), reduces the number of leguminous pods and seed abortions (Boutraa and Sanders 2001) and scanty pods ratio is increase (Beshir et al. 2016). "Produce shorter shoots and smaller leaves in beans crops and decrease the pods length due to water scarcity" (Durigon et al. 2019). Semi-leafless pea variety has decreased the leaf area considered to be low water consumption and has higher water use efficiency (WUE) than average leafy varieties (Baigorri et al. 1999). During the pod filling stage reduced the economic yield in leguminous crops under drought stress condition.

4 Salinity

A saline soil is generally classified as one in which the saturation extract (EC)'s electrical conductivity in the root zone exceeds 4 dS m − 1 (approximately 40 mM NaCl) at 25 °C and has 15% exchangeable sodium. The yield of several crops is the, though several crops display yield loss at lower EC (Munns 2005). Some vegetables' salinity sill (EC) is low with 1.0–2.5 dS m − 1 in saturated soil extracts, and the tolerance to vegetable salts decreases with salt water. The harmful effects of salt accumulation on agricultural fields have seriously affected production in large parts of the world's arable land. Salt contaminated land that has more than 6% of the total land area in the world. It was calculated that 20% of total crops worldwide and 33% of irrigated agricultural land are affected by high salinity. In addition, low precipitation, weathering of native rocks, high surface evaporation, poor cultural practices and irrigation with saline water the salinized areas are increasing at a rate of 10% each year. It was estimated that more than 50% of arable land would be salinized by 2050 (Jamil et al. 2011). The sensitivity of the plant to salt stress is reflected in the effects of growth reduction, turgor, epinasty and wilting, leaf curling, leaf abscission, respiratory changes, low photosynthesis rate, cellular integrity loss, potentially death of the plant and tissue necrosis.

Excessive soil salinity reduces the harvested yield of most vegetable crops, which are particularly sensitive throughout the plant's ontogeny. A 50% decrease in seed germination was observed in French bean due to an increase in salinity of 0–180 mM NaCl. (Bayuelo-Jiménez et al. 2002). Salinity reduces the economical yield with soil physicochemical properties results, and the area's ecological balance. The salinity consequences include — poor productivity, low economic returns with soil erosion (Hu and Schmidhalter 2002). Due to low osmotic ability of soil solution (osmotic stress), different ion effects, nutritional imbalances, or a combination of many factors, the saline growth medium induces many adverse effects on plant growth and development (Ashraf 2004). Muuns and James, 2003, are reported to cause adverse effects at the physiological and biochemical levels on plant growth and development. The consequences of salinity are the result of complex interactions between morphological, physiological and biochemical processes, as well as seed germination, plant growth and development (Akbarimoghaddam et al. 2011). Additionally, salt stress adversely affects nodulation and economic yield. Pulses are highly susceptible to salt stress, leading to loss of yield, Table 1. Early studies showed that salt stress interferes with the absorption of nitrogen and the biological fixation of nitrogen from soil, which is poor in legume crops. The fixation of nitrogen from the atmospheric composition of legume crops depends on nodules but the process of nodulation is very susceptible to salt stress. For example, earlier studies revealed that salt stress significantly influences the activity and compactness of faba bean nodules (Rabie and Almadini 2005).

Table 1 Loss of yield under different salinity levels (Muhammad et al. 2014)

Legume Crops (*Botanical Name*)	Concentration of salt (dSm^{-1})	Loss of yield (in per cent)
Glycine max (Galarsum)	14.4	50
Glycine max (Lee)	8.5	53
Glycine max (loam soil)	7.0	46
Glycine max (clay soil)	6.3	46
Vigna radiata (var. 245/7)	8.0	60
Vigna radiata (var. NM-51)	12	77
Vigna radiata (var. NM-92)	8.0	61
Vigna radiata (var. 6601)	12	72
Cicer arietinum (var. FLIP 87–59)	3.8	69
Cicer arietinum (var. FLIP 87–59)	2.5	43
Cicer arietinum (var. ILC 3279)	3.8	72
Vicia faba (loam soil)	6.6	50
Vicia faba (clay soil)	5.6	52
Vicia faba (loam soil)	4.9	28
Vicia faba (clay soil)	4.3	19
Lens culinaris (cv. 6796)	3.1	100
Lens culinaris (cv. 6796)	2.0	14
Lens culinaris (cv. 5582)	2.0	24
Lens culinaris (cv. 5582)	3.1	88

5 Flooding

The major abiotic stress has affected the growth and economic yield of vegetable crops due to flooding, which is widely called flood prone crops (Parent et al. 2008), Legume is yield of losses due waterlogged and accumulated in field condition around the crops. Stagnate water to decrease the levels of oxygen in the root zone which inhibit aerobic processes in crops. High temperatures coupled with waterlogged around the crops due to fast wilting and death of plants. Kuo et al. 2014 recorded waterlogged intensity in field conditions with increasing temperatures; "rapid wilting and death of tomato plants" is typically observed at high temperatures following a short period of flooding. Floods can promote the growth of waterborne pathogens, droughts and heat waves can influence plants to infection, and storms can increase sporal dispersal by wind (Pautasso et al. 2012).

The mean values for rainfall are not projected to change, there are likely to be more effective and extreme weather events that will reduce economic yield. Heavy rain, hail storms and flooding can physically and economically damage crops.

Physiology of effects on vegetable plants during floods. The loss of stomatal conductance is one of the earliest physiological reactions of plants to floods (Folzer et al. 2006). It results in the maximization of the water potential in the leaf, loss of stomatological conductivity resultant in a significant decrease in the rate of carbon exchange and an increase in the internal concentration of CO2 (Liao and Lin 2014). Leaf chlorosis due to flooding in sensitive crops reduces shoot and root growth, dry matter accumulation, and total plant yield (Malik et al. 2012).

6 Pests and Diseases Responses to Climatic Change

Insect-pest is one of the important factor that have a substantial effect on crop growth and development and losses of economic yield. All country clame that crop yield is day by day decreases of about 25% due to insect-pest. In changing environmental situations, low yield, poor quality and an increasing number of pest and disease problems are common and make vegetables unprofitable. Several insect pests damage these plants, of which pod borer, spiny pod borer, spotted pod borer, stem fly pea, pod fly, bean weevil, aphids, leafhoppers, white fly, blister beetles and thrips cause heavy yield losses due to variation of environmental parameters. Impact on the geographical spread of hosts and pathogens, changes in host physiology – pathogenic interactions, changes in pathogen growth rate due to climate change, e.g. increased over the summering and overwintering of pathogens, dispersal and transmission of pathogen, and emergence of new disease. The occurrence and more losses due to these pests differ over seasons as well as particular crops, cultivation pattern and method, increased temperature and locations, in some category of insects with short life cycles such as aphid and diamond back moth, earlier life cycle completion, increases fecundity. As a result, they can generate more generations each year than their usual rate (FAO 2009). Temperature causes insect-pest species to migrate to higher latitudes, whereas heat temperatures in tropics may adversely affect different species of plague. Insect growth and oviposition rates, insect outbreaks and introduction of invasive species during rise in atmospheric temperature; whereas it reduces the effectiveness of insect bio-control by fungi (Das et al. 2011).

The distribution of insect pests would be strongly influenced by changes in the host crop range, since the distribution of the pest also depends on the host's availability. However, whether or not a pest moves with a crop to a new area will based on certain environmental conditions such as the being of soil quality, and moisture, e.g. number of pod borers, may migrate to temperate regions and attain higher densities in tropics, resulting in most part of damage to food legumes and other crops (Sharma et al. 2010). In addition to the "direct effects of changes in temperature on development levels, rises in food quality due to plant stress can lead to drastic increases in the rate of development of pest populations" (White 1984). Higher temperatures below the upper lethal limit of the species could result in faster rates of development and therefore a more rapidly increase in number of pest populations as the time for reproductive maturity.

7 Adaptation Strategies and Mitigation to Climate Change

The only way to mitigate the adverse effects of climate change on vegetable production and in particular on its productivity of legume, yield and quality is by implementing effective and efficient steps. By integrating cultural traditions with genetic enhancement, it may mitigate the deleterious effects of abiotic stress on agricultural products. The new cultivar that superior result under heat temperatures leading to enhanced harvested yields due to genetic materials (Varshney et al. 2011). New varieties tolerant to salinity and heat stress and flood resistant, use of effective technologies, change in seeding date, like soil and measures of moisture conservations, fertilizers management through drip irrigation, fertigation, plant regulators, grafting techniques, pest management and conservation production are the effective adaptations technique for minimize the impact of climate change. Genetic material of major vegetable crops has been established which are resistant to heat temperatures, drought and flooding, and advanced lines of germplasm are being developed in many institutions Table 2.

The vegetable production is to depend on salinity levels in the root zone at values equal to or less than the EC of the crop. Management should involve saline soil reclamation to regulate salinity rates, and fertilization and irrigation practices should aim to reduce the effect of soil salinisation and the use of saline irrigation water on legume crop growth and production.

The micro irrigation technique with use fertilizer is called fertigation will decrease soil salinisation and mitigate the effects of salt stress as it improves fertilizer use performance, increases nutrient availability and timing of application, and the fertilizer concentration is easily regulated with higher yield.

Table 2 Varieties and advanced lines which are tolerant to abiotic stress

Sl. No.	Tolerant	Vegetables	Varieties	Advance lines
1.	Photo-insensitive	Dolichos	Arka Soumya, Arka Jay, Arka Sambram, Arka Amogh, Arka Vijay	–
		Cowpea	Arka Samrudhi, Arka Suman, Arka Garima	IIHR Sel.-16-2
2.	High temperature	French bean	–	IIHR-19-1
		Peas	–	IIHR-1 and IIHR-8

Source: Rai and Yadav (2005)

In addition to inorganic fertilizers, rivalry between ions is primarily dependent (one ion inhibits the absorption of another ion). Method of irrigation, control (irrigation plan and leaching fraction), and artificial drainage can avoid and reduce soil and water salinity effects. There's an urgent need therefore to grow salt-tolerant legume varieties through traditional breeding and transgenic approaches. Genes improve plant resistance to salinity by using many mechanisms such as reducing the absorption rate of salts from the soil and further transportation of salts within the plant system, controlling the growth of leaves and the onset of plant senescence, and modifying the ionic and osmotic cell balance in roots and shoots.

8 Conclusion

Vegetables are more susceptible to changes in the environment, as they lose of yield, productivity, quality and profitability. Temperature effects on crops caused by global warming are significant among all the effects of climate change. It is important to develop crop-based adaptation strategies combining all the options available for maintaining production and productivity, depending on the vulnerability of individual crops and the agro-ecological field. Nutrient and water management approaches need to consider the effect of salinity on growth, with variety of salt tolerances. There is an enormous need to cultivate new varieties of legumes with stable and maximum economic yield over adverse climate. The development of legume genomes could help classify genes that could be used to produce salt-tolerant legumes for use in transgenic or breeding approaches Use of biofertilizers for improve salt tolerance of legume crops and reduce salinisation of soil. The approaches such as farming system growth, improved varieties with increased productivity in water use, as well as tolerance to various temperatures, drought, flood, cold and insect-pest are taken to mitigate the impact of climate change on legume yield.

References

Abdelmageed AH, Gruda N, Geyer B (2014) Effects of temperature and grafting on the growth and development of tomato plants under controlled conditions. Rural Poverty Reduction through Research for Development and Transformation

Akbarimoghaddam H, Galavi M, Ghanbari A, Panjehkeh N (2011) Salinity effects on seed germination and seedling growth of bread wheat cultivars. Trakia J Sci 9(1):43–50

Ali M, Gupta S (2012) Carrying capacity of Indian agriculture: pulse crops. Curr Sci 102(6):874–881

Andrews M, Hodge S (2010) Climate change, a challenge for cool season grain legume crop production. In: Yadav S, Redden R (eds) Climate change and management of cool season grain legume crops. Springer, Dordrecht, pp 1–9

Anonymous (2007) The state of food and agriculture 2007. Paying farmers for environmental services, Rome

Anonymous (2009) Global agriculture towards 2050 issues brief. High Level Expert Forum, Rome, pp 12–13

Arora SK, Partap PS, Pandita ML, Jalal I (2010) Production problems and their possible remedies in vegetable crops. Indian Horticul 32:2–8

Ashraf M (2004) Some important physiological selection criteria for salt tolerance in plants. Flora 199:361–376

Asseng S, Ewert F, Martre P, Rotter RP, Lobell D, Cammarano D, Kimball B, Ottman M, Wall G, White JW (2015) Rising temperatures reduce global wheat production. Nat Climate Change 5:143

Baigorri H, Antolin., MC. Sanchez-Diaz M. (1999) Reproductive response of two morphologically different pea cultivars to drought. Eur J Agron 10:119–128

Bayuelo-Jimenez JS, Craig R, Lynch JP (2002) Salinity tolerance of *Phaseolus* species during germination and early seedling growth. Crop Sci 42:1584–1594

Beshir H, Bueckert R, Tar AB (2016) Effect of temporary drought at different growth stages on snap bean pod quality and yield. Afr Crop Sci J 24:317–330

Bhardwaj ML (2012) Effect of climate change on vegetable production in India. In: Bhardwaj et al. (eds) Vegetable production under changing climate scenario. Centre for Advanced Faculty Training in Horticulture (Vegetables), Dr. YS Parmar University of Horticulture and Forestry, Solan, India, pp. 1–12

Bita CE, Gerats T (2013) Plant tolerance to high temperature in a changing environment: scientific fundamentals and production of heat stress-tolerant crops. Front Plant Sci 4:273

Bonan GB, Doney SC (2018) Climate, ecosystems, and planetary futures: the challenge to predict life in Earth system models. Science 359:8328

Boutraa T, Sanders FE (2001) Influence of water stress on grain yield and vegetative growth of two cultivars of bean (*Phaseolus vulgaris* L.). J Agron Crop Sci 187:251–257

Boyer JS (1982) Plant productivity and environment. Science 218:443–448

Christophe S, Jean-Christophe A, Annabelle L, Alain O, Marion P, Anne-Sophie V (2011) Plant N fluxes and modulation by nitrogen, heat and water stresses: a review based on comparison of legumes and non legume plants. In: Shanker A, Venkateswarlu B (eds) Abiotic stress in plants–mechanisms and adaptations. Intech Open Access Publisher, Rijeka, pp 79–118

Das DK, Singh J, Vennila S (2011) Emerging crop pest scenario under the impact of climate change a brief review. J Agricul Phys 11:13–20

Durigon A, Evers J, Metselaar K, Lier QDJV (2019) Water stress permanently alters shoot architecture in common bean plants. Agronomy 9:160

Erickson AN, Markhart AH (2012) Flower development stage and organ sensitivity of bell pepper to elevated temperature. Plant Cell Environ 25:123–130

Fang X, Turner NC, Yan G, Li F, Siddique KHM (2010) Flower numbers, pod production, pollen viability, and pistil function are reduced and flower and pod abortion increased in chickpea (*Cicer arietinum* L.) under terminal drought. J Exp Bot 61:335–345

FAO (2009) Global agriculture towards 2050 Issues Brief. High level expert forum. Rome, 12–13

Folzer H, Dat JF, Capelli N, Rieffel D, Badot PM (2006) Response of sessile oak seedling to flooding: an integrated study. Tree Physiol 26:759–766

Goswami BN, Venugopal V, Sengupta D, Madhusoodan MS, Xavier PK (2006) Increasing trend of extreme events over India in a warming environment. Science 314:1442–1445

Harrington R, Fleming RA, Woiwod IP (2010) Climate change impacts on insect management and conservation in temperate regions: can they be predicted. Agric For Entomol 3:233–240

Harris D, Tripathi RS, Joshi A (2002) On-farm seed priming to improve crop establishment and yield in dry direct-seeded rice. In: Pandey S, Mortimer M, Wade L, Tuong TP, Lopes K, Hardy B (eds) Direct seeding: research strategies and opportunities. International Research Institute, Manila, pp 231–240

Hazra P, Samsul HA, Sikder D, Peter KV (2007) Breeding tomato (*Lycopersicon esculentum* Mill) resistant to high temperature stress. Int J Plant Breed 1:31–40

Hu Y, Schmidhalter U (2002) Limitation of salt stress to plant growth. In: Hock B, Elstner CF (eds) Plant toxicology. Marcel Dekker Inc., New York, pp 91–224

Hungate BA, Stiling PD, Dijkstra P, Johnson DW, Ketterer ME, Hymus GJ, Hinkle CR, Drake BG (2004) CO_2 elicits long-term decline in nitrogen fixation. Science 304:1291

IPCC (2007) Climate change 2007: the physical science basis. Contribution of working group I to the fourth assessment report of the intergovernmental panel on climate change. Cambridge University Press, Cambridge

IPCC (2014) Climate change 2014: synthesis report. Contribution of gorking groups I, II and III to the fifth assessment report of the intergovernmental panel on climate change [Core Writing Team, R.K. Pachauri and L.A. Meyer (eds.)]. IPCC, Geneva, Switzerland, 151

Jamil A, Riaz S, Ashraf M, Foolad MR (2011) Gene expression profiling of plants under salt stress. Crit Rev Plant Sci 30(5):435–458

Kaya MD, Okçub G, Ataka M, Çıkılıc Y, Kolsarıcıa O (2006) Seed treatments to overcome salt and drought stress during germination in sunflower (Helianthus annuus L.). Eur J Agron 24:291–295

Khan A, Ijaz M, Muhammad J, Goheer A, Akbar G, Adnan M (2016) Climate change implications for wheat crop in Dera Ismail Khan district of Khyber Pakhtunkhwa. Pak J Meteorol 13:17–27

Kirda C (2002) Deficit irrigation scheduling based on plant growth stages showing water stress tolerance; deficit irrigation practice water report 22. FAO, Rome, pp 3–10

Kuo DG, Tsay JS, Chen BW, Lin PY (2014) Screening for flooding tolerance in the genus Lycopersicon. Hort Sci 17:76–78

Liao CT, Lin CH (2014) Effect of flooding stress on photosynthetic activities of *Momordica charantia*. Plant Physiol Biochem 32:479–485

Malik AI, Colmer TD, Lambers H, Setter TL, Schortemeyer M (2012) Short-term waterlogging has long-term effects on the growth and physiology of wheat. New Phytol 153:225–236

Muhammad N, Li J, Yahya M, Wang M, Ali A, Cheng A, Wang X, Ma C (2014) Grain legumes and fear of salt stress: focus on mechanisms and management strategies. Int J Mol Sci 20(4):799

Munns R (2005) Genes and salt tolerance: bringing them together. New Phytol 167:645–663

Munns R, James RA (2003) Screening methods for salinity tolerance: a case study with tetraploid wheat. Plant Soil 253:201–218

Nam HN, Chauhan YS (2001) Effect of timing of drought stress on growth and grain yield of extra-short-duration pigeonpea lines. J Agric Sci 136(02):179–189

Nemeskeri E, Molnar K, Vígh R, Nagy J, Dobos A (2015) Relationships between stomatal behaviour, spectral traits and water use and productivity of green peas (*Pisum sativum* L.) in dry seasons. Acta Physiol Plant 37:1–16

Nemeskeri E, Molnar K, Helyes L (2018) Relationships of spectral traits with yield and nutritional quality of snap beans (*Phaseolus vulgaris* L.) in dry seasons. Arch Agron Soil Sci 64:1222–1239

Nemeskeri E, Molnar K, Racz C, Dobos AC, Helyes L (2019) Effect of water supply on spectral traits and their relationship with the productivity of sweet corns. Agronomy 9:63

Okcu G, Kaya MD, Atak M (2005) Effects of salt and drought stresses on germination and seedling growth of pea (*Pisum sativum* L.). Turk J Agr 29:237–242

Pachauri RK, Allen MR, Barros VR, Broome J, Cramer W, Christ R, Church JA, Clarke L, Dahe Q, Dasgupta P (2014) Synthesis report. IPCC; Geneva, Switzerland: Contribution of Working Groups I, II and III to the Fifth Assessment Report of the Intergovernmental Panel on Climate Change

Parent C, Capelli N, Berger A, Crevecoeur M, Dat JF (2008) An overview of plant responses to soil water logging. Plant Stress 2:20–27

Parry ML, Rosenzweig C, Iglesias A, Livermore M, Fischer G (2004) Effects of climate change on global food production under SRES emissions and socio-economic scenarios. Global Environ Change—Hum Policy Dimens 14:53–67

Pautasso M, Doring TF, Garbelotto M, Pellis L, Jeger MJ (2012) Impacts of climate change on plant diseases-opinions and trends. Eur J Plant Pathol 133:295–313

Poorter H, Nagel O (2000) The role of biomass allocation in the growth response of plants to different levels of light, CO_2, nutrients and water. Aust J Plant Physiol (27):595–607

Rabie GH, Almadini AM (2005) Role of bioinoculants in development of salt-tolerance of *Vicia faba* plants under salinity stress. Afr J Biotechnol 4:210–222

Rai N, Yadav DS (2005) Advances in vegetable production. Researchco Book Centre, New Delhi

Scoffoni C, Vuong C, Diep S, Cochard H, Sack L (2014) Leaf shrinkage with dehydration: coordination with hydraulic vulnerability and drought tolerance. Plant Physiol 164:1772–1788

Sharma M, Mangla UN, Krishnamurthy L, Vadez V, Pande S (2010) Drought and dry root rot of chickpea. [Abstract] Page 263 in 5th International Food Legumes Research Cionference (IFLRC V) & 7th European Conference on Grain Legumes (AEP II) April 26–30, Antalya, Turkey

Sivakumar R, Nandhitha GK, Boominathan P (2016) Impact of drought on growth characters and yield of contrasting tomato genotypes. Madras Agric J 103:78–82

Teixeira EI, Fischer G, Van Velthuizen H, Walter and Ewert F. (2013) Global hot-spots of heat stress on agricultural crops due to climate change. Agric For Meteorol 170:206–215

Thornton PK, Ericksen PJ, Herrero M, Challinor AJ (2014) Climate variability and vulnerability to climate change: a review. Glob Chang Biol 20:3313–3328

Varshney RK, Bansal KC, Aggarwal PK, Datta SK, Craufurd PQ (2011) Agricultural biotechnology for crop improvement in a variable climate: hope or hype? Trends Plant Sci 16:363–371

Wahid A, Gelani S, Ashraf M, Foolad M (2007) Heat tolerance in plants: an overview. Environ Exp Bot 61:199–223

White TCR (1984) The abundance of invertebrate herbivores in relation to the availability of nitrogen in stressed food plants. Oecologia 63:90–105

Zanetti S, Hartwig UA, Luscher A, Hebeisen T, Frehner M, Fischer BU, Hendrey GR, Blum H, Nosberger J (1996) Stimulation of symbiotic N2 fixation in Trifolium repens L. under elevated atmospheric CO_2 in a grassland ecosystem. Plant Physiol 112:575–583

Impact of Climate Change on Bulb Crops Production and Mitigation Strategies

Manoj Kumar, Meenakshi Kumari, and Shashank Shekhar Solankey

1 Introduction

Even though the earth's climate is relatively stable for the past 10,000 years or so in one or another way and it has been always shown few changes due to natural causes such as volcanic factors. But in the twenty-first century, the climate is changing at faster rate than normal due to more human intervention. In the past 100 years raise in mean global temperature is reported about 0.74 °C and best estimates predict that the global annual mean temperatures will increase in the range of 1.8–4 °C during the twenty-first century; resulted to increase variability in climatic phenomena. The report of Fourth Intergovernmental Panel on Climate Change (IPCC) conceived that the farming, water assets, common eco-framework and food healthful security will be profoundly influenced by the worldwide and territorial anticipated environmental change (IPCC 2007). Consistently, a few sorts of catastrophic events, for example, dry spell substantial downpour, hailstorm, twister, fire, ice, flood and other abiotic stresses are seen in one or different districts or provisions because of environmental change which brought about high and low-temperature system and furthermore expanded precipitation inconstancy and henceforth undermined agrarian profitability (Malhotra and Srivastava 2014; Eduardo et al. 2013). The most abundant gases which cause greenhouse effect are methane (CH_4), carbon dioxide (CO_2),

M. Kumar (✉)
Division of Vegetable Crops, ICAR-Indian Institute of Horticultural Research, Bengaluru, Karnataka, India

M. Kumari
Department of Vegetable Science, Chandra Shekhar Azad University of Agriculture & Technology, Kanpur, Uttar Pradesh, India

S. S. Solankey
Department of Horticulture (Vegetable and Floriculture), Bihar Agricultural University, Sabour, Bhagalpur, Bihar, India

S. S. Solankey et al. (eds.), *Advances in Research on Vegetable Production Under a Changing Climate Vol. 1*, Advances in Olericulture,
https://doi.org/10.1007/978-3-030-63497-1_8

nitrous oxide (N2O), ozone (O_3), water vapor (H_2O) and chlorofluorocarbon (CFCs). The figure shows the sector-wise emission of greenhouse gases (GHGs) (Fig. 1).

"Vegetables are any sort of vegetation or plant item (Ward) which incorporates products of the soil as tomato, cucumber, watermelon, peas; root and tuber/root vegetables, for example, carrot, potato, yam, radish, elephant foot sweet potato; green verdant vegetables, for example, Amaranthus, celery, cabbage, curry leaf and bulb vegetables, for example, onion and garlic. These vegetables are a rich source of nutrition and minerals which plays an important role in overcoming malnutrition. Along with this, it also helps to generate higher incomc and employment to small and marginal land holding farmers. Among the vegetables, onion is more important vegetable and its production is estimated at around 23.28 million tonnes which are found slightly higher production than in 2017–2018. The expansion in the creation and utilization of vegetables could be an incredible pathway to improve dietary decent variety and quality at the time where the weight control plans are ruled by high-vitality nourishments that are poor in micronutrients. Be that as it may, the creation of numerous vegetables, particularly bulb crops are restricted because of natural boundaries, for example, high temperatures and constrained soil dampness in light of the fact that these components greatly affect different physiological and biochemical procedures including photosynthesis, digestion, enzymatic movement, makes warm injury the tissues, influence fertilization and organic product set, and so on., which will be additionally amplified by environmental change (Afroza et al. 2010).

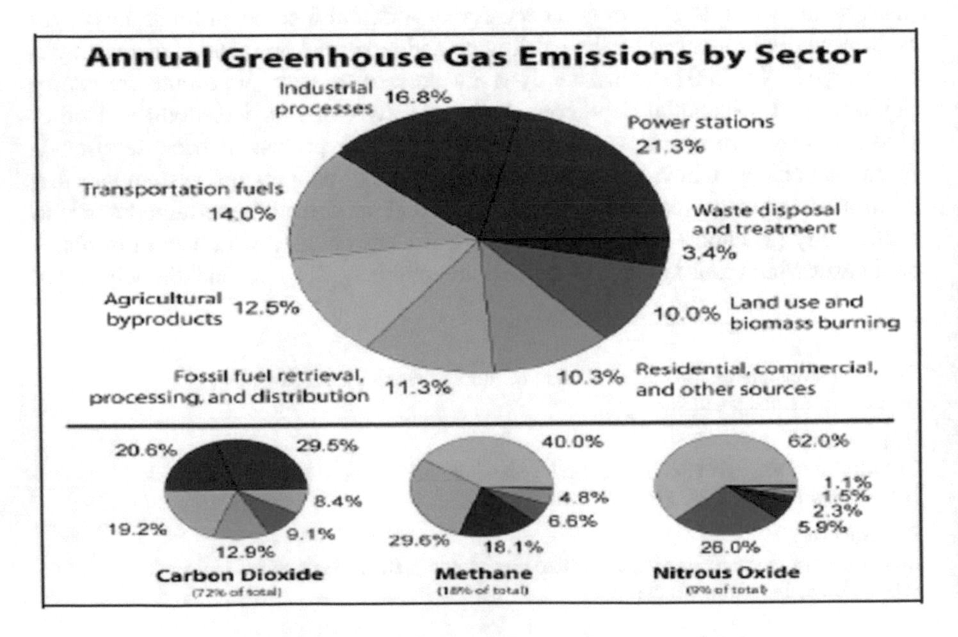

Fig. 1 Annual greenhouse gas emission by sector. (Source: Ranveer et al. 2015)

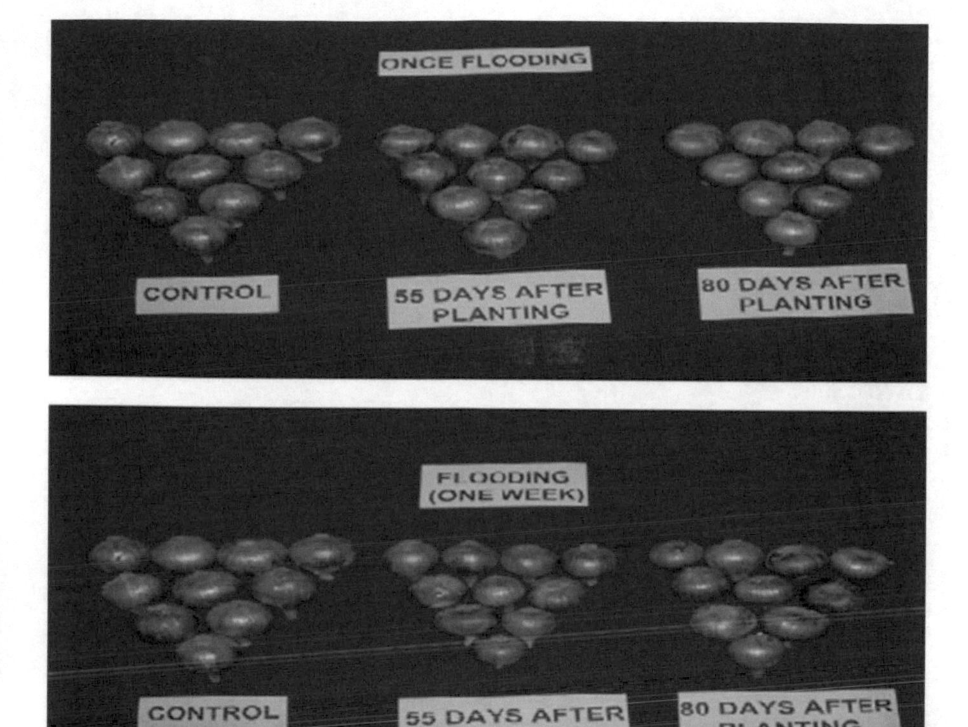

Plate 1 Impact of flooding on bulb size during bulb inception stage (55 DAP) and dynamic development stages (85 DAP)

"Environmental change is characterized as an adjustment in the mean of the different climatic boundaries, for example, temperature, precipitation, relative moistness and air gases organization, and so forth and in properties over a more extended period and a bigger geological zone. The major climatic change parameters like unequal and untimely precipitation, higher in atmospheric temperature excess exposer to UV radiation and incidence of Natural hazardous like drought and flood are the major threats of bulb crop production in tropical areas (Tirado et al. 2010). The bulb crop productivity is threatened by the major consequence of climate change such as a shift in weather pattern and leading to uncertainty in rainfall distribution and high and low-temperature regimes (Somarribaa et al. 2013; Malhotra and Srinivasa 2014). The price of bulb crops like onion may hike due to climate change as most of the onion cultivation is under rainfed conditions. Other than these, change in climatic parameters leads to an outbreak of new strains of insect pest and disease-causing fungus, bacteria, and viruses and also increases the rate of spread of prevailing disease. In this changing climatic conditions and declining cultivable land and available waters as well, the situation is very much challenging to sustain and accomplish the focused on creation to fulfill the developing needs as the populace episode. This problem can be solved by improving production of vegetables by using climate-smart vegetable interventions, which requires intensive knowledge and are highly in this challenging environment (Malhotra and Srinivasa 2015).

2 Stresses Occur During Climate Change and Their Impact on Bulb Crop Production

A brief explanation of environment stresses that affects the bulb production is given below:

Temperature The primary effect of climate change is fluctuation in daily mean maximum and minimum temperature that have adverse effects on bulb crop production because several physiological, biochemical, and metabolic activities are dependent on temperature. The impact of variable temperature on bulb crops happens all the more as often as possible in tropical and parched regions (Abdelmageed et al. 2014). High temperature leads to significant shortening of crop duration which is responsible for reduced yield (Daymond et al. 1997a, b). In another study, the size of bulb reduced as the temperature reaches above 40 °C, if the temperature increases about 3.5 °C above 38 °C reduced yield (Lawande et al. 2010). The high temperature makes a decreased span of development and driving lower in crop yield in onion (Wheeler et al. 1996). Photoperiod has a major impact on the growth of onion but once growth induced it occurs more rapidly at higher temperatures (Brewster 1997). The plants show growth and development to some extent even at a temperature above 25 °C through heat adaptation. In such plants, temperature raise causes to change in the lipid composition of chloroplast membranes and responsible for disruption of the photosynthetic electron transport system (Fitter and Hay 1987). A few plants demonstrated diminished shoot dry weight, development, and net osmosis rate because of higher temperature (Wahid and Close 2007). The potential for summer heat stress reduces due to larger increase in winter temperature than in summer temperature which affects the flower induction and dormancy in bulb crops. The flowering in onion is collectively affected by seedling age nutritional status, day length, and winter temperature and their interactions. Cultivar selection and planting date directly influence the flower initiation therefore it is suggested to plant the seedling sufficiently large to help the development of a huge bulb. Thus, the selection of cultivar and date of planting could be different in mild winters. Daymond et al. (1997a, b) reported the negative correlation between elevated temperature and crop yields as a warmer temperature to shorten the duration of growth. Fruit quality also affected by high temperature. Bulb splitting in onion is a common problem under high-temperature conditions (Fig. 2).

Drought Climate change is the major cause of the shortage of water availability and the drought condition severely affects crop production, particularly bulb crops.

Fig. 2 Bulb splitting in onion

Drought stress is a primary cause of crop loss at a global level in most of the arid and semi-arid region where average yield loss of >50% have been observed in many crop plants (Sivakumar et al. 2016). Drought stress causes the insufficient availability of moisture and water needed for proper functioning of the metabolic process that resulted in the abnormal health of crop plants and restricts crop development (Vadez et al. 2012). Seed germination is restricted in crops like onion due to the occurrence of drought (Arora et al. 2010). Drought condition causes an expansion of relative solute focus in soil condition in this condition the water moves out of the plant cell due to osmotic flow of water and makes the plant dehydrated which hamper various physiological and biochemical activities and collectively reduces the efficiency of most vegetables (De la Peña and Hughes 2007). It also reduces the stomatal conductance that hinder the photosynthetic rate (Yordanov et al. 2013). A yield reduction of 26% was reported under water stress at early growth stages.

Flooding This is also important abiotic stress affects the growth and responsible for yield reduction in flood susceptible crops like an onion (Parent et al. 2008). Oxygen deficiency is the major problem under the flooded condition as this oxygen emerges from a moderate dissemination of gases in water and O2 utilization by microorganisms and plant roots. Many vegetables including onion and garlic are highly sensitive to flooding, also having very limited variations for this trait. Onion shows severe sensitive during bulb development and causes yield losses of about 30–40%. A research was conducted to study the effect of flooding on onion cv. of Arka Kalyan at ICAR-IIHR, Bengaluru and results showed 75–80% reduction in photosynthesis, 28–46% loss in fresh weight and 26–47% loss in dry large scale manufacturing under flooding condition in an onion field. The decrease in leaf zone was seen about 25% and 51% in once and ceaselessly overflowed plants separately (Fig. 3). More leaf senescence was seen in overflowed plants. In continuous flooding condition, the bulb initiation stage was affected and this stage brought about greatest decrease in bulb size (27.2%) and bulb yield (48.3%) (Fig. 4).

During flooding, the physiology of vegetable plants is affected. In the hypoxic condition the stomatal conductance reduces and this condition emerges from a moderate dispersion of gases in water and O_2 utilization by microorganisms and plant roots (Folzer et al. 2006). Therefore, elevation in internal CO_2 due to significant reduction in the carbon exchange rate which took place due to increase leaf water potential and decreased stomatal conductance (Liao and Lin 2014). Leaf chlorosis, reduction in shoot and root development, dry issue gathering, and all out plant yield are the common losses of sensitive plants to flooding (Malik et al. 2012). Flood acts as simplest mean for spreading waterborne pathogen as compare to other like wind, storm, etc. (Pautasso et al. 2012).

Salinity The production and productivity of vegetables are severely affected in salt-affected areas. Physiologically, an initial water deficiency occurs due to salinity as a relatively higher concentration of solute in the soil and altered ion balance and K^+/Na^+ ratio ultimately resulting in higher concentration of Na^+ and Cl^- in plants which are detrimental to them. Under salt stress condition plants shows loss of turgor, decrease in development, shriveling, leaf abscission, diminished photosynthesis and breath, loss of cell trustworthiness, tissue rot lastly demise of the plant (Cheeseman 2008). Among vegetables, onion is highly sensitive to saline soil (Jamil and Rha 2014).

Effect of Carbon Dioxide Reduction in bulbing duration occurs with the increasing CO_2 concentration in cultivars of onion (*Allium cepa* L. Cepa group), but the time required for bulb maturation from bulbing was increased (Daymond et al. 1997a, b). In the presence of higher CO_2 apart from the longer duration of bulbing the bulb yield increases of about 28.9–51% because of an expansion in the pace of leaf territory extension and pace of photosynthesis during the pre bulbing period.

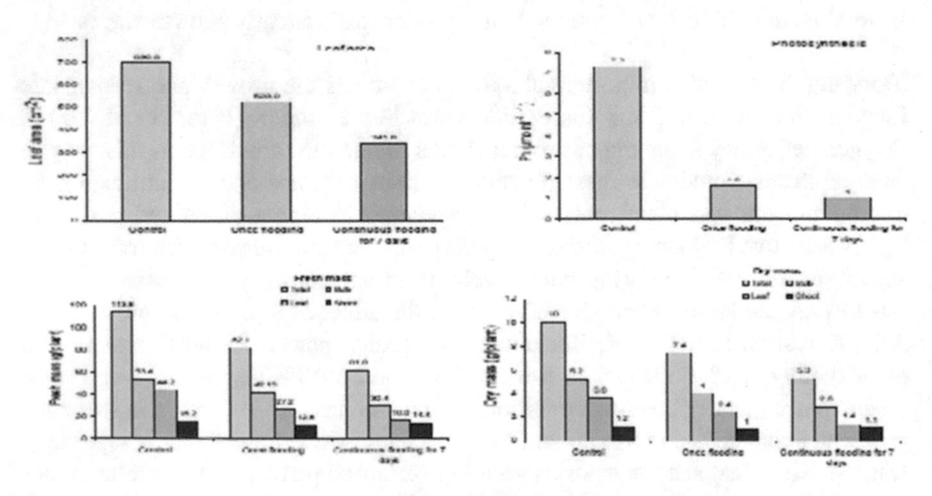

Fig. 3 Consequences of flooding on leaf zone, photosynthesis, new and dry issue gathering in Arka Kalyan variety of Onion. (Source: Yearly Report (2008–2009) of ICAR organize venture on effect, adjustment and weakness of Indian Agriculture to Climate Change)

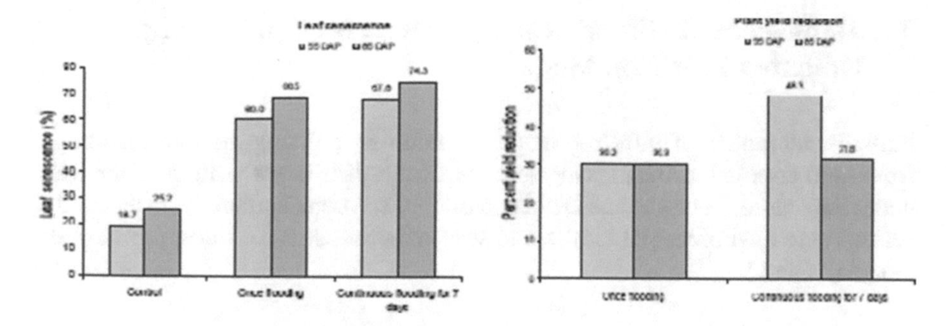

Fig. 4 Impact of flooding on leaf senescence and plant yield of Arka Kalyan variety of Onion. (Source: Yearly Report (2008–2009) of ICAR organize venture on effect, adjustment and weakness of Indian Agriculture to Climate Change)

Pest and Disease Activity in Bulb Crops The ecology and biology of insect Pests are influenced by climate change (Jat and Tetarwal 2012). Migration of insect species towards higher latitudes occurs due to an increase in temperature, while in the tropics areas the higher temperatures may unfavorably influence explicit nuisance species. As the temperature expands the creepy crawly populace, oviposition rates, bug flare-ups, and intrusive species presentations is increases along with the higher temperature decrease the viability of creepy crawly bio-control by organisms, dependability of monetary edge levels, bug decent variety in biological systems and parasitism (Das et al. 2011). The stenotherm (cold blooded) nature of insect make them more sensitive to temperature. Normally, insects have a faster development rate with less time between generations at a higher temperature. High temperature fastens the development of onion maggot (Newton et al. 2011). High temperatures also enhance the reproductive rate of insects by providing prolong favourable conditions for the breeding season. Several researchers concluded that an increase in temperatures can suite to creepy crawlies to arrive at their base flight temperature early, supporting in higher dispersal capacities in aphids and moths. High temperature leads to a shorter length of bug diapauses because of the fast exhaustion of put away supplement assets as it accelerates the metabolic rate (Hahn and Denlinger 2007). The warmer winter season may cause delays in the beginning and late-spring may prompt quicker end of diapauses in creepy crawlies, which would then be able to continue their dynamic development and improvement. This proved an important implication that the increase in the temperature range from 1 °C to 5 °C would lead to increased insect survival as the winter mortality reduced and favours higher population build-up, early infestation, and ultimately crop harm under a worldwide temperature alteration situation (Harrington et al. 2010).

3 Assessment of Climate Change Effect on Onion Using Validated InfoCrop Model

Further calibrations of InfoCrop model in onion by utilizing the information got from field tests led at IIHR, Bangalore, and from distributed writing of the trials under Bangalore, Dharwad and Delhi conditions. The contributions from tests led at various agro-environmental locales and seasons were taken for the approval of the model (Table 1).

4 Adaptation and Mitigation Strategies to Climate Change

Climate change is a reality and on based on much evidence, it is proved that the outflow of ozone depleting substances is liable for an Earth-wide temperature boost and environmental change. Hence, the only and successful way to mitigate the adverse effect of climate change is to adopt an effective and efficient measure which includes the incorporation of the new gene with tolerance or resistance to changeable parameters like temperature, salinity, excess moisture, etc. (Table 2). Different measures remember a change for the planting date, utilization of effective advances like trickle water system, soil and dampness preservations measures, manures the board through fertigation, use of grafting techniques, use of plant regulators, protected cultivation, improving pest management are the effective adaptations strategies for reducing the impact of climate change. Identification of germplasm with resilience to abiotic stresses need to conserve and maintain for future use. Adopt the climate-smart production technologies which requires site-explicit evaluations to recognize the appropriate approach and practices to address the various difficulties confronted during bulb crop cultivation (Malhotra 2014). Bulb crops are thermosensitive and were produce under only long day conditions in a temperate climate but using adjusted production system and development of heat tolerance cultivars made them grow in subtropical and tropical climate with good productivity (Malhotra and

Table 1 Validation of the model for different locations

			Days to bulb initiation		Days to maturity		Yield/ha	
Location	Year/season	Date of planting	M	O	S	O	S	O
Bengaluru	1985 Rabi	15th Nov	50	48	135	140	43	40
Nasik	2000 Rabi	1st Jan	52	50	125	130	22.5	25
Delhi	2000 Rabi	1st Jan	70	75	130	135	37	35
Hyderabad	1994 Rabi	2nd Dec	60	55	122	125	26	25
Dharwad	1999 Kharif	23rd July	50	47	120	125	30	26

S = Simulated, O = Observed

Source: Yearly Report (2008–2009) of ICAR organize venture on effect, adjustment and weakness of Indian Agriculture to Climate Change

Table 2 Rundown of some assortment and propelled line tolerant to abiotic/biotic pressure

Crop	Tolerance/resistance	Variety/line
Onion	Soil moisture stress	ST 154, MST 690–58, Satara local, AFLR, AK & PBR-140
	Purple blotch diseased	Arka Kalyan, IHR 25, IHR-56-1, VL-1, PBR-1
	Basal rot	IHR 14, IHR 506, Sel-13-1, Italian silver queen and red Bermuda
	Thrips tabaci	White Persian, Udaipur 103, N-2-4 1, IHR-141, Poona red.

Srivastava 2014). In this situations, the danger of environmental change could be changed over into circumstance by utilizing the experience and innovative research, to do so it will require representation of likely change, its effect and intending to relieve its terrible effect. The accessible biotechnological apparatuses could add to the speedier conveyance of examination results.

Enhanced Production System The yield of bulb crops can be enhanced by adopting various management practices under tropical conditions. A few advancements to relieve creation challenges because of restricted water system water and flooding has been developed by AICRP on bulb crops, to cope-up with the adverse effect of salinity and fitting accessibility of supplement to the plants.

Water Saving Irrigation Management The yield and quality of bulb crops determined by the quality and effectiveness of water the executives. Climate and weather conditions are the deciding factor for optimum frequency and amount of water to be applied along with variety used, phase of development and establishing attributes of plants, soil water maintenance limit and surface, and furthermore water system framework and management factors.

Cultural Practices That Conserve Water and Protect Crops Soil moisture conservation, prevention of soil degradation, protection of vegetable from heavy rains, high temperature, and flooding can be possible by following mulching, use of shelters, and raised bed like crop management practices.

Development of Climate-Resilient Cultivars Used of improved and well adapted germplasm of bulb crops is the most efficient way to overcome the challenge of climate change. But, most of the modern cultivars did not have an ample amount of genetic variability for tolerance to environmental stresses. Therefore, it is necessary to develop the new varieties, which can be suitable for the intensive and high production system in developing countries, under optimum growth condition having traits that can provide adaptation or tolerance to unfavourable environmental conditions and with low input requirements.

Tolerance to High Temperatures Expanding of the hereditary base through making intra-varietal crosses and mutations in potential tolerant tropical lines and having disease-resistant characters in temperate or winter varieties of bulb crops are the main key to achieve the cultivars with high yield and heat tolerance. Furthermore, there is a need to develop a large number of varieties to fulfil the requirement of changing climate without compromising the yield under stress conditions. Therefore, it is necessary to explore the wide scope of genotypic variety to distinguish more potential source for heat tolerance.

Drought Resistance and Water Use Proficiency There are several phenomena in plants to resist drought stress. The shortening of the life cycle is the mean of slowly developing water deficient in which plants may escape drought stress (Chaves and Oliveira 2004). The exchange and usage of qualities from drought-resistant species could be used to develop bulb crops tolerance to drought stresses.

Tolerance to Salinity in Soil and Irrigation Water Developing salt tolerance cultivars has limited success through conventional breeding approach because of the hereditary and physiologic multifaceted nature of this attribute (Flowers 2004). Other than this, it is not compulsory that tolerance to salt at one developmental stage confirms the tolerance at all the stages because it is a developmentally regulated and stage-specific phenomenon (Foolad 2004). Salt tolerance breeding also depends on the existence of genetic variability, the effectiveness of screening methods, and the capacity to move the qualities to the types of intrigue. And it is also not advisable to screen the plants for salt tolerance under open filed because of variable degrees of saltiness in field soils.

5 Conclusion

The major outcomes of environmental change, for example, a shift in the weather pattern, uncertainty in rainfall distribution, high and low-temperature regimes are the major threats of bulb crop productivity. The discharge of ozone depleting substances is liable for an unnatural weather change and environmental change. Therefore, the only successful way to mitigate the adverse effect of climate change is to adopt effective and efficient measure like the use of tolerance or resistance variety which are developed through using a foreign gene to changeable parameters like temperature, salinity, excess moisture etc.

References

Abdelmageed AH, Gruda N, Geyer B (2014) Effects of temperature and grafting on the growth and development of tomato plants under controlled conditions. Rural poverty reduction through research for development and transformation. J Hortic Sci 14:102–112

Afroza B, Wani KP, Khan SH, Jabeen N, Hussain K (2010) Various technological interventions to meet vegetable production challenges in view of climate change. Asian J Horticul 5:523–529

Arora SK, Partap PS, Pandita ML, Jalal I (2010) Production problems and their possible remedies in vegetable crops. Indian Horticul 32:2–8

Brewster JL (1997) Onions and garlic. In: Wien HC (ed) The physiology of vegetable crops. CAB International, Wallingford, pp 581–620

Chaves MM, Oliveira (2004) Mechanisms underlying plant resilience to water deficits: prospects for water-saving agriculture. J Exp Bot 55:2365–2384

Cheeseman JM (2008) Mechanisms of salinity tolerance in plants. Plant Physiol 87:547–550

Das DK, Singh J, Vennila S (2011) Emerging crop pest scenario under the impact of climate change – a brief review. J Agric Phys 11:13–20

Daymond AJ, Wheeler TR, Hadley P, Ellis RH, Morison JIL (1997a) The growth, development and yield of onion (*Allium cepa* L.) in response to temperature and CO2. J Horticul Sci 72:135–145

Daymond AJ, Wheeler TR, Hadley P, Ellis RH, Morison JIL (1997b) Effects of temperature, CO2 and their interaction on the growth, development and yield of two varieties of onion (*Allium cepa* L.). J Exp Bot 30:108–118

De la Peña R, Hughes J (2007) Improving vegetable productivity in a variable and changing climate. J SAT Agric Res 4:1–22

Eduardo S, Rolando C, Luis O, Miguel C, Hector D (2013) Carbon stocks and cocoa yields in agroforestry systems of Central America. Agric Ecosyst Environ 173:46–57

Fitter AH, Hay RKM (1987) Environmental physiology of plants, 2nd edn. Academic, London, p 423

Flowers TJ (2004) Improving crop salt tolerance. J Exp Bot 55:307–319

Folzer H, Dat JF, Capelli N, Rieffel D, Badot PM (2006) Response of sessile oak seedlings (Quercus petraea) to flooding: an integrated study. Tree Physiol 26:759–766

Foolad MR (2004) Recent advances in genetics of salt tolerance in tomato. Plant Cell Tiss Org Cult 76:101–119

Hahn DA, Denlinger DL (2007) Meeting the energetic demands of insect diapause: nutrient storage and utilization. J Insect Physiol 53:760–773

Harrington R, Fleming RA, Woiwod IP (2010) Climate change impacts on insect management and conservation in temperate regions: can they be predicted? Agric For Entomol 3:233–240

IPCC (2007) Climate change 2007: fourth assessment report of the intergovernmental panel on climate change (IPCC), WMO, UNEP

Jamil M, Rha ES (2014) The effect of salinity (NaCl) on the germination and seedling of sugar beet (Beta vulgaris L.) and cabbage (*Brassica oleracea var capitata* L.). Korean J Plant Res 7:226–232

Jat MK, Tetarwal AS (2012) Effect of changing climate on the insect pest population national seminar on sustainable agriculture and food security: challenges in changing climate. Indian Horticul 3:41–49

Lawande KE (2010) Impact of climate change on onion and garlic production. In: Singh HP, Singh JP, Lal SS (eds) Challenges of climate change in Indian horticulture. Westville Publishing House, New Delhi, pp 100–103

Liao CT, Lin CH (2014) Effect of flooding stress on photosynthetic activities of *Momordica charantia*. Plant Physiol Biochem 32:479–485

Malhotra SK (2014) Development strategies for climate smart horticulture. In: *Global Conference* held at Navsari on 28–30 May, 2014

Malhotra SK, Srivastava AK (2015) Fertiliser requirement of Indian horticulture. Indian J Fertiliser 11:16–25

Malhotra SK, Srivastva AK (2014) Climate smart horticulture for addressing food, nutritional security and climate challenges. In: Srivastava AK et al (eds) Shodh Chintan – scientific articles. ASM Foundation, New Delhi, pp 83–97

Malik AI, Colmer TD, Lambers H, Setter TL, Schortemeyer M (2012) Short term waterlogging has long term effects on the growth and physiology of wheat. New Phytol 153:225–236

Newton AC, Johnson SN, Gregory PJ (2011) Implications of climate change for diseases, crop yields and food security. Euphytica 179:3–18

Parent C, Capelli N, Berger A, Crèvecoeur M, Dat JF (2008) An overview of plant responses to soil waterlogging. Plant Stress 2:20–27

Pautasso M, Doring TF, Garbelotto M, Pellis L, Jeger MJ (2012) Impacts of climate change on plant diseases-opinions and trends. Eur J Plant Pathol 133:295–313

Ranveer CA, Latake TP, Pawar P (2015) The greenhouse effect and its impacts on environment. Int J Innovat Res Creative Technol 1(3):1–7

Sivakumar R, Nandhitha GK, Boominathan P (2016) Impact of drought on growth characters and yield of contrasting tomato genotypes. Madras Agric J 103:78–82

Somarribaa E, Cerdaa R, Orozcoa L, Cifuentesa M, Dávila H (2013) Carbon stocks and cocoa yields in agroforestry systems of Central America. Agric Ecosyst Environ 173:46–57

Tirado MC, Clarke R, Jaykus LA, McQuatters GA, Frank JM (2010) Climate change and food safety: a review. Food Res Int 43:1745–1765

Vadez V, Berger JD, Warkentin T, Asseng S, Ratnakumar P (2012) Adaptation of grain legumes to climate change: a review. Agron Sustain Dev 32:31–44

Wahid A, Close TJ (2007) Expression of dehydrins under heat stress and their relationship with water relations of sugarcane leaves. *Biol Plant* 51:104–109

Wheeler TR, Ellis RH, Hadley P, Morison JIL, Batts GR, Daymond AJ (1996) Assessing the effects of climate change on field crop production aspects. Appl Biol 45:49–54

Yordanov I, Velikova V, Tsonev T (2013) Plant responses to drought, acclimation, and stress tolerance. Photosynthetica 38:171–186

Impact of Climate Change on Cucurbitaceous Vegetables in Relation to Increasing Temperature and Drought

Randhir Kumar and K. Madhusudhan Reddy

1 Introduction

Being a warm season crop cucumber cannot withstand severe frost. In cucumber, short days and high humidity promotes pistillate flowers. For seed germination minimum temperature required is 18 and 20 °C for growth and development. The most favorable temperature for cucumber is 18–24 °C. The economic sex ratio in cucumber is 15:1. Ridge gourd needs extensive and tropical climate with preference to moist warm climate. It develops well in monsoon season. It requires ideal temperature of 24–37 °C for its cultivation. Bottle gourd is warm season crop and prefers a hot humid climate for the best growth. Short day and humid climate produce female flowers. It requires 25–30 °C for seed germination with minimum of 18 °C temperature. Both sub-tropical and tropical climates with higher percentage of humidity are suitable for the cultivation ash gourd. Optimum temperature requirement for ash gourd cultivation is 24–30 °C. Both sub-tropical and tropical climates with higher percentage of humidity are suitable for the cultivation of snake gourd but the problem is at above 1500 m altitude its cultivation is may not be efficacious. Both tropical and sub-tropical climates are suitable for the bitter gourd cultivation. But warm weather conditions reflected best. Being a warm season crop bitter gourd cannot withstand the chilling temperatures. Bitter gourd requires 18 °C as a minimum temperature for germination of seed and 30 °C for its development and growth. The optimum temperature for its cultivation is 24–27 °C. Short day (8–10 hr) situations will aid in enhancing pistillate flower production.

R. Kumar (✉) · K.M. Reddy
Department of Horticulture (Vegetable & Floriculture), Bihar Agricultural University, Sabour, Bhagalpur, Bihar, India

S. S. Solankey et al. (eds.), *Advances in Research on Vegetable Production Under a Changing Climate Vol. 1*, Advances in Olericulture,
https://doi.org/10.1007/978-3-030-63497-1_9

Being a tropical season crop watermelon also cannot withstand the chilling temperature. Watermelon needs 18 °C as a minimum temperature for its germination of seed and 24–27 °C for its crop physiology. In watermelon, higher temperatures at maturity or ripening stage leads to the production of good and best quality fruits. Humid warm climate is not suitable for the cultivation of musk melon. It needs only hot and dry climatic conditions. Temperature range of 27–30 °C is the ideal for muskmelon cultivation. Short days promote pistillate flowers. For proper ripening and high sugar content in muskmelon the conditions like high temperature, plenty of sun shine, low humidity are necessary. Muskmelon plants are susceptible to chilling temperatures and frost. Cool night & warm weather is ideal for sugar accumulation.

Atmosphere variation might be an adjustment in the methods for a few climatic factors, for example, precipitation, temperature, relative moistness and air gases alignment etc. and in assets over an all-inclusive time frame and a more noteworthy geological zone. It can likewise be expressed as any fine-tuning in sky later certain period, nonetheless of wheather as of typical variation or as of social exploit. As per Schneider et al. (2007) powerlessness of some framework to atmosphere variety is the amount to which these frameworks are defenseless and incapable to make due with the unfavorable effects of environmental modification. At current-day, because of man-made achievements viz., mechanization, removal of forest trees and autos and so forth deviations in the atmosphere are existence occurred, these all are once more go unsafe to natural life (Rakshit et al. 2009). Fluctuations in atmosphere may involve varieties in temperatures, increment in salt content of the soil (salinity), stagnation of water, higher air carbon dioxide (CO_2) fixation and ultra violet (UV) radiation. In elevation temperature is as of the lengthened degree of ozone depleting substances like carbon dioxide (CO_2) and CH_4 in environment (Table 1), which is regularly known as a dangerous atmospheric deviation or greenhouse effect. In India mean yearly temperature is expanded by 0.46 °C above a period of maximum latest an extensive period from 1901 (24.23 °C) to 2012 (24.69 °C) (Source-Data Portal India 2013).

Reproductive development can be enhanced by high temperatures. High temperature can reduce the spell for photosynthesis to give to fruit formation. The

Table 1 Worldwide yearly exterior average bounties (2016) and patterns of main GHGs as the WMO GAW worldwide GHG inspectional system

	Carbon dioxide (CO_2)	Methane (CH_4)	Nitrous oxide (N_2O)
Universal richness in 2016	403.3 ± 0.1 ppm	1853 ± 2 ppb	328.9 ± 0.1 ppb
2016 richness comparative to year 1750a	145%	257%	122%
2015–2016 complete upsurge	3.3 ppm	9 ppb	0.8 ppb
2015–2016 comparative upsurge	0.83%	0.49%	0.24%
Average yearly total upsurge during last 10 years	2.21 ppm $yr.^{-1}$	6.8 ppb $yr.^{-1}$	0.90 ppb $yr.^{-1}$

Source: Global Climate Report (WMO), 2018

degree of heat stress that can arise in a definite climatic region be influenced by the possibility of high temperatures working on and their length throughout the day or night. Where universal climate dissimilarity is happening these forecasts may not be anticipated well based only on historic archives for definite localities. Heat stress is a multifarious task of strength (temperature degrees), period and rate of temperature augmentation.

May of 2018 is collective common temperature over the worldwide soil and sea exteriors were 0.80 °C beyond the twentieth century usual of 14.8 °C. Five years from 2014 to 2018 position the five (5) hottest Mays on best ever, with 2016 is the hottest May at +0.88 °C. Recent May of 2018 similarly denotes the 42nd successive May and the 401th back to back month with high temperature, at littlest evidently, past the twentieth century mean data (NOAA 2018). The Universal mean temperature for the dated January month to October month of 2018 was 0.98 ± 0.12 °C past the pre-modern standard (1850–1900). Year 2018 is on movement to be the fourth hottest year on best ever. It would awful that the previous 4 years viz., 2015, 2016, 2017 and 2018 – are likewise the four hottest years in the arrangement. Year 2018 is the chilling (coolest) among four. In dissimilarity to the 2 hottest years, year 2018 started with feeble La Niña situations, normally connected with inferior worldwide temperatures. The 20 hottest years ensure all happened in the previous 22 years (WMO 2018). This temperature upsurge will alter the planning and proportion of precipitation, openness of water, wind models and causes recurrence of atmosphere limits, for instance, droughts, heat waves, floods or whirlwinds, assortments in ocean streams, maturation, forest area flares and surges pace of ozone utilization (Minaxi et al. 2011; Kumar 2012).

Ozone weariness in the region of stratosphere, acknowledged by follow gases, for instance, CFCs (chlorofluorocarbons) and N_2O (nitrogen oxides) realizes extended components of splendid ultra violet – B radiation (280–315 nm) accomplishing the Universe exterior, which is dangerous to life (Zajac and Kubis 2010).

2 Impact of Increasing Temperatures on Cucurbitaceous Vegetables (Temperature Stress)

Remarkably in elevation or little temperatures may source disturbances in ordinary physiological and metabolic procedures. Higher temperatures can influence cell development, synthesis of cell wall, plant hormonal connections, amalgamation of proteins, stomatal opening (respiration) and photosynthesis i.e. carbon dioxide (CO_2) assimilation (Hasanuzzaman et al. 2012).

Cucurbits need warm & cordial circumstances for development. For instance, cucumber necessitates beyond 16 °C temperature for germination of seed, emergence and seedling rise, development and advancement of plant. Ideal temperature for some cucurbit species is 32 °C, above which the development is seized. On higher temperatures (38–45 °C), development of plants in 3–6 leaf stage might be eased back, and leaf edges may seem yellow, contingent on the cultivar, species, duration of acquaintance, and other natural components. At very higher temperatures (42–45 °C) newly emerged leaves may show up light green to yellowish after generally tiny contacts (24–48 hr.). High temperatures during fruit augmentation frequently brings about diminished yield and fruit quality, contingent upon the degree and term of pressure scene. Blossoms and fruits prematurely end, sex articulation is changes from female to male if the temperature is ascends over 38 °C for any obvious time.

Low or chilling temperatures i.e.10–17 °C can postpone seedling emergence and rise and source reduction in development of plant (Lower 1974). For the most part, there are no sores on the leaves anyway stems are shorter and leaves are humbler. Acquaintance of plants with low or chilling temperatures (<17 °C) not long previously or during flower opening stage may provoke a move in sex enunciation (to a higher repeat of female blooms). Variations in sex enunciation are dependent on genetic constitution (for instance the effect of modifying characteristics on critical sex-choosing characteristics).

In numerous cucurbit species, temperatures <10 °C prompts chilling injury (Paul et al. 1979). The underlying indications are the white regions on cotyledons and white or light earthy colored edges on completely extended leaves. Fundamental proof propose that squash and water melon are increasingly safe (more unaffected), sponge gourd and melons are moderate and cucumber is progressively helpless (more vulnerable). Serious chilling treatments bring about necrotic sores over huge territories of the leaves at last causing demise of the plants.

Chilling or low temperature injury is (<10 °C) is expanded by the accompanying conditions; (1) Longer length of chilling, (2) Low temperature, (3) In elevation light force at particular time of chilling, (4) Higher wind speed at the time of chilling and (5) High development temperature earlier to chilling happens. This injury is more terrible in powerless (prone) cultivars.

Control measures to forestall chilling injury incorporate cultivating safe varieties, choosing a sheltered planting date and time (planting after ice peril is finished), utilizing plastic mulch and line spreads to give security from chilling. Likewise overhead water system gives some proportion of insurance during frost.

2.1 Impact of High Temperatures on Germination and Emergence of Cucurbitaceous Vegetables

Melon and cucumber seed germination is enormously smothered at 42 and 45 °C, respectively but in case of winter squash, watermelon, pumpkin and summer squash the seed germination is won't happen at 42 °C (Kurtar 2010).

Seeds of cucurbits comprise of an incipient organism (embryo) and two cotyledons secured with a protected seed coat. Vitality stores are put away in the heavy cotyledons. The procedures of seedling emergence are run of the mill of dicots with above surface (epigeal) development. Intake of water (imbibition) is trailed by

Table 2 Soil temperature required for seed germination of different cucurbitaceous vegetables

	Soil temperature required for germination (°C)			
Crop	Minimum	Optimum range	Optimum	Maximum
Muskmelon	15.5	24.0–35.0	30.0	37.8
Cucumber	15.5	15.5–35.0	30.0	40.5
Watermelon	15.5	21.0–35.0	30.0	40.5
Summer squash	15.5	21.0–35.0	30.0	37.8
Pumpkin	15.5	21.0–32.0	30.0	37.8
Bitter gourd	15.5	20.0–35.5	32.0	37.8
Ridge gourd	15.5	21.0–35.0	30.0	40.5

biological action, prolongation and development of the root (radicle). The hypocotyl rises up out of the seed coat and protracts to drive the hypocotyl snare over the dirt apparent. Presented to sunlit, the snare rectifies, hauling the seed leaves out of the dirt where they grow and start CO_2 assimilation i.e. photosynthesis. The seed coat may stay in the dirt, or may cling to either of the cotyledons, here and there avoiding their ordinary development.

The period of time required for germination in a clammy situation relies upon the temperature. Somewhere in the range of 15 °C and 25 °C, muskmelons and cucumbers need 194 degree days over a vile temperature of 12.2 °C (Taylor 1997). Table 3 demonstrates the opportunity to development for a few cucurbitaceous vegetables at various soil temperatures. Germination of summer squash may takes place at as low as 4.4–10 °C temperatures, and cucumber at 11.6 °C, yet watermelon and muskmelon need in any event 15.5 °C (Wien 1997).

Triploid watermelon seeds are highly inclined to sprouting issues than different cucurbitaceous vegetables. Regularly, the seed coat adheres to the cotyledons. Putting the seed on a level plane, or with the extreme termination jagged despondent at a 45° or 90° edge from even decreases the quantity of seed coats adhering to the seed leaves (cotyledons) (Maynard 1989). Seedling emergence might be expanded by evacuating or cutting the seed coat, treating the seed with hydrogen peroxide (H_2O_2), or giving a more-oxygen condition (Duval and Nesmith 2000; Grange et al. 2003). Exorbitantly misty media decreases seedling emergence process (Grange et al. 2003) (Table 2).

2.2 *Impact of High Temperatures on Crop Growth and Development of Cucurbitaceous Vegetables*

The temperature vacillations postponement the maturing of fruits and diminish the sugariness in the melons. Little dampness content in the dirt is inconvenient to natural fruit superiority and advancement in gourds and melons (Arora et al. 1987). Hot and warm muggy atmosphere increment the somatic development and effect the deprived generation of female blooms in cucurbit crops like bottle gourd, pumpkin, wax gourd which causes little yield (Singh 2010).

Ideal normal temperatures for development are 18.3–23.8 °C for squash, pumpkin, muskmelon and cucumber, and 23.8–29.4 °C for watermelon. Outrageous normal temperatures are (Indiana CCA Conference Proceedings, 2007) 35 °C for watermelon and 32 °C for different yields. (Maynard and Hochmuth 2007). Forecasts of yield improvement dependent on developing commencement days or degree-days have been created for a portion of these harvests in unmistakable conditions. Vile temperatures of various cucurbits are viz., 7.2 °C for squash, 10 °C for muskmelon, and 12.7 °C for watermelon and cucumber are expressed for North Carolina (Maynard and Hochmuth 2007). Anticipating crop development can be intense for those yields reaped in excess of a couple of days after fertilization and when various natural fruits are available, perhaps on account of the organic fruit burden impacts on organic fruit development rate (Perry and Wehner 1990).

Cucurbitaceous vegetables were vulnerable to low temperature or chilling damage at temperatures underneath 7.2–10 °C. Lesser the temperature and the more drew out the span lead to more prominent the damage. Side effects of chilling incorporate water splashed acnes on shrubberies, withering, and demise of roots. Frost effected vegetation can endure, however they are interfered with being developed. The produce of these particular plants additionally are inclined to chilling damage. Manifestations comprise indented recognizes that are frequently attacked by bacteria and fungi. The harm may not be promptly evident on fruit; however the natural product won't keep going as long. Dynamic research on cucurbits keeps on expanding comprehension of these harvests. In the United States, significant territories of research incorporate insect and pest management, organic production, grafting, decreased culturing frameworks, fruit quality, enhancing supplement and water controlling, and reproducing and sub-atomic hereditary qualities (Tables 3 and 4).

Table 3 Temperature requirement for successful cultivation of cucurbitaceous vegetables:

Monthly mean temperature (°C)			
Minimum	Maximum	Optimum	Cucurbitaceous crops
18.0	35.0	25.0–27.0	Watermelon, muskmelon,
10.0	35.0	20.0–25.0	Cucumber, pumpkin, pointed gourd, bottle gourd, bitter gourd, ridge gourd, snake gourd, summer squash,

Source: Hazra and Som (2006)

Table 4 Impact of temperature on seedling development for some cucurbits

	Number of days prerequisite for development of seedlings at different soil heats from seed embedded ½ in deep				
Crop	15 °C	20 °C	25 °C	30 °C	35 °C
Rock melon	–	8	4	3	–
Cucumber	13	6	4	3	3
Watermelon	–	12	5	4	3

Source: Maynard and Hochmuth (2007). Adapted from J.F. Harrington and P.A. Minges, "Vegetable Seed Germination," California Agricultural Extension Mimeo Leaflet (1954)

2.3 *Impact of High Temperatures on Pollination and Fruit Set of Cucurbitaceous Vegetables*

Anthesis is occur at temperatures beyond 10 °C (squashes and pumpkins), 15.5 °C (watermelon and cucumber), or 18.3 °C (muskmelon). Blossoms are remains open for multi day by virtue of watermelons, muskmelons, and cucumbers, or a huge segment of multi day or less for *Cucurbita* spp. A great part of the time, fruit set requires the activity of pollinators, for instance, honey bees or neighborhood squash bumble bees. Sufficient feasible pollen dust must be passed on to the shame so that there is one grain of pollen dust for each making seed in the trademark common natural fruit product. Exactly when pollen dust is on the shame, syngamy is up 'til now not assured. The pollen dust should create and raise a pollen tube dejected the shame to circulate spermatozoa to the ovule. If there is no plentiful pollen dust, or circumstances doesn't sensible for pollen tube advancement, simply the ovules neighboring to the flower head may be effectively pollinated and fertilized. Seeds ascending close to the blossom end empower improvement in that bit of the fruit, anyway the rest of the fruit remains pretty much nothing. It leads to a distorted fruit.

By virtue of triploid (3x) watermelon, preparation with reasonable pollen is crucial to strengthen fruit improvement regardless of the way that seeds don't develop. Triploid plants don't make enough sensible pollen themselves hence a pollinizer cultivar must be planted adjoining. Expectedly, seeded cultivars that produce fruit ostensibly not equivalent to result of the seedless cultivars are used as pollinizers. Even more starting late, certain cultivars have been developed that are advanced solely as pollinizers; they don't yield appealing or attractive fruit. For compelling fruit formation in triploid (3x) watermelons it is significant that pollinizer cultivars yield workable pollen dust during female blossoms are uncluttered on the triploid (3x) plant.

Certain female flower producing (gynoecious) cucumbers are parthenocarpic (formation of fruits without fertilization) thus don't need pollinators. Normal parthenocarpy is referred to happen in further cucurbits also, prominently summer squash, yet is regularly not depended upon for business generation (Wien 1997).

Natural environments and the state of the plant can meddle with fertilization and fruit formation. Climate situations impact movement of pollinator. For instance, bumble bees are inactive when climatic situations are warm and dry. Applications or deposits of various pesticides can annihilate or deflect honey bees. Fruit officially emerging on the plant frustrate powerful fruit set in more youthful blossoms, particularly those on a similar branch or stem (Table 5).

Table 5 Temperature prerequisite for blossoming and fruit formation in different cucurbitaceous vegetables

Crop	Temperature prerequisite for fruit set (°C)	Impact of variation from optimum temperature
Bitter gourd, watermelon, muskmelon, pumpkin, summer squash, cucumber	12.3–18.3	Higher temperatures encourage more male blossoms and thus harmfully disturb fruit formation.
Bottle gourd, ridge gourd	24–28	Cool night temperature augments sugar levels in the fruits and thus enhances sweetness.

2.4 *Impact of High Temperatures on Fruit Development of Cucurbitaceous Vegetables*

Fruit of cucurbit develops exponentially for a period after fruit formation, and a short time later the advancement rate moderates. The increasing speed in fruit measure after treatment is generally an eventual outcome of cell advancement instead of a development in the amount of cells. Cucurbits can be portrayed into two critical social events subject to whether the characteristic natural fruits are gathered when adolescent (juvenile) – summer squash and cucumbers – or created (develop) – a wide scope of gourds, melons, pumpkins and winter squash. Summer squash and cucumbers are gathered during the hour of snappy development. These might be set up for procure when 3 days after fertilization, dependent upon the market necessities. In the further yields, fruit ordinarily accomplish their optimum mass around 2–3 weeks after fertilization, and proceeds alternative in any event 3 weeks to develop a horticultural maturity. Through this time seeds make to advancement and pleasantness, flavor and concealing (shading) make in the natural fruit. The skin hardness, ends up being less vulnerable to water, and by virtue of muskmelon, makes airy netting. A shading adjustment may arise, whichever straightforward as in the conversion from light green colour to yellowish colour in muskmelon; or just with respect to the fruit adjacent the the soil, yellowish colour spot on fruit surface of a watermelon where it touches the ground; or over the whole natural fruit surface, as in case of pumpkin. Muskmelons and watermelons normally develop 42–46 days later fertilization, while pumpkins and winter squash need 50–90 days to accomplish accumulate advancement. Propelling natural item puts extensive solicitations on the plant, lessening the advancement of fresh shrubberies, roots, and some extra natural fruits emerging meanwhile. (Loy 2004; Wien 1997).

The dimension of the developed fruit is biased by hereditary qualities, condition, and plant surroundings throughout improvement of the female bloom and fruit. Circumstances that diminish the measure of integrate accessible will in general abatement the dimension of distinct fruit. Improved plant thickness, bigger quantities of fruits per plant, and dense water supply will in general lessen fruit measure. In watermelon and muskmelon, the dissolvable solids substance of the fruit is a basic proportion of value. Like fruit measure, dissolvable solids slope to be lesser under situations that decrease level of assimilate. Under high night temperatures fruits per plant and enlarged plant thickness would all be able to lessen solvent solids (Wien 1997). Rather than its effect on fruit estimate, diminished water source can build fruit solvent solids (Bhella 1985; Fabeiro et al. 2002) (Figs. 1, 2, and 3).

3 Impact of Increasing Drought (Moisture Stress) on Cucurbitaceous Vegetables

What Is Drought?

Dry spell or dampness stress is one of the most groundbreaking ecological anxieties making tremendous misfortune the agriculture and horticulture around the world. Vegetables are exceptionally delicate to dry spell as contrast with numerous different crops. Edifying yield under dry season is a significant objective of plant breeding.

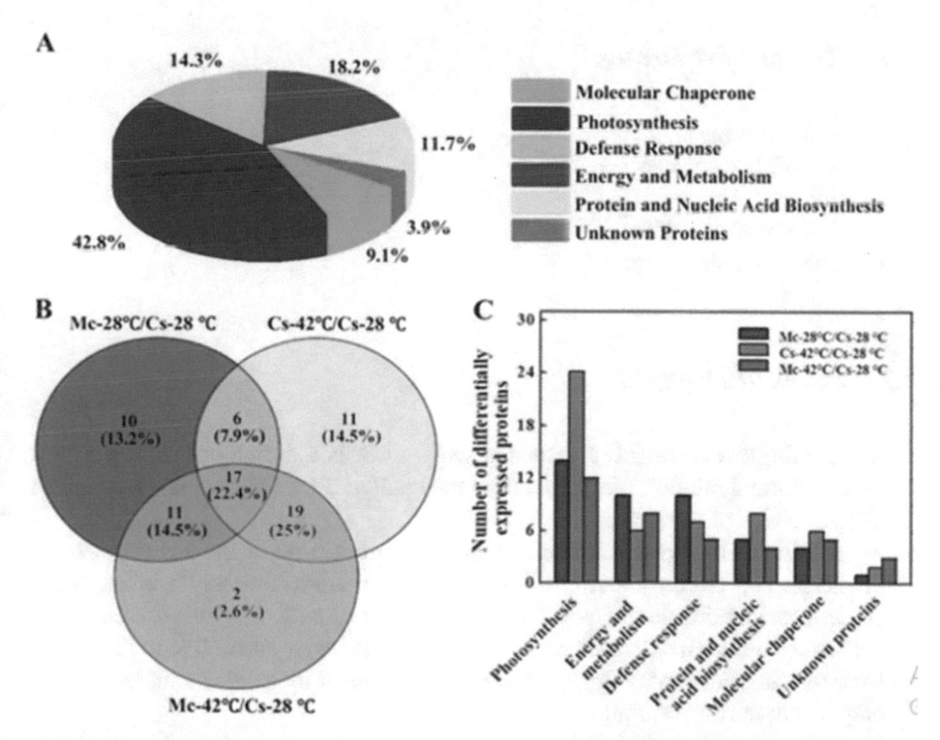

Fig. 1 Dissemination of differentially gathered proteins by *Momordica* rootstock and/or temperature stress in cucumber leaves. (Ye Xu et al. 2018):
(**a**) Functional cataloguing and scattering of all 77 differentially amassed proteins. (**b**) Venn diagram presenting total of overlying proteins differentially controlled by Momordica rootstock and/or heat stress compared to control. (**c**) Functional protein dissemination in associated groups (changes ≥1.5-fold or ≤0.67-fold)

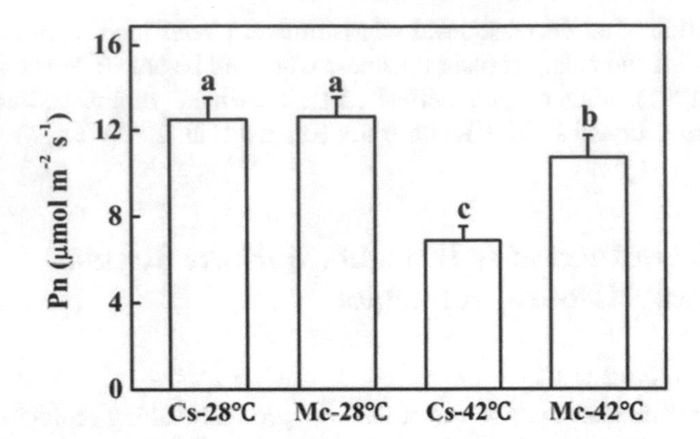

Fig. 2 The effects of temperature stress on net CO_2 assimilation rates (Pn) of self-grafted and rootstock-grafted cucumber scions. (Ye Xu et al. 2018):
Means of three biological replicates (±SE). Means with same letter are not significantly differ at $P < 0.05$ conferring to Duncan multiple range test. Three self-determining trials were accomplished with parallel consequences

3.1 Reasons for Drought

- Rainfall Shortage
- Waterless Time of year
- El Nino
- Erosion Social activities
- Climatic Vagaries

3.2 Kinds of Drought

- **Meteorological Drought:** It happens when there is a drawn out time with not exactly normal rainfall. Meteorological dry spell as often as possible goes before different sorts of dry season.
- **Agricultural Drought:** It influences the crop production or the environment of the range. This circumstance can likewise emerge self-sufficiently from any variety in precipitation levels when soil conditions and disintegration created by half-baked agrarian exercises cause a deficiency in water accessible to the crops. However, in a customary dry spell, it is brought about by an all-inclusive time of underneath normal rainfall.
- **Hydrological Drought:** It is achieved when the water holds accessible in sources, for example, springs, lakes and supplies fall underneath the measurable normal.

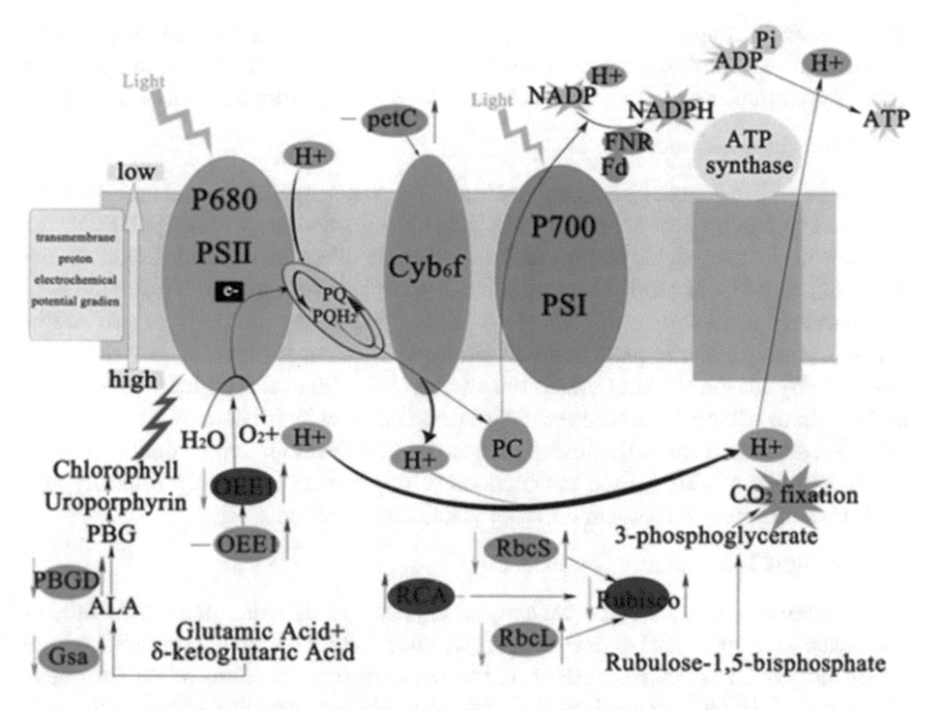

Fig. 3 Schematic presentation of effects of temperature stress and Momordica rootstock on CO_2 assimilation metabolism in cucumber leaves. (Ye Xu et al. 2018):
Variations in protein (noticeable in red ellipses) and gene appearance (marked in blue ellipses) were combined. Arrows at left of ellipses indicate variations brought by heat stress and arrows at right specify variations brought by *Momordica* rootstock under heat stress. Red or green arrows display up regulation or down regulation, respectively, while black short lines specify no variation. OEE1 oxygen-evolving enhancer protein 1, Rubisco ribulose-l,5-bisphosphate carboxylase/oxygenase, RCA ribulose-l,5-bisphosphate carboxylase/oxygenase activase, petC cytochrome b6-f complex iron–sulfur subunit, PBGD porphobilinogen deaminase, Gsa glutamate-1-semialdehyde 2,1-aminomutase, RbcS rubisco small subunit, RbcL rubisco large subunit

3.3 *Mechanism of Drought Tolerance*

Dry season or dampness stress is one of the most significant ecological anxieties making enormous misfortune the agriculture and horticulture around the world. Vegetables are exceptionally delicate to dry spell as contrast with numerous different harvests. There are various kinds of dry season get away from components are examined beneath,

(A) **Drought Escape:**

The capacity of a yield of plant to finish its natural life cycle before improvement of genuine earth and plant water shortages is called as dry spell escape. This strategy includes fast phenological improvement for example early blooming and development, variety in term of development time relying upon the degree of water shortage. For example, in Black eyed pea early erect cultivars, for example, 'Ein El Gazal' and 'Melakh', have accomplished healthy when the precipitation spell was small yet particular because of their capacity to get away from late-season dry spell.

(B) **Drought Avoidance:**

It alludes to the capacity of plant to suffer stages deprived of critical precipitation even as keeping up a in elevation plant standing at great crop water potential, for example desiccation delay or dry season evasion. In other manner, dry season shirking is the capacity of crops to keep up moderately great tissue water potential notwithstanding a lack of soil dampness. Enlightening the components of water take-up, putting away in plant cell and diminishing water misfortune give dry spell evasion. Dry season shirking systems are connected with useful entire plant components viz., overhang resilience and leaf zone decay (which decreased adsorption, radiation and transpiration), closing of stomata and development of cuticular wax, and alterations of sink-source connections through changing root profundity and thickness, progress in root hair and root water driven conductivity.

(C) **Drought Tolerance:**

Dry season resilience is the capacity of a yield to suffer dampness shortfalls at low tissue water potential or drying out resistance. Under dry spell condition, plants get by through an exercise in careful control between support of turgor with decrease of water misfortune. Dry spell resilience systems are orchestrating of turgor through osmotic modification (solute development in cell), increment in versatility in cell however shrinkage in cell mass and parching resistance by protoplasmic resistance. In lab investigation of tomato cultivar PS-10 disclosed little osmotic potential at all polyethylene glycol (PEG), medicines and in this way it went to be a superior dry season open minded cultivar than another cultivar Roma while cultivars Peto and Nora indicated normal dry season resilience (Table 6).

Table 6 Basic phases of dry spell (drought) pressure and its effect on cucurbitaceous vegetables crops

Cucurbit	Acute time of irrigation	Influence of drought on cucurbits
Cucumber	Flowering stage and fruit development stage	Misshapen and non-viable pollen grains, acrimony and irregularity in fruits, deprived seed viability
Melons	Flowering stage and evenly all over fruit development stage	Reduced fruit superiority in muskmelon due to decline in ascorbic acid (vit-C), TSS and reducing sugar, upsurge nitrate levels in watermelon fruit, meager seed viability
Summer squash	At the time of bud development and flower formation	Distorted and non-viable pollen grains, malformed fruits

3.4 Consequences of Drought Stress in Cucurbitaceous Vegetables

Dry season pressure diminishes leaf size, stems augmentation and root multiplication, interferes with plant water relations and diminishes water-use productivity. Plants show a better than average assortment of physiological and organic come backs at cell and entire substance powers toward extraordinary dry spell (dry season) pressure, along these lines making it an inconvenient sign. CO_2 assimilation by leaves is reduced generally by stomatal end, layer damage and upset development of various proteins, especially those of CO_2 fixation and adenosine tri-phosphate blend. Improved metabolite movement through the photograph respiratory pathway grows the oxidative weight on the tissues as the two methodology produce receptive oxygen species. Injury enacted by receptive oxygen species to natural macromolecules under dry season tension is the noteworthy deterrents to improvement.

Plants show an extent of frameworks to withstand drought pressure. The superior segments incorporates hardened water disaster by improved diffusive restriction, redesigned water take-up with profitable and progressively significant root structures and its able use, and slighter and heavenly leaves to diminish the transpiration mishap. Among the principal segments, potassium particles help in osmotic modification; silicon fabricates root endodermal silicification and grows the cell water balance (Hessini et al. 2009). Low-sub-nuclear weight osmolytes, alongside glycine betaine, proline and other amino acids, regular acids, and polyols, are fundamental to withstand cell limits under drought. Plant advancement controllers, for instance, salicylic destructive, auxins, gibberellins, cytokinins and abscisic corrosive control the plant responses toward dry season.

Polyamines, citrulline and different impetuses go about as cancer prevention agents and reducing the antagonistic effects of water shortage. At sub-nuclear level different dry season responsive qualities and interpretation viewpoints have been seen, for instance, the absence of hydration responsive part – restricting quality, aquaporin, late embryogenesis plentiful proteins and dehydrins. Plant dry season versatility can be overall by completing procedures, for instance, mass screening and raising, marker-helped assurance and exogenous usage of hormones and osmoprotectants to seed or creating plants, similarly as structuring for drought impediment.

The dry season pressure altogether decreased cucumber fruit weight, length, yield, leaf region, and number of fruits. The tallness, dry, and new loads of plant diminished with the expansion in the dry spell feeling of anxiety (Najarian et al. 2018). Dry season pressure is a central point answerable for the constrained development of cropss and CO_2 assimilation decrease and subsequently may prompt the decrease of crop yields (Efeoglu et al. 2009). This perspective is generally hard to handle due to the significant connection among photosynthesis and transpiration (Posch and Bennett 2009). Under the dry season pressure circumstance, crop administration strategies that advance the crop protection from dry spell pressure might be advantage for an appropriate plant development (Egilla et al. 2001). Water inadequacy diminishes development rate, net CO_2 assimilation and transpiration paces of cucumber crop plants (Li et al. 2009).

Dry season is the most extreme crucial angle that cause starvation and upset the world food security. Being delicious in nature vegetables are profoundly influenced by water pressure. High temperature combined with low precipitation coming about because of environmental change will decreased the accessibility of water system water and simultaneously evapo-transpiration will be expanded. In this way, this will prompts serious harvest water pressure coming about low yield and nature of vegetables. Dry season extends the salt obsession in the soil and impacts the transform osmosis of loss of water from plant cells. This prompts an extended water hardship in plant cells and deterrent of a couple of physiological and biochemical systems, for instance, photosynthesis, breath, etc., as such lessen gainfulness of most vegetables (Pena and Hughes 2007).

Drought pressure also reduces the openness, take-up, move and assimilation of enhancements (Farooq et al. 2009). Introduction of drought stress in plants for the most part reduces both the take-up of enhancements by roots and translocation from roots to shoots (Hu and Schmidhalter 2005). Dry season impelled declines in take-up and translocation of full scale supplements (N, P, and K) have been represented in various plant species (Kuchenbuch et al. 1986; Subramanian et al. 2006; Asrar and Elhindi 2011; Suriyagoda et al. 2014). Low soil sogginess availability under drought tension lessened the root improvement and the pace of supplement inflow with respect to both per unit of root length and root biomass (Kuchenbuch et al. 1986). Moreover, drought pressure causes the qualification in dynamic vehicle and film permeability of cations (KC, Ca_2C, and Mg_2C), thusly realizing lessened digestion of these cations through roots (Hu and Schmidhalter 2005; Farooq et al. 2009). Drought pressure compels the activities of synthetic concoctions related with supplement assimilation (Figs. 4, 5, and 6).

3.5 Moisture Stress Leads to Formation of Reactive Oxygen Species (ROS)

Dry season pressure improves the generation of ROS in cell compartments, for example, chloroplasts, peroxisome and mitochondria. ROS are in part decreased types of climatic oxygen. They characteristically consequence from the excitation of O_2 to outline singlet oxygen ($O_2$1) or from the trading of 1, 2 or 3 electrons to O_2, for superoxide radical (O_2−), hydrogen peroxide H_2O_2 or a hydroxyl radical (OH), independently. The cells are commonly guaranteed in contradiction of Reactive Oxygen Species (ROS) by the undertaking of the malignancy anticipation operator boundary structure including enzymatic (APX, POD, SOD, CAT, GR, PPO) and non-enzymatic (ascorbate, alpha-tocopherol, glutathione, carotenoids) portions. The happenings of impetuses of the cell fortification structure in crops underneath tension are ordinarily seen as a pointer of the flexibility of genotypes in

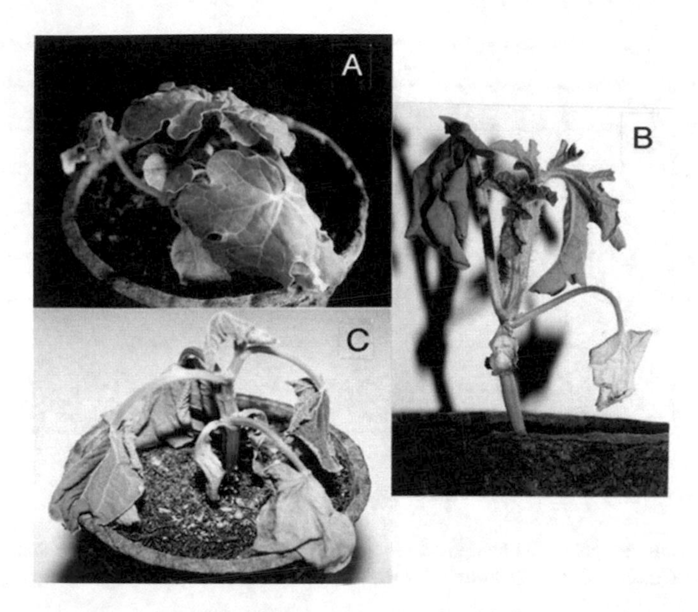

Fig. 4 Distinctive case of wild watermelon subjected to drought for 8 d (**a**). Domesticated watermelon (**b**) and cucumber plants I (**c**) treated for 3 and 2 d, respectively, are also shown as controls. (Shinji Kawasaki et al. 2000)

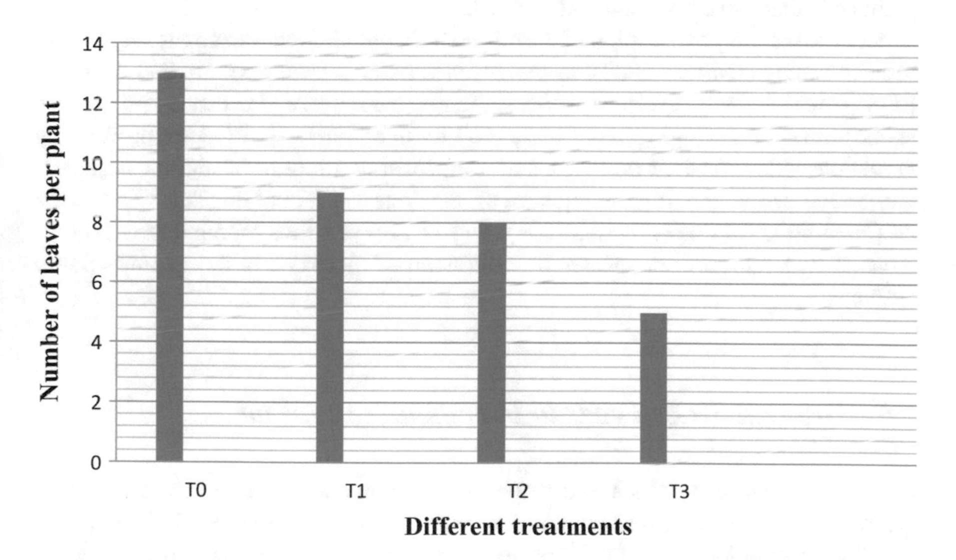

Fig. 5 Effect of moisture stress on number of leaves/plant of bitter gourd. (Ayesha Shahbaz et al. 2015)

Where, T_0 = Control, T_1 = 25% drought, T_2 = 50% drought, T_3 = 75% drought

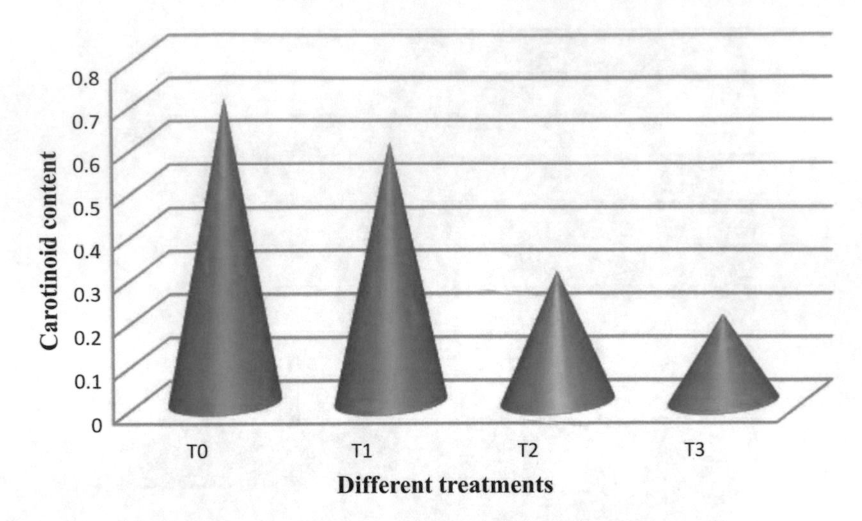

Fig. 6 Effect of drought on carotenoid content (mg/g) of bitter gourd. (Ayesha Shahbaz et al. 2015) Where, T_0 = Control, T_1 = 25% drought, T_2 = 50% drought, T_3 = 75% drought

contradiction of hassle conditions (Iturbe-Ormaetxe et al. 1998). All around, the consideration of Reactive Oxygen Species (ROS) in numerous metabolic systems in plant cells may have common consequences.

ROS bases the peroxidation of film lipids, the breakdown of proteins and harm to nucleic acids (Mittler 2002). In the event that dry season pressure is drawn out, ROS generation will overpower the searching activity of the cancer prevention agent framework, bringing about broad cell harm and demise. ROS are profoundly pernicious side-effects of pressure, and are probably going to be significant auxiliary errand people that trigger adjustment reactions to the evolving condition (Cruz de Carvalho and Contour-Ansel 2008). Dry season pressure prompts the development of dynamic oxygen species by confusion of electrons in the genuine photo systems.

3.6 *Drought Stress Leads to Formation of Proline*

Proline aggregation in leaves of dry spell focused on plants and its job as an osmolyte or osmoprotectant has been plentifully archived (Pagter et al. 2005). Despite the fact that proline has for some time been estimated as a good osmolyte, ongoing outcomes feature its different capacities in stress adjustment, recuperation and flagging. Proline aggregation because of dry spell pressure results from an animated blend, hindered debasement or an impeded fuse of proline into proteins (Heuer 1999). Likewise, proline expect a more brilliant movement in giving dry spell resistance than in going about as a basic osmolyte (Szabados and Savoure 2009). It might ensure about proteins structure by keeping up their assistant valiance

(Rajendrakumar et al. 1994), go about as free unbelievable forager (Reddy et al. 2004), and be connected with the reusing of NADPH+ H+ through its blend from the glutamate pathway (Hare and Cress 1997) including glutamyl kinase, glutamyl phosphate reductase and D–pyrroline-5-carboxylate synthetase in tomato. Proline amalgamation may give some insurance against photograph limitation under unsavory conditions by reestablishing the pool of the terminal electron acceptor of the photosynthetic electron transport chain (Szabados and Savoure 2009). In Laurus, close by its proposition in osmotic change, the weight incited complete of proline in certain masses could in addition be identified with dry season began adjustment of cell divider proteome (Maatallah et al. 2010). It might in like way work as a protein-impeccable hydrotrope, encouraging cytoplasmic acidosis, and keeping up fitting NADP/NADPH degrees extraordinary with ingestion (Hare and Cress 1997). Furthermore, rapid breakdown of unending flexibly of weight may give satisfactory diminishing pros that help mitochondrial oxidative phosphorylation and time of ATP for recovery from stress and fixing of weight impelled damages.

3.7 Drought Stress Leads to Formation of LEA Proteins

Late embryogenesis abundant (LEA) proteins may amass in response to dry spell (drought) pressure in plants and assume a significant job in plant assurance against the antagonistic impacts brought about by dry season pressure (Gosal et al. 2009). The putative job of LEA proteins in plant dry season resistance has been recommended to be because of their association in the upkeep of cell film structure and particle balance, official of water, and their activity as sub-atomic chaperones.

3.8 Drought Stress Leads to Formation of Abscisic Acid (ABA)

Plant hormones oversee pressure responses not by methods for straight pathways, yet through complex sub-nuclear frameworks. Abscisic corrosive (ABA), a terpenoid phyto-hormone, is locked in with the rule of various pieces of plant improvement and headway consolidating into seed advancement structures, getting of drying obstruction and torpidity, senescence of leaf and opening of stomata (Wasilewska et al. 2008). ABA is similarly the key hormone that presents versatility to characteristic tensions, for instance, drought, as needs be permitting plants to create where water accessibility is compelled or inconsistent. ABA center was represented to upsurge up to 30-fold for the duration of drought pressure (Outlaw 2003). ABA organized in light of drought pressure, is known to provoke stomatal end and to decrease transpirational water disaster. Measurements of endogenous ABA upsurge in tissues presented to osmotic concern as a result of evaporating. ABA starts the mix of ROS in guardian cells by a layer bound NADPH oxidase, and ROS mediate stomatal end by inciting (through hyperpolarization) plasma film Ca2+

Table 7 Numerous roles of proline in plants

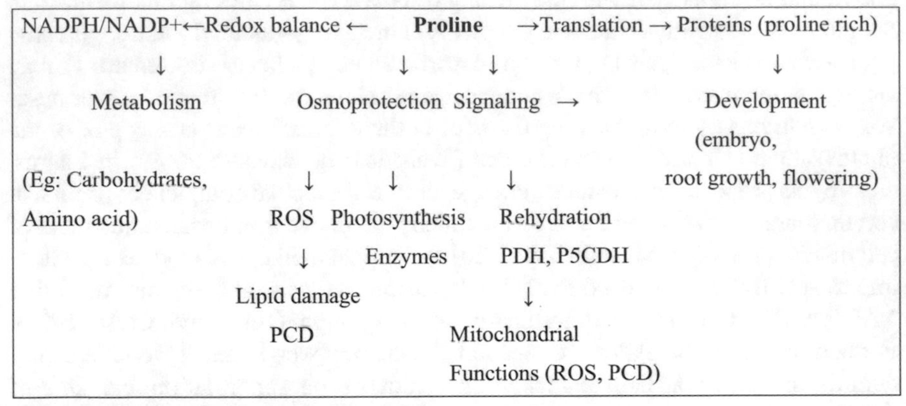

Adopted from Szabados and Savoure (2009)

Table 8 Genes responsive exclusively to drought (Hafiz A. Hussain et al. 2018)

General process	Annotation	Functions
Signaling and transcriptional control	bZIP	Basic leucine zipper (bZIP) proteins are ABA reactive transcription factors energetic in osmotic tension
	AP2/ DREB2A/ DRF1	DEHYDRATION-RESPONSIVE ELEMENT BINDING PROTEIN 2 family is involved in heat acclimation and positive regulation of transcription
	HD-ZIP	HD-ZIP is related to genes reactive to dryness and involved in cell differentiation
	ERF1	Eukaryotic Release Factor is involved in regulation of growth, translational termination

channels. It can bring about the extended period of dynamic oxygen species (AOS), improves the activities of malignant growth avoidance specialist impetuses, for instance, SOD, CAT, APX and GR. In any case, bits of knowledge concerning the relationship between ABA, AOS and disease counteraction operator response remain to be settled (Tables 7 and 8).

4 Conclusion

Vegetable crops are more susceptible to climate alteration, as it obstructs yield, productivity and quality. Impacts of temperature produced by an Earth-wide temperature boost (global warming) on yield plants are the most significant among all the environmental transformation impacts. It is again in charge of different burdens like dampness stress (dry spell), saltiness and atmospheric gases like CO_2 concentration. Evangelist approaches like advancement of creation framework, improved assortments with upgraded water use productivity just as resistance to various abiotic and biotic anxieties are taken up to lighten the impacts of environmental change on vegetable creation.

References

Arora SK, Partap PS, Pandita ML, Jalal I (1987) Production problems and their possible remedies in vegetable crops. Indian Hortic 32(2):2–8

Asrar AWA, Elhindi KM (2011) Alleviation of drought stress of marigold (*Tageteserecta*) plants by using arbuscularmycorrhizal fungi. S J Biol Sci 18:93–98

Ayesha S, Khalid H, Muhmmad QA, Khalid N, Abdul M, Syeda MB (2015) Changes in growth, morphology and photosynthetic attributes by drought in bitter gourd (*Momordica charantia* L.). Bot Res Int 8(3):54–58

Bhella HS (1985) Muskmelon growth, yield, and nutrition as influenced by planting method and trickle irrigation. J Am Soc Hortic Sci 110(6):793–796

Cruz de Carvalho MH, Contour-Ansel D (2008) Drought stress and reactive oxygen species: production, scavenging and signaling. Plant Signal Behav 3:156–165

Data Portal India (2013) Annual and seasonal mean temperature of India, National Informatics Centre of Govt. of India. Downloaded from http://data.gov.in/dataset/annual-andseasonal-mean-temperature-india

Duval JR, Nesmith DS (2000) Treatment with hydrogen peroxide and seedcoat removal or clipping improve germination of 'Genesis' triploid watermelon. HortScience 35(1):85–86

Efeoglu B, Ekmeki Y, Cicek N (2009) Physiological responses of three maize cultivars to drought stress and recovery. S Afr J Bot 75:34–42

Egilla JN, Davies FT, Drew MC (2001) Effect of potassium on drought resistance of Hibiscus rosasinensis cv. Leprechaum: plant growth, leaf macro and micronutrient content and root longevity. Plant Soil 229:213–224

Fabeiro C, Martin de Santa Olalla F, DeJuan JA (2002) Production of muskmelon (*Cucumis melo* L.) under controlled deficit irrigation in a semi-arid climate. Agric Water Manag 54(2):93–105

Farooq M, Wahid A, Kobayashi N, Fujita D, Basra SMA (2009) Plant drought stress: effects, mechanisms and management. Agronomy for sustainable development. Springer/EDP Sciences/INRA 29(1):185–212

Gosal SS, Wani SH, Kang MS (2009) Biotechnology and drought tolerance. J Crop Improv 23:19–54

Grange S, Leskovar DI, Pike LM, Cobb BG (2003) Seed coat structure and oxygen-enhanced environments affect germination of triploid watermelon. J Am Soc Hortic Sci 128(2):253–259

Hare PD, Cress WA (1997) Metabolic implications of stress induced proline accumulation in plants. Plant Growth Reg 21:79–102

Hasanuzzaman M, Hossain MA, da Silva JAT, Fujita M (2012) Plant responses and tolerance to abiotic oxidative stress: antioxidant defense is a key factor. In: Bandi V, Shanker AK, Shanker C, Mandapaka M (eds) Crop stress and its management: perspectives and strategies. Springer, Berlin, pp 261–316

Hazra P, Som MG (2006) Environmental influences on growth, development and yield of vegetable crops. In: Vegetable science. Kalyani Publishers, pp 37–108

Hessini K, Martínez JP, Gandour M, Albouchi A, Soltani A, Abdelly C (2009) Effect of water stress on growth, osmotic adjustment, cell wall elasticity and water-use efficiency in *Spartinaalterniflora*. Environ Exp Bot 67:312–319

Heuer B (1999) Osmoregulatory role of proline in plants exposed to environmental stressed. In: Pessarakli M (ed), Handbook of Plant and Crop Stress. Marcel Dekker, New York, pp 675–695

Hu Y, Schmidhalter U (2005) Drought and salinity: a comparison of their effects on mineral nutrition of plants. J Plant Nutr Soil Sci 168:541–549

Hussain HA, Hussain S, Khaliq A, Ashraf U, Anjum SA, Men S, Wang L (2018) Chilling and drought stresses in crop plants: implications, cross talk, and potential management opportunities. Front Plant Sci 9:393

Iturbe-Ormaetxe I, Escuredo PR, Arrese IC, Becana M (1998) Oxidative damage in pea plants exposed to water deficit or paraquat. Plant Physiol 116:173–181

Kawasaki S, Chikahiro M, Takayuki K, Shinichiro F, Masato U, Akhiho Y (2000) Responses of wild watermelon to drought stress: Accumilation of an argE homologue and Citrulline in leaves during water deficits. Plant Cell Physiol 41(7):864–873

Kuchenbuch R, Claassen N, Jungk A (1986) Potassium availability in relation to soil-moisture. Effect of soil-moisture on potassium diffusion, root-growth and potassium uptake of onion plants. Plant Soil 95:221–231

Kumar SV (2012) Climate change and its impact on agriculture: a review. Int J Agric Environ Biotechnol 4(2):297–302

Kurtar ES (2010) Modelling the effect of temperature on seed germination in some cucurbits. Afr J Biotechnol 9(9):1343–1353

Li W, Sui XL, Gao LH, Ren HZ, Zhang ZX (2009) Effect of rapid dehydration on photosynthetic and fluorescent properties of cucumber leaves detached from low light treated seedlings. Eur J Hortic Sci 74:210–217

Lower RL (1974) Measurement and selection for cold tolerance in cucumber pickle. Pak J Plant Sci 4:8–11

Loy JB (2004) Morpho-physiological aspects of productivity and quality in squash and pumpkins (*Cucurbita* spp.). Crit Rev Plant Sci 23(4):337–363

Maatallah S, Ghanem ME, Albouchi A, Bizid E, Lutts E (2010) A greenhouse investigation of responses to different water stress regimes of *Laurus nobilis* trees from two climatic regions. J Arid Environ 74:327–337

Maynard DN (1989) Triploid watermelon seed orientation affects seed coat adherence on emerged cotyledons. Hortic Sci 24(4):603–604

Maynard DN, Hochmuth GJ (2007) Knott's handbook for vegetable growers. Wiley, Hoboken

Minaxi RP, Acharya KO, Nawale S (2011) Impact of climate change on food security. Int J Agric Environ Biotechnol 4(2):125–127

Mittler R (2002) Oxidative stress, antioxidants and stress tolerance. Trends Plant Sci 7:405–410

Najarian M, Mohammadi-Ghehsareh A, Fallahzade J, Peykanpour E (2018) Responses of cucumber (*Cucumis sativus* L.) to ozonated water under varying drought stress intensities. J Plant Nutr 41(1):1–9

NOAA (2018) Global climate report. National Oceanic and Atmospheric Administration, Silver Spring, Maryland, United States

Outlaw WH (2003) Integration of cellular and physiological functions of guard cells. Crit Rev Plant Sci 22:503–529

Pagter M, Bragato C, Brix H (2005) Tolerance and physiological responses of *Phragmites australis* to water deficit. Aquat Bot 81:285–299

Paul RE, Patterson BD, Graham D (1979) Chilling injury assays for plant breeding. In: Lyons JM, Graham D, Raison JK (eds) Low temperature stress in crop plants – the role of the membrane. Academic, New York, pp 507–519

Pena R, Hughes J (2007) Improving vegetable productivity in a variable and changing climate. SAT e J 4(1):1–22

Perry KB, Wehner TC (1990) Prediction of cucumber harvest date using a heat unit model. HortScience 25:405–406

Posch S, Bennett LT (2009) Photosynthesis, photochemistry and antioxidative defence in response to two drought varieties and with re-watering in *Allocasuarina luehmannii*. Plant Biol 11:83–93

Rajendrakumar CSV, Reddy BVB, Reddy AR (1994) Proline-protein interactions: protection of structural and functional integrity of M4 lactate dehydrogenase. Biochem Biophys Res Commun (2):957–963

Rakshit A, Sarkar NC, Pathak H, Maiti RK, Makar AK, Singh PL (2009) Agriculture: a potential source of greenhouse gases and their mitigation strategies. IOP Conf Ser Earth Environ Sci 6(24):242033

Reddy AR, Chaitanya KV, Vivekanandan M (2004) Droughtinduced responses of photosynthesis and antioxidant metabolism in higher plants. J Plant Physiol 161:1189–1202

Schneider SH, Semenov S, Patwardhan A, Burton I, Magadza CHD, Oppenheimer M, Pittock AB, Rahman A, Smith JB, Suarez A, Yamin F (2007) Assessing key vulnerabilities and the risk from climate change. Climate Change 2007: impacts, adaptation and vulnerability. In: Parry et al (eds) Contribution of Working Group II to the fourth assessment report of the Intergovernmental Panel on Climate Change. Cambridge University Press, Cambridge, pp 779–810

Singh AK (2010) Climate change sensitivity of Indian horticulture – role of technological interventions, Souvenir of Fourth Indian Horticultural Congress. HSI, New Delhi, pp 85–95

Subramanian K, Santhanakrishnan P, Balasubramanian P (2006) Responses of field grown tomato plants to arbuscular mycorrhizal fungal colonization under varying intensities of drought stress. Sci Hortic 107:245–253

Suriyagoda L, De Costa WAJM, Lambers H (2014) Growth and phosphorus nutrition of rice when inorganic fertilizer application is partly replaced by straw under varying moisture availability in sandy and clay soils. Plant Soil 384:53–68

Szabados L, Savoure A (2009) Proline: a multifunctional amino acid. Trends Plant Sci 2:89–97

Taylor AG (1997) Seed storage, germination and quality. In: Wien HC (ed) The physiology of vegetable crops. CAB International, New York, pp 1–36

Wasilewska A, Vlad F, Sirichandra C, Redko Y, Jammes F, Valon C, Frey NFD, Leung J (2008) An update on abscisic acid signaling in plants and more. Mol Plant 1:198–217

Wien HC (1997) The cucurbits: cucumber, melon, squash and pumpkin. In: Wien HC (ed) The physiology of vegetable crops. CAB International, New York, pp 345–386

WMO (2018) Global climate report. World Meteorological Organization, Geneva

Ye X, Yuan Y, Dy N, Wang Y, Shu S, Jin s, Guo S (2018) Proteomic analysis of heat stress resistance of cucumber leaves when grafted onto Momordica rootstock. Hortic Res 5:53

Zajac MR, Kubis J (2010) Effect of UV-B radiation on antioxidative enzyme activity in cucumber cotyledons. Acta Biol Cracov Ser Bot 52(2):97–102

Impact of Carbon Sequestration and Greenhouse Gasses on Soil

D. K. Verma and Shashank Shekhar Solankey

1 Introduction

Climate change is an emerging challenge and serious issue facing the world. Human activities during the past two centuries have elevated to unprecedented levels the atmospheric attentions of carbon dioxide gas (CO_2) and other global warming creating gases (GHGs) are play important role for enhance the temperature. Climate mitigation targets must involve the agricultural sector, which contributes 10%–14% of global anthropogenic greenhouse gas (GHG) emissions (Jantke et al. 2020). The total contribution of green houses in the atmospheric CO_2 is accounts are approximately 60%. In the atmospheric carbon dioxide concentration (CO_2) enhance from about 280.0 part per million in the year 2014. Concentration of carbon dioxide has been developing into the atmosphere as a suitable level in the upcoming future. We need to adopt proper planning for control the carbon sequestration. In addition of cellulose into the soil its reduce the carbon mineralization process. Combination of cellulose and other nutrient affected comparable responses of crop leaf, through substantial effects of nitrogen and phosphorus fertilizers. Well developed and proper managed soils, has the capacity to stock carbon and play important role for the mitigation of the releases greenhouse gas (Erickson 2003). Improve the organic carbon concentration into the soils is not only minimise the greenhouse gas productions, however, it's also promote the soil health, reducing soil erosion and endure ecology. Carbon present into the soil was found in two forms: organic and inorganic carbon. It is a complex chemistry of the carbon mixes were found in the terms of humus. It

D. K. Verma (✉)
Department of Soil Science and Agricultural Chemistry, Dr. Kalam Agricultural College, Kishanganj, Bihar Agricultural University, Sabour, Bhagalpur, Bihar, India

S. S. Solankey
Department of Horticulture (Vegetable and Floriculture), Bihar Agricultural University, Sabour, Bhagalpur, Bihar, India

S. S. Solankey et al. (eds.), *Advances in Research on Vegetable Production Under a Changing Climate Vol. 1*, Advances in Olericulture,
https://doi.org/10.1007/978-3-030-63497-1_10

contains everything used to plant nutrition and biological decomposition (Baldock and Skjemstad 1999). Soil organic carbon originally comes from atmospheric in the form of CO_2 it is taken by different green plant through the photosynthesis process (Burke et al. 1989). The total quantity of the organic carbon is a equilibrium of carbon incoming and outgoing of carbon into different humus form. Weathering of the different type of rock and minerals produces the inorganic carbon in the form of carbonic acid (CO_2 dissolved in water) into the soil and lead to as different carbon minerals like, aragonite, dolomite and calcite (Lal 2007). Liming is an important process in agricultural its can be stored significant more carbon it is depending on the source of the carbon, soil pH and calcium present into the soil (Rasse et al. 2006). Worldwide, soils hold around three time's higher carbon than the atmosphere. If we increase fairly lesser amount of carbon content into the soils its can give a significant result to reducing carbon dioxide content in to the atmospheric.

2 Soil and Soil Carbon

Carbon in soils were found in different forms like (1) organically bound like, complex mixture of organic matter (2) in-organic form as carbonate and bicarbonate minerals (Calcite or dolomite, gypsum). Parent's materials like in-organic materials and organic matter components like soil, water, air and all living thing. Creation of carbonates is an secondary important components and it's generally occurs in arid and semi-arid climates (Buckman and Brady 1970). The balance between carbon incoming and outgoing is important for soil organic carbon change. Contributions include belowground and aboveground crop west materials, compost, animal product decomposition etc., whereas degradation of the organic materials through air and water erosion, movement of the gas by the microbial and plant respiration and leaching process. The process of the decomposition its produce humus as a fluid arrangement and it's highly impervious to degradation (Schlesinger 1982). Various organic composites in the soil are nearly linked with inorganic particles of the soil. In well-known fact that about 5% of the organic carbon were found in our agricultural land (West and McBride 2005) esimated the net arbon by the application of limestone and dolomite.

Decrease organic carbon stocks connected with transformation of inherent agricultural production system is attributable to:

- Burning the harvest crop and stubble inputs reduce organic matter.
- Continuous cultivation increases the rates decomposition.
- Topsoil organic matter content decreases due to soil erosion.
- Depletion availability of plant nutrients.
- Decreased crop yields.
- Diminished cation-exchange capacity.
- Improved bulk density of the soil.
- Defeat soil structure of the soil.

- Decreased soil moisture and hydraulic conductivity of the soil.
- Increased leaching losses of fertilizers, agrochemicals and heavy metals into the soil system.
- Degradation of biological activity and bio-diversity of the soil.

3 Greenhouse Gasses and Their Source

Increase the different type of greenhouse gases concentration into the atmosphere causes a decrease outward infrared radiation, hence the climate of Earth's need to change somehow to reinstate and balance the incoming and outgoing radiation. While the majority of farmers perceive climate change as an important issue and see GHG reduction potential in the agricultural sector, only a few stated that they estimate GHG emissions on their farm. This suggests a lack of knowledge about adequate tools to calculate on-farm emissions at regular intervals. Available GHG tools to assess agricultural and forest practices include calculators, protocols, guidelines, and models (Denef et al. 2012; Colomb et al. 2012). Improvements of such tools may, however, be needed, as several existing GHG calculators show limited agreement on the magnitude of GHG emissions (Green et al. 2017; Lewis et al. 2013). Technology-intensive changes such as the adoption of precision agriculture yet incur high investment cost, which could hamper their implementation (Barnes et al. 2019). The most dangerous and dominant greenhouse gases in atmosphere like water vapour, carbon dioxide, methane, nitrous oxide and Chlorofluorocarbons (CFC). The concentrations of greenhouse gases into the atmosphere are strong-minded through the stability amongst sources and sinks. The percentage contribution to the different greenhouse gasses and their effect the major gases are: water vapour 36–70%, carbon dioxide 9–26%, methane 4–9%. These are important funders to the atmospheric greenhouse effect; clouds also release infrared energy and thus, have effect on radioactive substances of the atmosphere.

The net effect of global warming gasses on agriculture systems it's not only carbon dioxide (CO_2) and methane (CH_4) productions is important but due to their high precise greenhouse potential also the site and management related N_2O emissions. Here are fallowing factors affect the step to which any greenhouse gas will Impact global warming, as follows:

- It's plenty present in the air.
- Capacity to hold into the atmosphere.
- It has potential for warming global.

4 Climate

Climate represent the cumulative effect of air, heat and water vapour affects both gain and losses of organic substances present into the soil thus, quantity of carbon impacting the deposited the soil system since, these affecting factor controls the respiration rate of organic matter. Availability of the enough quantity of water and optimum temperature increase the rate of soil organic matter decomposition, passive pools hold slow carbon its result greater loss of carbon through respiration. In hot climates soil usually contains low organic carbon compare to the cold climates (Lal 2007).

5 Factor Affecting of Carbon Sequestration

Atmospheric is the primary source of the carbon its comes from CO_2 that is taken by plants through the photosynthesis process. Rainfall, air and temperature are main factors impact the plant biomass contribution and subsequent rate of soil organic carbon decomposition as a certain type of soil. Some soil controlling factors influence the equilibrium of carbon into the soil system (Burke et al. 1989). Organic carbon equilibrium into the soil always equal due to the additions and removal process of the soil organic carbon. Organic carbon storage is affected by different soil factors these are given below.

5.1 *Soil Type and Depth of Soil*

Atmospheric removal of the carbon dioxide is only one advantage its enhanced carbon storage into the soils profile. Its improved soil quality, enhance water holding capacity of the soil, reduces the nutrient losses, control the soil degradation, improve aeration of the soil system and improved crop productivity may result from increasing the amount of carbon stock in agricultural soils.

5.2 *Conservation Tillage*

Minimizes tillage operation of the cultivated cropland. Practice of the soil cover like, use of the different type of mulching materials plant leaves, crop residues, paddy straw on the soil surface. These types of agronomical practices usually reduce soil erosion, enhance soil moisture and improve carbon stock into the soil. In conservation tillage exploit the crop residue. It has potential to sequester the significant quantity of carbon dioxide. The amount of organic matter present into the soil was measured by passive pools it's also improved through the accumulation reprocessed organic substances like, plant west materials, compost, manure and bio -solids into the soil and by slowing down through the decomposition process.

Introduction of the helping tools to mitigate the effect of the climate change and it's resulted to enhance soil carbon loading. Practise of used things into the soil it's very imperative environmental advantage (Lal 2002, 2004; Smith 2008).

- Its Improve soil health, water holding capacity and crop yield.
- Its decrease requirement of agro-chemicals and fertilizers.
- Reduction of soil degradation by climatic factors.
- Upgraded physical and biological condition of the soil.

5.3 Cover Cropping

Crop residues are also a main resource for a number of competing off-farm uses. The use of crops such as leguminous and cover for improvement of the biological activity and fertility status of the soil, it's also helpful to minimise the loss of soil. Its deliver many benefits to agricultural activity including weed control, soil aggregation and water storage and are also promote soil carbon formation. Its increase carbon sequestration and adding organic matter to the soil.

5.4 Crop Rotation

Different type of land use pattern and crop growing practices can greatly upset the quantity of carbon sequestered into the soil profile. According to the USDA-NRCS (2014) we need to be developing a framework for demonstrating and analyse the effects of cropping variations on soil carbon sequestration. It is a represent the sequence of crops grown in regularly in same or confined area of land. It is a simulator of natural variation in the ecologies, it has extremely diligently than intensive mono-cropping system. Different type of the crops cultivation practices can also increase the level of organic matter into the soil. However, crop rotational dominancy depends on the type of crops. Different results show that the dynamics of cover crop can also encourage carbon sequestration for a very huge portion. In the same crop rotations carbon sequestration were found highest in clay soils and lowest on sandy soils. Maize, winter wheat followed by catch crops residues donate higher amount to the total carbon sequestered.

6 Potential Costs of Soil Sequestration of Carbon

Different types of crop cultivation practices that have been recommended a method for carbon sequestering these are the following.

6.1 Nitrogen Fertilizer

The nitrogen fertilizer has also raise organic matter content into the soil because nitrogen was present in a limited quantity in the agro-ecosystems. Added N in soil its reduced carbon emissions from the soil surface, respired carbon outcome into the soil are not much clear. Nitrogen application into the soil can also reduce the carbon mineralization process as an average of 2110 mg carbon per gram of soil completed the path of the incubation, an amount of soil carbon equal to approximately 25% of the microbial carbon group. The application of mineral fertilizers is a major source of agricultural GHG emissions (Singh et al. 2019). However, carbon dioxide released from organic matter incineration at the time of the production, transport and application of nitrogen containing fertilizer can reduce the net amount of carbon sequestered. Nitrogenous fertilizers can be losses due to the leaching and run off agricultural lands into adjacent water body where it may have severe environmental significances (Fig. 1).

6.2 Growing Plants on Semiarid Lands

Cultivation of crop in semiarid areas has been mentioned as a way to enhance the carbon storage in soils profile. Different type of plant roots are playing vital role of subsoil to increase the organic matter. The worldwide assessment of root distributions for the different crop species like, grasses were found shallow in depth, trees were found intermediate root system and shrubs were found deepest into the soil profiles but, the organic matter was fixed into the soil in long time these type lands may costs of irrigating and surpass any net advance in carbon sequestration Plant roots play the importance for soil carbon sequestration that was emphasized by the fact they have a high prospective to become stable in soil. Moreover, in many parts of the semi-arid regions groundwater was available high level of soluble calcium, magnesium, carbonate and bicarbonate ions. These type elements are directly deposited into the soil and they can release carbon dioxide into the atmosphere.

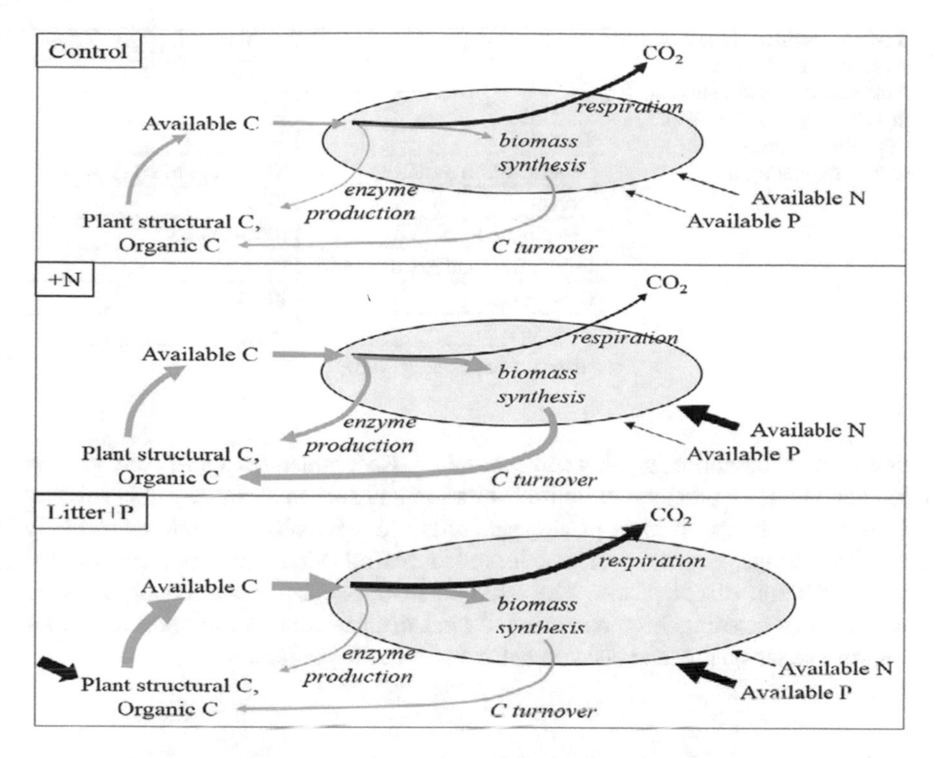

Fig. 1 Movement of C into and out of microbial biomass (shaded). Line thickness represents relative differences in C movement related to addition of N and addition of leaf litter þ P. Black lines indicate variables that were measured and blue lines represent hypothesized responses to treatments (Fisk et al. 2015)

7 Connection of Land Management, Carbon Sequestration and Different Environmental Factors

We use effective land management practices, entirely soil and water conservation, deforestation, Agroforestry, conservation tillage, Wetland Reserve Program, can key to both its improved soil surface carbon sequestration and to improved soil organic carbon. Tillage operation affects the system of the micro flora it can be demonstrated by the situation of mycorrhizal fungi. Numerous studies on soils acquisition soil organic carbon are also usually adding different type of nutritional qualities that can improve the crop productivity and secure environmental safety. The amount, place and type of organic matter affect the relative sizes of the numerous ecological groups in the community. Deep tillage was found not harmful for all species. Epigeic and anecic species are poorly affected by the mixture of crop residues into the soil. Increases in soil organic carbon generally developed good soil structure; increase soil aeration and water holding capacity it's also improve biological activity of the soil.

In common here is a favourable interrelation between carbon sequestration and much type of suggested land management practices it's linked to soil fertility,

Table 1 estimate of carbon reservoirs of different biosphere compartments and order of magnitude estimates of potential capacities for carbon sequestration

Reservoir size	Gt (billion tonnes) carbon
Oceans	44,000
Atmosphere	750
Terrestrial	2200
Sequestration potential	**Gt (billion tonnes) carbon**
Oceans	1000s
Deep saline formations	100s–1000s
Depleted oil and gas reservoirs	100s
Coal seams	10s–100s
Terrestrial	10s

Adapted from Herzog (2001)

agronomical operation, grazing and forestation. Recommended agronomic, grazing land and forestry practices, wildlife habitation and soil-water quality also enhance land sustainability. In most places, generally we were observed environmentally sensitive things, such type of practices also control wind and water erosion that reduce the soil carbon stocks. The same positive association that occurs between carbon sequestration and recommended land management can in some situations improve water quality and encouragement of wildlife habitation.

8 Soils Include as Carbon Sequestration

One meter of top soil in the comprises soil organic carbon near 3/4 of the total earth's carbon but, here is incredible capacity to sequester supplementary carbon into the soil i.e. different cultivated areas soils of the US have vanished as much as 50% of their usual soil carbon by the belongings of land clearing and cultivation operation. Present farming practices destroy soil carbon due to the burning of crop residue. Recent advanced management practices of crops, forest and grazing lands is assessed to potentially balance 30–60 thousand million metric tons of carbon free from the remains burning materials over the subsequent fifty (50) years (Table 1).

9 Greenhouse Technology

On an average 95% of crops are produce under open condition and only 5% are grown under protected cultivation. The horticultural crops are predominantly grown under protected structure i.e. poly houses, glass houses, net houses, etc. In earliest condition, people have cultured how to cultivate crop below natural condition. Mostly in temperate regions where the climatic conditions are extremely cold i.e. hard/adverse for survival of any crop plant, where various protected structure are very much helpful for growing of few high value crops. We can say that a green house is a protected structure structures, made by glass/net/polythene that provides favourable environmental condition for growth, development to the plant crop. Thus, on the basis of above facts, the greenhouse technology is a technique of providing favourable environment condition to the plants for complete its optimum life cycle. Its protect the plants from the adverse climatic conditions such as cold, heat, precipitation, wind, excessive radiation, insects and disease infections. It is also dynamic importance to generate an ideal micro-climate around the plants canopy. This is possible by building a greenhouse or glass house, where the environmental conditions are so modified that can grow any plant in any place at any time by providing suitable environmental conditions with minimum men power. Greenhouses are framed or inflated structures covered with clear or translucent UV stabilised polythene/glass material, large enough to grow crops under partial or fully controlled environmental conditions to get optimum growth and development as well as the growers can obtain a better productivity under these structures.

9.1 Advantages of Greenhouses

- Disease-free, healthy planting materials can be produced under these houses.
- The productivity in more than 10 times as compare to the outdoor cultivation, however, it's also depending upon the types of greenhouse, crop, variety and environmental management facilities.
- This technique generally suitable for flower and vegetables crops.
- Under greenhouse farming increases crop dependability.
- It's time to need increase production of floricultural crops.
- This technique helps to produce off-season vegetable and fruit crops.
- Minimise the water requirement of the crops and it's maintain easily under this condition.
- Application of chemicals, pesticides to manage the pest and diseases in efficient manner.
- Preservation of plants root stock, cultivating grafted plant and micro propagated plant.
- It is a very useful in controlling and monitoring the variability of much ecological system.

10 Mitigation of Climate Change Through Soil Carbon Sequestration

- Organic matter into the soil is differently removed by action of the wind and water. Several enriched carbon residues are reallocated of the land surface, many of the put into the depressed soil sites and more reserved into the moist condition. Whereas a portion of carbon trans-located through the soil erosion it may be supressed and reallocated into the atmosphere in the form of carbon dioxide by mineralization or as methane by methanogens process.

- Induced erosion of the soil and its deposition a large quantity carbon released into the atmosphere. Estimation of carbon emission against burial is a high significance. However, effective control of soil erosion is important for cultivated land and enhancing water quality.
- Degradation rate of the soil nutrients per year of sub- Saharan Africa affected by low input/subsistence agricultural is assessed to be apply 40 kg of NPK/ha of cultivated land since the mid- 1960s. Removal of nutrients from the organic carbon degradation it's also damages the atmosphere through organic matter decomposition. Thus, deplete organic carbon pool and soil fertility status, increase and decrease the productivity of the crop by application of fertilizer in per unit area and recover the damaged soil fertility status.
- Availability of the good quality water and proper managed land is the primary requirement for crop production. Association with the water and carbon cycles over the use conservation practices of water; in dry land areas it's an essential to enhance the crop productivity and carbon sequestration. The low soil carbon stock in rain fed areas can be enhanced through the practice of soil and water conservation, rain water harvesting and efficient use of water in agriculture. Through, the minimum tillage farming, drought management and apply good management of the agronomical practices improve the soil carbon stock in dry land ecosystems.
- Global warming is an emerging challenge and a global issue. Carbon sequestration is an associated this issues but separate with its own qualities of increasing crop productivity, improving quality of water and renovating the degraded soils, unrelatedly of the global warming deliberation. Balancing vestige energy emissions by possible soil carbon potential provides numerous physical, biological and social benefits. Besides, soil carbon sequestration is a connection through three global issues like; desertification, climate change and biodiversity.

10.1 Disadvantages of Greenhouse Gasses

10.1.1 Global Warming

The continuous rise in atmospheric temperature is truly disappointing. Greenhouse gases concentration of greenhouse into the atmosphere was artificially increased. Water vapour, soil particles, cloud, water body and surface of the soil are playing major role to reflect the sunlight into the atmosphere. Different greenhouse gasses like carbon dioxide, CFC and methane other trace gases present into the atmosphere and its exuded reverse to the earth surface and increase the atmosphere temperature. This type of environment variation has comprised a warming situation on the surface of the earths. Rise the atmospheric temperature is increase the rate of breakdown of the organic matter in to the soil and the decomposition process rate of carbon sequestration.

10.1.2 Agricultural Impact

Different agricultural practices like crop cultivation, tillage, fertilizer application, green manuring and composting release the carbon dioxide, methane and other greenhouse gasses. These free gasses increase the atmosphere temperature and its result improves the rate of the degradation of the soil organic carbon. The higher concentrations of carbon dioxide into the atmosphere can increase the growth of plants superior and faster growing. Still, the global warming may also affect the global pattern of the rainfall and soil moisture content.

10.1.3 Rise Sea Level

The global rise the atmospheric temperature its result sea level will rise due to firstly, warmer temperature of the atmosphere cause thermal expansion of seawater and increase sea level. Secondly, melting of glaciers and the icecap of Greenland and the Antarctica would also add water to the ocean.

10.1.4 Unbalance of Water Cycle

Rate of precipitation on worldwide is probable to rise. Still, this is not currently recognized that, how our local forms of precipitation will change. Different areas may were found higher rate of rainfall, while others places were observed less rainfall. Besides, rate of evaporation also increase due to the higher temperatures. All these type of climatic changes would possibly generate new pressures for water management practices.

10.1.5 Economic Impact

Maximum population of the world lives within 100 km of the coastal areas. Most of this urban population sea as a picnic spot. An assessable that time rise the level of sea will have an unfavourable impact on economy and low lying coastal areas of the islands, like if increasing the rates of soil erosion on beach along with seaboards and moving fresh groundwater its result contaminate the fresh ground water through intrusion process for an extensive distance.

11 Agronomical Practices Using as a Tools for the Management of Carbon Storage

Balance of organic carbon represents the net soil carbon inputs and losses. Consequently, different agronomical practices that enhance the carbon through improve production of the crop, over the incorporation of various type of exterior organic sources like, compost, green manuring, crop residue and bio-solids etc. The effective management practices of the soil will increase the soil carbon.

- Crop yield will increase by adopting good management practices.
- Apply conservation agriculture principal and practices.
- Adopt cropping system, grazing, agroforestry and pasture management.
- Apply different type of organic matter sources.
- Adopt alternative farming practices.
- Use organic farming practices.

12 Conclusion

Carbon sequestration is a natural phenomenon and we minimise the carbon sequestration and achieve the goal of food production by application of the different type of cultural agronomical and social awareness. If we minimise the carbon losses by our food habit and living lifestyle. If you reduce the carbon dioxide concentration into the atmosphere than carbon concentration of the soil will increase and enhance the soil quality, environment condition and improve the crop productivity.

References

Baldock JA, Skjemstad JO (1999) Organic soil C/soil organic matter. In: Prveril KI, Sparrow LA, Reuter DJ (eds) Soil analysis: an interpretation manual. CSIRO Publishing, Collingwood, pp 159–170

Barnes A, De Soto I, Eory V, Beck B, Balafoutis A, Sanchez B, Vangeyte J, Fountas S, van der Wal T, Gomez-Barbero M (2019) Influencing factors and incentives on the intention to adopt precision agricultural technologies within arable farming systems. Environ Sci Policy 93:66–74

Buckman HO, Brady NC (1970) The nature and properties of soils. The Macmillan Company, London, p 653

Burke IC, Yonker CM, Parton WJ, Cole CV, Flach K, Schimel DS (1989) Texture, climate, and cultivation effects on soil organic matter content in U.S. grassland soils. Soil Sci Soc Am J 53(3):800–805

Colomb V, Bernoux M, Bockel L, Chotte JL, Martin S, Martin-Phipps C, Mousset J, Tinlot M, Touchemoulin O (2012) Review of GHG calculators in agriculture and forestry sectors: a guideline for appropriate choice and use of landscape based tools; French agency for environment and energy management. French Research Institute for Development, Food and Agricultural

Organization, Rome, Italy. Available online:http://www.fao.org/fileadmin/templates/ex_act/pdf/ADEME/Review_existingGHGtool_VF_UK4.pdf

Denef K, Paustian K, Archibeque S, Biggar S, Pape D (2012) Report of greenhouse gas accounting tools for agriculture and forestry sectors; interim report to USDA under contract No. GS23F8182H; ICF International, Fairfax, VA, USA

Erickson C (2003) Historical ecology and future explorations. In: Lehmann J, Kern DC, Glaser B, Woods WI (eds) Amazonian dark earths: origin, properties, management. Kluwer Academic Publishers, Dordrecht, pp 455–500

Fisk M, Sharon S, Kevan M (2015) Carbon mineralization is promoted by phosphorus and reduced by nitrogen addition in the organic horizon of northern hardwood forests. Soil Biol Biochem 81:212–218. https://doi.org/10.1016/j.soilbio.2014.11.022

Green A, Lewis KA, Tzilivakis J, Warner DJ (2017) Agricultural climate change mitigation: carbon calculators as a guide for decision making. Int J Agric Sustain 15:645–661

Herzog H (2001) What future for carbon capture and sequestration? Environ Sci Technol 35(7):149A–153A

Jantke K, Hartmann MJ, Rasche L, Blanz B, Schneider UA (2020) Agricultural greenhouse gas emissions: knowledge and positions of German farmers. Land 9:130. https://doi.org/10.3390/land905013

Lal R (2002) C sequestration in dry land ecosystems of West Asia and North Africa. Land Degrad Dev 13:45–59

Lal R (2004) C emissions from farm operations. Environ Int 30:981–990

Lal R (2007) C management in agricultural soils. Mitig Adapt Strateg Global Change 12:303–322

Lewis KA, Green A, Warner DJ, Tzilivakis J (2013) Carbon accounting tools: are they fit for purpose in the context of arable cropping? Int J Agric Sustain 11:159–175

Rasse DP, Mulder J, Moni C, Chenu C (2006) Carbon turnover kinetics with depth in a french loamy soil. Soil Sci Soc Am J 70(6):2097–2105

Schlesinger WH (1982) Carbon storage in the caliche of arid soils: a case study from Arizona. Soil Sci 133:247–255

Singh H, Northup BK, Baath GS, Gowda PP, Kakani VG (2019) Greenhouse mitigation strategies for agronomic and grazing lands of the US Southern Great Plains. Mitig Adapt Strateg Glob Chang

Smith P (2008) Land use change and soil organic C dynamics. Nutr Cycl Agroecosyst 81:169–178

USDA-NRCS (2014) Code 590 guidelines

West TO, McBride AC (2005) The contribution of agricultural lime to carbon dioxide emissions in the United States: dissolution, transport and net emissions. Agric Ecosyst Environ 108:145–154

Effect of Green House Gases on Vegetable Production

Meenakshi Kumari, Manoj Kumar, Shashank Shekhar Solankey, and Saurabh Tomar

1 Introduction

The climate of earth is regular changing and evolving due to excess human intervention. Human activities like deforestation, emissions of different gases from industry and transport are major cause of climate change led to stored gases and aerosols in the atmosphere which are turning detrimental to life (Rakshit et al. 2009). Climate change defined as a change in mean of the various climatic factors and composition of different gases in the atmosphere etc. along with their properties in a larger geographical area over a longer period. Climate change is also known as any change in climate due to natural variability or by human activity. The degree at which systems are susceptible and unable to survive with the adverse impacts of climate change is known as Vulnerability of any system (Schneider et al. 2007). Other than these, change in other parameters like flooding, water logging, drought, soil salinity, high concentration of CO_2 in the atmosphere and UV radiations also occurs due to climate change (Bates et al. 2008). The increased concentration of greenhouse gases like CO_2 and CH_4 in atmosphere is responsible for increased temperature (Table 1) which is known as global warming or greenhouse effect because these gases act as trapping agent for heat and raise the temperatures on the surface of the planet.

M. Kumari (✉) · S. Tomar
Department of Vegetable Science, Chandra Shekhar Azad University of Agriculture & Technology, Kanpur, Uttar Pradesh, India

M. Kumar
Division of Vegetable Crops, ICAR-Indian Institute of Horticultural Research, Bengaluru, Karnataka, India

S. S. Solankey
Department of Horticulture (Vegetable and Floriculture), Bihar Agricultural University, Sabour, Bhagalpur, Bihar, India

S. S. Solankey et al. (eds.), *Advances in Research on Vegetable Production Under a Changing Climate Vol. 1*, Advances in Olericulture,
https://doi.org/10.1007/978-3-030-63497-1_11

Table 1 Increase in atmospheric concentration of greenhouse gases since pre industrial times

Greenhouse gas	Conc. in 2010	Increase since pre industrial time
Carbon dioxide 389 ppm 39%	Carbon dioxide 389 ppm 39%	Carbon dioxide 389 ppm 39%
Methane 1808 ppb 158%	Methane 1808 ppb 158%	Methane 1808 ppb 158%
Nitrous oxide 323.2 ppb 20%	Nitrous oxide 323.2 ppb 20%	Nitrous oxide 323.2 ppb 20%

Source: WMO (2013)

According to Minaxi et al. (2011) at global level the averaged surface temperature by the last decade of the twenty-first century expected to rise by between 1.1 °C up to 6.4 °C. Another study conducted by Houghton et al. (2001) reported that concentration of atmospheric CO_2 increases by increase in air temperature of 1.4–6.4 °C along with significant changes in rainfall pattern. From 1881–90 to 2001–10, the global combined surface temperatures over land and sea have been increased from 13.68 °C to 14.47 °C (WMO 2013). However, in India over a period of last 111 years since 1901 (24.23 °C) to 2012 (24.69 °C) the mean annual temperature has been increased by 0.46 °C (Data Portal India 2013). Altering in timing and amount of rainfall, water availability, wind patterns and different weather patterns such as drought, heat waves, floods, storms, change in ocean currents, acidification, forest fires are the result of increasing temperature in the environment (Kumar 2012).

2 Consequences of Climate Change on Vegetable Crops

Major activity of agriculture production depends on climate; therefore, it is largely going to affected by climate change. Several consequences like excessive rainfall, flood, rising sea level, global warming, change in weather pattern, drought etc. are due to such rapid change which leads to extremity of all kinds. These changes will lead in delay and uncertainty of summer monsoon along with more intense temperature during the winter season in South Asian (Lal et al. 2001). In most of vegetable crops, reduction of potential yield and quality is mainly due to shortening of the growing period, unavailability of water and poor vernalization. Other than these, occurrence of new pests and diseases along with re-emergence of existed pest and disease with new and more sever strains and increase the vectors that carry the diseases are the other consequences.

In India, primary stage of crop production is the main stage of greenhouse gas emission (GHG) (Pathak et al. 2014). In environment green house gases generated through cultivation of different cop, use of agricultural inputs and farm machinery, soil residue management and irrigation (Table 2). Main contribution to climate change among agriculture commodities is made by rice production and livestock rising through significant consumption of fossil fuels and methane emission (Ahmad et al. 2011). Different climate related factors could decline India's GDP upto 9% according to the estimation of Indira Gandhi Institute of Development Research if

Table 2 List of major vegetable crops with per cent GHG emission from crop

Crop	Yield (tonnes/ha)	GHG (kg/ha)	GHG (kg/kg product)	GHG (kg/kcal)
Tomato	130	3000.00	0.15	0.88
Potato	23.83	3406.00	0.22	0.33
Onion	19.55	1599.65	0.10	0.39
Pea	1.39	540.09	0.42	0.81

Source: Vetter et al. (2017)

the predictions related to global warming made by the inter-governmental panel on climate change comes to fruition (Priyadarshini 2009).

3 Implication of Climate Change on Vegetables

Two major climatic parameters that affect the vegetable production and responsible for reduced productivity of vegetable are erratic rainfall patterns and unpredictable high temperature spells. The preconditions like shifting of latitude and altitude in ecological and agro-economic zones, degradation of land, unavailability of water, extreme geophysical events, rise in sea level and salinization are of major consideration for future prospects (FAO 2004).

On the basis of experimental studies conducted by various researchers, climate change impact on different vegetable is given below:

- The area suitable for cultivation of major vegetable crops may become unsuitable in another 25 years or the non-traditional area may be more suitable for their cultivation.
- Rise in temperature will leads to change in production timing. Much more variation may not be observed in photoperiod due to excess rise in temperature which will responsible for faster maturity of photosensitive vegetables.
- Reduction in winter regime and chilling period in temperate regions will affect the temperate vegetable crops.
- Pollination mechanism, floral abortion, flower and fruit drop have adversely affected by fluctuation in temperature.
- It also leads to increase the annual irrigation requirement along with reduced duration for achieving the particular heat unit requirement.
- Temperature above optimum requirement affect pollination process in many crops, tuber initiation in potato and quality of tomato as tip burn and blossom end rot is common phenomenon in tomato.
- High temperature leads to bolting in crucifers and affect anthocyanin production in capsicum.

4 Quality and Production of Vegetable Affected by Climate Change

Vegetable are rich in essential nutrients, vitamins and minerals which are major components of human diet and needed for overcoming micronutrient deficiencies, hence these are regarded as protective food. Nutritional value of any produce is depends on the quality of the produce and the quality of produce is dependent on a minor change of several parameters like soil conditions, available temperature, light and CO_2. Any change in climatic factors affect the nutritional value of produce like in tomato Vitamin C, sugars, acid and carotenoids content increases with increase in the level of CO_2 while in many vegetables vitamin C, starch, sugars and many anti-oxidants especially anthocyanins and volatile flavour compounds decreases with increase in temperature. Several reports are available on the effect of temperature fluctuation during cultivation and affects of storage on the post-harvest quality of many vegetables (Cotty and Jaime-Garcia 2007).

The vegetable production is doubled at global level over the past quarter century and value of trade in vegetable is more than the cereals. In Asian continent, due to temperate and sub-temperate climatic condition in east yield of vegetable is highest. In vegetable production, India ranks second next to china with total production of 187.36 mt (Anonymous 2019) which is 1.6% is higher than previous year. In the past three decades, production of vegetables in India increased about 2.75 times from 58.5 mt in 1991–92 to 187.36 mt in 2018–19 (Anonymous 2019). Up to some extent variation in climatic factors can be adopted by plants but under extreme conditions most of plants especially vegetables shows sensitive reactions. Therefore, very low yield is obtained due to change in various physiological and biochemical processes like photosynthesis, several enzymatic activities, altered metabolism, thermal injury to the tissues, failure in pollination and fruit set which is result of change in climatic condition. According to Bray et al. (2000), major losses of crop production at global level is mainly due to environmental factor which is responsible for yield reduction of more than 50%.

The effect of environmental stress severity on vegetable crops is influenced by climate change and similarly plants responses to these stress depends on the plant developmental stage and the length and severity of the stress (Bray 2002). Plant protects themselves by modifying the biochemical and morphological mechanisms (Capiati et al. 2006). Tomato productivity is significantly reducing at high temperature because higher temperature results in poor pollination, abortion of flower and fruits and smaller fruit size.

Before anthesis, exposure of plants to temperature stress results in changes in development of anthers, irregularities in the epidermis and endothecium, problem in stomium and ultimately results in poor pollen formation (Sato Peet and Thomas 2002). In tomato plants, higher temperature stress express symptoms like poor production and viability of pollen grains, pollen dehiscence, abortion of ovule, bud drop, abnormal flower development, reduced availability of carbohydrate, and other reproductive abnormalities (Hazra et al. 2007). In case of capsicum, fertilization is affected by high temperature stress, therefore fruit set inhibited at high temperature during post pollination but at pre-anthesis stage it has no determinantal effects on fruit set (Erickson and Markhart 2002).

Plants exposed to salt stresses shows the symptoms like decrease in photosynthesis rate, abscission, curling and wilting of leaves, changes in respiratory rate, loss of cellular integrity, turgor and growth, necrosis of tissue and finally death of plant.

Flooding or water logging is another factor which affects both quality and production of most of vegetable crops and with respect to these characters information of genetic variations are limited. The damage in flooded plants (*e.g.* tomato) is mainly due to accumulation of endogenous ethylene (Drew 1979) because production of ethylene precursor in such plants root (1-aminocyclopropane-1-carboxylic acid or ACC) under low oxygen level is increased. The combined effect of water logging and high temperature leads to rapid wilting and death of tomato plants (Kuo et al. 1982). In developing countries from the last 40–50 years the air pollution is increasing at alarming rate and causing great loss on crop production due to presence of major pollutants like sulphur dioxide, nitrogen oxide, hydro fluride, ozone and acid rain. Major losses in term of reducing growth, yield and quality of vegetable produce are mainly caused by ozone. Air pollution cause significant reduction in the yield (>50%) of vegetable crops like cauliflower, lettuce and radish. It cause more damage in highly susceptible vegetables like in solanaceous family (tomato, potato), root crops (carrot, beet, turnip), Cucurbitaceae crop (water melon, squash, cantaloupe), soyabean and peas. Daily ozone causes 5–15% reduction in crop yield when concentrations reach to greater than 50 ppb (Raj Narayan 2009). Frost is detrimental in most of vegetables like cucumber, melons and fenugreek which is responsible for chilling injury in these vegetables. But, in case of leafy vegetables growth stage plays an important role to mitigate the frost injury therefore, no source of resistance is identified in available germplasm against low temperature injury (Tables 3 and 4).

Table 3 List of tolerance vegetable with respect to stress is given below

Sl. No.	Tolerance	Crop
1	Tolerance to drought	*Capsicum annuum, Citrullus lanatus, Solanum lycopersicum, Allium cepa*
2	Tolerance to heat	*Pisum sativum,, Solanum lycopersicum, Phaseolus limensis,*
3	Tolerance to salinity	*Brassica oleracea var. acephala, Momordica charantia, Beta vulgaris var. Bengalensis, Lactuca sativa*
4	Tolerance to flooding/ excess moisture	*Solanum lycopersicum, Allium cepa, Capsicum annuum*
5	Tolerance to soil acidity	*Solanum tuberosum, Ipomea batatus, Rheum rhaponticum*

Source: Rai and Yadav (2005)

Table 4 Indian varieties and advanced lines tolerant to abiotic stress

Sl. No.	Tolerant	Crop	Variety	Advance line
1	Drought/rainfed	Tomato	Arka Vikas	RF- 4A
		Onion	Arka Kalyan	MST-42 and MST-46
		Chilli	Arka Lohit	IIHR Sel.-132
2	Photo insensitive	Dolichos	Arka Jay, Arka Vijay, Arka Sambram, Arka Amogh, Arka Soumya	IIHR-16-2
		Cow pea	Arka garima, Arka Suman, Arka Samrudhi	
		French bean	Arka Anoop, Arka Bold, Arka Suvidha, Arka Komal, Kashi Kanchan	
3	High temperature	Capsicum		IIHR Sel.-3
		French bean		IIHR-19-1
		Peas		IIHR-1 and IIHR-8
		Cauliflower	Pusa Meghna, Arka Kanti, Pusa Early Synthetic	IIHR 316–1 and IIHR-371-1
		Cabbage	Pusa Ageti, Green Express, KK Cros	
		Radish	Pusa Chetki, Pujab Safed	
		Carrot	Pusa Vrishti, Pusa Kesar	
		Tomato	Pusa Hybrid-1	
		Indian bean	Kashi Khushal	
4	Tolerant to high and low temperatures	Tomato Chilli	Pusa Sadabahar Kashi Abha	
5	Tolerant to low temperatures	Tomato Indian bean Pea Cucumber	Pusa Sheetal Kashi Sheetal Alderman, Thomas Laxton Japanese long green, Straight-8, Pusa Sanyog	
6	Tolerant to low moisture stress	Dolichous bean Cow pea	Arka Vijay Arka Jay Arka Garima	
7	Salt tolerance	Tomato Pea Palak Cabbage	Sabour Suphala Newline Perfection, Market Prize Jobner green, HS-23, Pusa Harit Golden Acre, Pusa Synthetic, Pride of India	

Source: Hazra and Som (1999) and Rai and Yadav (2005)

5 Strategies for Reducing and Managing the Risks of Climate Change

Very less data is available regarding impact of climate change on vegetable crops. Therefore, it's very difficult to explain the problems of climatic change on vegetable production as compare to other food crops. Hence, accurate report and advance management practice is very important to solve the issue of climate change and problems arises from it. Agriculture can play also a major role in mitigating the climate change strategy by direct emissions of green house gases from livestock such as cows, agricultural soils and rice production (Smith et al. 2008, 2013). The effective adaption strategies to overcome the harmful effects of climate change are development of cultivars which are tolerant to heat, drought, salinity and flood. Some of the other techniques are use of drip irrigation, management of fertilizers with fertigation, conservation of soil and moisture, use of grafting techniques, use of plant regulators, protected cultivation, improved pest management practices.

Other than these, following the organic farming and adoption of resource conservation techniques are the other mitigation techniques to overcome the problems arise from climate change. Most of agricultural crops have any carbon sequestration potential but most of annual vegetables crop do not have these potential hence, there is limited scope for reducing emissions in their cultivation and information regarding these aspects is also lacking. Several new varieties and advanced breeding lines with tolerance to climate resilience have been developed in many vegetable crops. Efforts are also in progress to identify and develop the germplasm with efficient nitrogen use efficiency.

6 Conclusion

Climate change and agriculture are related to each other but became global problem from the last few years due to negative effect on crop production. Hence, vegetables are highly susceptible to climate change which hinders the production, quality and productivity of the crop. Therefore, olericulturist play major role in increasing the protection of crops to cope with climate change and reduce the impact of these changes on vegetable production. The most efficient strategy to mitigate the effect of climate change on vegetable production is adoption of new technologies, reduce the source of green house gases *i.e*, burning of fossil fuels for electricity, heat or transport, conservation of forest and water, sustainable use of land *i.e*, conservation agriculture etc. for sustainable vegetable production.

References

Ahmad J, Alam D, Haseen MS (2011) Impact of climate change on agriculture and food security in India. Int J Agric Environ Biotechnol 4(2):129–137

Anonymous (2019) Indian horticulture database. National Horticulture Database, Gurgon, India

Bates BC, Kundzewicz ZW, Wu S, Palutikof JP (2008) Climate change and water. IPCC Technical Paper VI, Geneva, p 210

Bray EA (2002) Abscisic acid regulation of gene expression during water-deficit stress in the era of the *Arabidopsis* genome. Plant Cell Environ 25(2):153–161

Bray EA, Bailey-Serres J, Weretilnyk E (2000) Responses to abiotic stresses. In: Gruissem W, Buchannan B, Jones R (eds) Biochemistry and molecular biology of plants. ASPP, Rockville, pp 1158–1249

Capiati DA, País SM, Téllez-Iñón MT (2006) Wounding increases salt tolerance in tomato plants: evidence on the participation of calmodulin-like activities in cross tolerance signalling. J Exp Bot 57:2391–2400

Cotty PJ, Jamie-Garcia R (2007) Influences of climate on aflatoxin producing fungi and aflatoxin contamination. Int J Food Microbiol 119(1–2):109–115

Data Portal India (2013) Annual and seasonal mean temperature of India, National Informatics Centre of Govt. of India

Drew MC (1979) Plant responses to anaerobic conditions in soil and solution culture. Curr Adv Plant Sci 36:1–14

Erickson AN, Markhart AH (2002) Flower developmental stage and organ sensitivity of bell pepper (*Capsicum annuum* L.) to elevated temperature. Plant Cell Environ 25:123–130

FAO (2004) Impact of climate change on agriculture in Asia and the Pacific. Twenty seventh FAO regional conference for Asia and the Pacific. Beijing: China, 17–21 May 2004

Hazra P, Som MG (1999) Technology for Vegetable Production and Improvement. Naya Prokash, Kolkata

Hazra P, Samsul HA, Sikder D, Peter KV (2007) Breeding tomato (*Lycopersicon esculentum* Mill) resistant to high temperature stress. Int J Plant Breed 1:1

Houghton J, Ding Y, Griggs D, Noguer M, Van der Linden P (2001) Climate change 2001: the scientific basis. Published for the Intergovernmental Panel on Climate Change. Cambridge University Press, Cambridge/New York, p 881

Kumar SV (2012) Climate change and its impact on agriculture: a review. Int J Agric Environ Biotechnol 4(2):297–302

Kuo DG, Tsay JS, Chen BW, Lin PY (1982) Screening for flooding tolerance in the genus *Lycopersicon*. Hortic Science 17(1):6–78

Lal M, Nozawa T, Emori S, Harasawa H, Takahashi K, Kimoto M, Abe-Ouchi A, Nakajima T, Takemura T, Numaguti A (2001) Future climate change: implications for Indian summer monsoon and its variability. Curr Sci 81:1196–1207

Minaxi RP, Acharya KO, Nawale S (2011) Impact of climate change on food security. Int J Agric Environ Biotechnol 4(2):125–127

Narayan R (2009) Air pollution – a threat in vegetable production. In: Sulladmath UV, Swamy KRM (eds) International conference on horticulture (ICH-2009). Horticulture for livelihood security and economic growth, pp 158–159

Pathak S, Bhatia A, Jain N (2014) Greenhouse gas emission from Indian agriculture: trends, mitigation and policy needs. Indian Agricultural Research Institute, New Delhi, p 39

Priyadarshini S (2009) Protected farming can reduce impact of climate change, News Paper Article published in The Assam Tribune on December 27, 2009

Rai N, Yadav DS (2005) Advances in vegetable production. Research co Book Centre, New Delhi

Rakshit A, Sarkar NC, Pathak H, Maiti RK, Makar AK, Singh PL (2009) Agriculture: a potential source of greenhouse gases and their mitigation strategies. IOP Conf Ser Earth Environ Sci 6(24):242033

Sato Peet MM, Thomas JF (2002) Determining critical pre- and post-anthesis periods and physiological process in *Lycopersicon esculentum* Mill. exposed to moderately elevated temperatures. J Exp Bot 53:1187–1195

Schneider SH, Semenov S, Patwardhan A, Burton I, Magadza CHD, Oppenheimer M, Pittock AB, Rahman A, Smith JB, Suarez A, Yamin F (2007) Assessing key vulnerabilities and the risk from climate change. Climate change 2007: impacts, adaptation and vulnerability. In: Parry et al (eds) Contribution of Working Group II to the fourth assessment report of the Intergovernmental Panel on Climate Change. Cambridge University Press, Cambridge, pp 779–810

Smith P, Martino D, Cai Z, Gwary D, Janzen H, Kumar P, McCarl B, Ogle S, O'Mara F, Rice C, Scholes B, Sirotenko O, Howden M, McAllister T, Pan G, Romanenkov V, Schneider U, Towprayoon S, Wattenbach M, Smith J (2008) Greenhouse gas mitigation in agriculture. Philos Trans R Soc B Biol Sci 1492:789–813

Smith P, Haberl H, Popp A, Erb K-H, Lauk C, Harper R, Tubiello F, de Siqueira Pinto A, Jafari M, Sohi S, Masera O, Böttcher H, Berndes G, Bustamante M, Ahammad H, Clark H, Dong HM, Elsiddig EA, Mbow C, Ravindranath NH, Rice CW, Robledo Abad C, Romanovskaya A, Sperling F, Herrero M, House JI, Rose S (2013) How much land-based greenhouse gas mitigation can be achieved without compromising food security and environmental goals? GCB Bioenergy 19:2285–2302

Vetter HS, Sapkota BT, Hillier J, Stirling MC, Macdiarmid IJ, Aleksandrowicz L, Green R, Joy JME, Dangour AD, Smith P (2017) Greenhouse gas emissions from agricultural food production to supply Indian diets: implications for climate change mitigation. Agric Ecosyst Environ 237:234–241

WMO (2013) The Global Climate 2001–2010 – a decade of climate extremes summary report. World Meteorological Organization, Geneva, Switzerland. Downloaded from http://library.wmo.int/pmb_ged/wmo_1119_en.pdf

Impact of Heat on Vegetable Crops and Mitigation Strategies

Pankaj Kumar Ray, Hemant Kumar Singh, Shashank Shekhar Solankey, R. N. Singh, and Anjani Kumar

1 Introduction

Mechanical changes continuously between the atmosphere, the ambient temperature continuously increasing a measurement of extremely harmful stress. The air temperature in the world, each predicted from 0.2 °C, which will result in a temperature of from 1.8 to 4.0 °C, to increase the ratio of 2100 (IPCC 2007) this level. This estimate is generated concern among researchers because heat stress has the effect or influence on the development of the life cycle of a known organism, unwavering through the adjacent ecological mechanisms of change. Vegetable crops, in specific, as sessile organisms, cannot change additional beneficial environment; then, crop growth and development method significantly affect, habitually lethal, high temperature stress.

P. K. Ray (✉)
Krishi Vigyan Kendra, Saharsa, Bihar, India

Bihar Agricultural University, Sabour, Bhagalpur, Bihar, India

H. K. Singh
Krishi Vigyan Kendra, Kishanganj, Bihar, India

Bihar Agricultural University, Sabour, Bhagalpur, Bihar, India

S. S. Solankey
Department of Horticulture (Vegetable and Floriculture), Bihar Agricultural University, Sabour, Bhagalpur, Bihar, India

R. N. Singh
Associate Director Extension Education, Bihar Agricultural University, Sabour, Bhagalpur, Bihar, India

A. Kumar
ICAR-Agricultural Technology Application Research Institute (ATARI), Patna, Bihar, India

S. S. Solankey et al. (eds.), *Advances in Research on Vegetable Production Under a Changing Climate Vol. 1*, Advances in Olericulture,
https://doi.org/10.1007/978-3-030-63497-1_12

Thermal stress in different plants and often unfavorable modifications to the growth, development, functional programs, and yield (Hasanuzzaman et al. 2012, 2013). The main significance of a HT is an additional stress generation of reactive oxygen, oxidative stress leads towards. The case of vegetable plants continuously fights for survival at several ecological pressure together with HT. Vegetable plants through plant body by generating a physical change inside alter metabolism and normal indications subjected to heat stress. Vegetable plants in various ways to change its orientation HT metabolic response, most solutes by constructing very suitable to establish the structure of proteins and cells, cells expanded by maintaining the osmotic pressure and stability and steady-state adaptation redox antioxidants remodeling cells (Valliyodan and Nguyen 2006). At the molecular level, the thermal stress results in a modification (Shinozaki and Yamaguchi-Shinozaki 2007) relates to the direct stress from the HT gene secure appearance. These include an osmoprotectant, detoxification enzymes, transport proteins and regulatory proteins appearance (Semenov and Halford 2009) gene response. As in the case of the HT, through physiological and biochemical changes in the appearance of methods of gene thermal resistance in the case of variations or domesticated perfect form, in order to adapt to the dynamic changes of the stabilized (Hasanuzzaman et al. 2010a). In the current period, HT effectively reduce the osmotic stress in a plant protection agent, a plant hormone, a signaling segment, polyamines, trace elements and nutrients in the form of protective agent has been brought exogenous application damage (Hasanuzzaman et al. 2010b).

Enhancement of novel plant varieties resistant to high temperature plant is the most important task for researchers (Moreno and Orellana 2011). Depending extreme period, but also depends on the plant species and surrounded by additional ecological aspects, plant vegetables on HT responded positively, but the deliberate character verification documents and tolerant HT still cannot tell (Wahid et al. 2007). HT participated in the survey pressure plant researchers are trying to find the heat of the plant response; they hate what way the plant can be successfully carried out a search in the atmosphere with HT. At present, generally considered the molecular method involves genomics and transgenic plants through management development target gene (Kosova et al. 2011). This basic molecular research method can be used to provide stress resistance varieties development, and agriculture is essential to cultivate crops in HT.

2 Temperature Stress

Greaves (1996) describes a suboptimal metabolic stress; such as temperature-induced growth or yield potential of the cells or tissue damage results hereditary slightly reduced within a predetermined range determined as a result of inspiration of unremitting temperature is higher than or lowers optimum thermal threshold biochemical and physiological movement or morphological generation progress.

2.1 Heat Stress

Levitt (1980) Plant Taxonomy of psychrophilic, mesophilic, thermophilic, and depending on whether they are tolerant lowest, intermediate, or highest temperatures. Psychrotrophic that the high temperature threshold value of 15–20 °C, mesophilic bacteria plants those plants, the high temperature threshold 35–45 °C of thermophilic, further these plants, the high temperature threshold range of from 45 to 100 °C. Is made directly from the reversible strain Levitt (1980) is a high temperature, i.e., more than respiration under elevated temperatures due to the indirect photosynthesis strain, i.e., losses, or direct or indirect damage, i.e. damage hunger. High temperatures may be experienced on a daily or seasonal basis device. There are long-term climate change leading to both higher average temperatures, expanding the geographical range where the high temperature limit crop yields become routine, increase the severity and frequency of evidence of an increasing number of extreme temperature events. Plants can be by exposure to a suitable elevated temperature, for example, temperature extremes very short time such as an extended, though different mechanisms to deal with these stresses. Thermal stress on quality and yield of grain.

2.2 Low Temperature Stress

By the cooling effect, it can damage both plants, and leads to physiological and developmental abnormalities by freezing and low temperature by direct cell damage caused by dehydration of the cell. Lyons (1973), many of the symptoms described in chilling injury. Many physiological processes, such as the rice is flowering at temperatures up to 20 °C extremely sensitive to low temperatures and damage may occur. Chilling injury leaves are usually visible symptoms include wilting, photo-oxidation of the pigment, waterlogged cell gap bleach maturity, browning and final leaf necrosis and plant death (Levitt 1980). Dudal (1976) estimates that 15% of arable land is frozen influence of pressure. Low temperatures reduce crop yields in many areas. Chilling and freezing damage can be damaged by physical or by interfering with the normal biochemical and physiological functions, thereby reducing the yield to obtain, directly affect crop growth. More subtly, to reduce potential productivity/agricultural varieties, often those who limit the maximum potential yield of crops or varieties can be grown in a particular area at a low temperature cold species. Low temperature exposure can be both of which factories must adapt, including seasonal factors, low temperature many months long day in some areas.

2.3 *Heat Stress in Vegetable Plants*

- Seedling establishment is hindered
- Drying of leaf margin and leaf burning effect
- Reducing plant growth
- Pollen development will be affected
- Repair in photosynthesis
- Total biomass is reduced
- Affect the development and quality of fruit

2.4 *High Temperature Stress Mitigation Strategies*

- Plants need light blocking under culture conditions.
- Overhead irrigation to avoid sunburn.
- It stimulates the production of α- amylase seed germination gibberellin applications.
- BAP lower leaf senescence & lipid peroxidation
- Salicylic acid enhances the ability of the heat resistance.
- Betaine reduces leakage ions.
- Improve seed germination ethylene application

2.5 *Low Temperature Stress Mitigation Strategies*

- Spraying of 0.15% ammonium reduces the effect of low temperature stress.
- Pre-soak gibberellic acid and proline to increase seed germination.
- MET application to increase the activity of scavenging enzymes.
- Electrolyte leakage is uniconazole (50 ppm) is applied is reduced.
- Cryoprotectants also for reducing the effect of stress.
- Induction of freezing tolerance in ABA.

3 Vegetable Plants Response to Heat Stress

HT response to the temperature of the vegetable plant, the duration and extent of plant type changes. In extreme HT, available in a few minutes, which could lead to a catastrophic collapse of the cellular tissue (Ahuja et al. 2010) cell damage or cell death occurring within. Thermal stress can affect the germination process and other plant growth, all aspects of the development, reproduction and production (Mittler and Blumwald 2010). Thermal stress differences affect the stability of various

Table 1 Effects of temperature stress in different crop species

Crops	Heat treatment	development stage	The main impact
Chili pepper (*Capsicum annuum*)	38/30 °C (day/night)	Reproductive, maturity and harvesting stage	If the reduced width and fruit weight, increasing the proportion of abnormal seeds per fruit.
Okra (*Abelmoschus esculentus*)	32 and 34 °C	The entire growth period	Reduced yield, pod quality parameters, such as fiber damage and destruction content of calcium pectate down.

proteins, membranes, RNA and cytoskeletal structure type and amount of altered cellular enzyme reaction efficiency, the main obstacle to physiological processes and creates metabolic imbalance (Pagamas and Nawata 2008). Some common effects of heat stress have been summarized in Table 1.

3.1 Growth

In the vegetable plant germination and growth stage, the first to be affected. Thermal stress is applied to the negative impact on seed germination various crops, although the temperature range is large crop species (Kumar et al. 2011). Low germination rate, seedling vigor abnormal plants, seedlings difference, to reduce the impact of the radicle and significant growth in various plant species geminated sprouts cultivated vegetables recording thermal stress. Inhibition of seed germination are also well described, which typically occurs HT induced by ABA (Essemine et al. 2010).

High temperature so that the amount of water content in cell size, growth, and final reduction of loss. In another net assimilation rate reduction reasons (NAR) is reduced relative growth rate (RGR) HT. Heat stress symptoms include burning morphology and sunburn leaves and twigs, branches and stems, leaf senescence and off, root and shoot, fruit discoloration and damage of the growth inhibition (Rodriguez et al. 2005). Hinder beans (*Phaseolus vulgaris*) morphological and physiological characteristics, such as climate, growth and extended partition, plant water relations and shoots severe thermal stress (Koini et al. 2009). Reason more plant species of vegetables in high-temperature superconducting (28/29 °C) substantially elongated stems and leaves entended and reduce total biomass. Can be displayed in a particular cell or minutes or even seconds because the protein may occur, on the other hand, to gather tissue degeneration or programmed cell death extended period of time of death gradually moderately high thermal stress in the extreme vegetable plants the high-temperature superconducting guide; both types of injury or death may result in leaf loss, miscarriage or the entire plant flowers and fruits, and even death (Rodriguez et al. 2005).

3.2 Photosynthesis

Photosynthesis is the most heat-sensitive physiological processes in plants and vegetables in one. High temperature, especially C_3 plants have a greater impact than the C_4 photosynthetic ability of plants to vegetables. In the chloroplast thylakoid substrate sheet photochemical reactions and carbon metabolism are considered to be the primary site of damage HTS (Wang et al. 2009). Thylakoid membrane is highly sensitive to HT. The main changes in the thylakoid as swelling grana in the chloroplast structure of the tissue changes the deposited particles and the loss of heat stress group (Rodriguez et al. 2005). Again, Photosystem II (PSII) activity is significantly reduced or even stopped at high temperature superconductivity. Reduce heat shock photosynthetic pigments amount.

Vegetables ability to maintain the thermal stress gas exchange and carbon dioxide assimilation rate is directly related to heat resistance (Kumar et al. 2005). Significantly affect heat Leaf Water Status, Stomatal conductance (GS) and intercellular CO_2 concentration. Under HT stomatal closure is another cause of impaired photosynthesis, intercellular CO_2. Thermal imposed as reduce the negative effects of plant leaf water potential of leaves, leaf area and reduce pre-mature overall performance of this photosynthesis of plants (Greer and Weedon 2012) the negative impact of leaf senescence. And plant carbohydrate reserve under prolonged thermal stress hunger depletion observed.

3.3 Reproductive Development

While all of the plant tissues easily in almost all stages of growth and development to heat stress, in the most sensitive reproductive tissue temperature a few degrees and the elevation angle during flowering time can cause loss of the entire crop cycle. During reproduction, the thermal stress can result in a short period a large although there is a significant reduction of the sensitivity change between plant species and within species and vegetables bud and abortion. Heating plant spells even in the reproductive stage of development may not produce flowers or flower may not produce fruit or seeds (Sato et al. 2006). Abiotic stress situations, with increasing the HT causes infertility in both male and female meiosis damaged organs, impaired pollen germination and pollen tube growth, reducing the stigma and style ovule position, abnormality, reducing the number of pollen fertilization to keep interference, obstacles in the endosperm, germ and not the original fertilized embryos by the stigma.

3.4 *Yield*

Elevated temperature increases worry about crop productivity and food security. The impact is terrible, even a small (1.5 °C) temperature increase of crop yield vegetables significant negative impact. Heat stress is a main loss of productivity and reduces the capacity of assimilation by changing the film due to the enhanced stability and maintenance respiration cost reduction in the radiation efficiency (Hay and Porter 2006) reduced photosynthesis. In the pod quality parameters, such as fiber damage and destruction content of calcium pectate downward stress is found in the HT okra. On the main stem Brassica seed yield was decreased from 89%, but all contribute to the branch 52% overall yield loss HT exceeds 30 °C. The reason for this reduction in yield is due to the pod infertility due to heating and reduces seed weight per seed pod (Sinsawat et al. 2004).

3.5 *Oxidative Stress*

Different metabolic pathway enzymes are sensitive to different degrees of dependency of HTS. Like other abiotic stress, heat stress, power is not connected enzymes and metabolic pathways that lead to undesirable and harmful most common ROS singlet oxygen, superoxide radicals, hydrogen peroxide and hydroxyl radicals, which it is responsible for the accumulation of oxidative stress. Physiological changes in a plant in a variety of thermal stress damage occurring in a horizontal exposure. Hydroxyl radical can be, such as cytochrome, proteins, lipids, and DNA and all potential biological molecules, and almost all the components of the reaction (Moller et al. 2007). Singlet oxygen can be directly oxidized proteins, polyunsaturated fatty acids, and DNA. By thermal stress can destroy the stability of the cell membrane protein denaturation peroxidation induced by oxidative stress and lipid membrane. Even in HTS functional photosynthetic light reaction medium is reduced is generated from the recorded thylakoid membranes increase caused by electron leakage of ROS, to induce oxidative stress (Camejo et al. 2006). While the damaging effects of ROS on huge plant metabolic processes, they have also speculated that the behavior of the trigger signal to the heat resistance of plants this development is confusing and should be leaked heat shock response.

4 Antioxidant Response to Heat-Induced Oxidative Stress

Factories must heat-induced oxidative stress protection; enabling them to survive in the HT. HT vegetable crops increased stress resistance antioxidant capacity has been associated. Tolerant plants of the damaging effects of reactive oxygen synthesizing apparatus, various enzymatic and nonenzymatic ROS-scavenging and detoxification systems are often protected. Different temperature sensitive Antioxidant Enzyme Activity and activation occur in different temperature ranges, but the enzyme activity with increasing temperature. Chakraborty and Pradhan (2011) observed that catalase (CAT), ascorbate peroxidase (in APX) and displays the initial decrease of superoxide dismutase (SOD), increases up to at 50 °C, over oxidase (the POX), glutathione reductase (GR) activity, of from 20 to 50 °C in all temperature ranges. Moreover, the total antioxidant activity maximum at 30 °C in susceptible and tolerant of 35–40 °C. Their activities also depend on the tolerance or sensitivity to different crop varieties, growth stage, different growing seasons (Almeselmani et al. 2006).

5 Mechanism of Signal Transduction and Development of Heat Tolerance

Since the regulation of many genes have been reported to help vegetable plants to withstand stress, resulting in plant adaptation. To adjust its response to the development of tolerance (Kaur and Gupta 2005), plant various sensing signals of different channels and independent of external or internal stress interrelated. Plant stress response is a complex integrated circuit, in which multiple pathways are involved. To stimulate the cells to produce a compartment or tissue response required cofactor interaction and signal transduction molecules. Genes involved in stress signaling molecules of activation reactions. Stress response gene activation, depending on the type and variety of plant types of stress associated with signal transduction molecules. When help detoxify ROS stress response gene (by activating detoxification enzymes, free radical scavengers) activation; re-activated structural proteins and enzymes, and processes all the above statements needed to help maintain cell homeostasis. This can be said by the heat resistance of plant development or typical model of tolerance.

6 Use of Exogenous Protectants in Mitigating Heat-Induced Damages

Trading adverse effects of heat stress may involve some exploration potential harmful effects of the molecular pathways of plants with HT protection. In recent decades, such as protection of exogenous application osmoprotectant, plant hormones, signaling molecules, and other trace elements, has been shown to reduce the growth of HT, and as growth-promoting antioxidant capacity of plant protection agents benefits. Protective Effect in the biosphere was also subjected to thermal stress observed in the beans. El-Bassiony and so on. Who, in one experiment (2012) bean plants sprayed with various concentrations of BR (25, 50 and 100 mg/l). They observed that the spray bean plants BR 25 and 50 mg/l increased vegetative growth, the concentration of total yield of lower quality and HT pods. However, there is no difference between treatments. Spray 25 mg/l BR as compared to control and increased leaf nacelle total phenols total free amino acids (FAA).

7 Molecular and Biotechnological Strategies for Development of Heat Stress Tolerance in Plants

With the different physiological and biochemical mechanisms, molecular biology methods to raise awareness concept of heat stress tolerance, it is clear in plants. This stress tolerance of plants by expression of a gene coordinated way by adjusting a plurality of different genes and.

7.1 Heat-Shock Proteins (HSPs): Master Players for Heat Stress Tolerance

In general, heat stress is several heat-induced genes responsible, often referred to as "heat shock gene" (HSGS), which encodes a heat shock protein and the active products, is essential to survival in the deadly HT plants increase. The majority of these proteins protect cells temperature induced constitutive expression of the protein is denatured and preserved by the protein folding and stability of their functions; it as a molecular chaperone. The heat shock proteins in nature very uneven. Heat shock protein is limited to as seed germination, embryogenesis, small spores and certain developmental stages of fruit ripening plant (Baniwal et al. 2004). Due to their thermal properties tolerance expression, HSP can be produced by heat treatment HSGS, which triggers transcriptional response to the present heat promoter region conserved heat shock element (HSE element) caused.

7.2 Genetic Engineering and Transgenic Approaches in Conferring Heat Stress Tolerance in Plants

Adverse effects of thermal stress and can use a variety of genetic engineering by a method for improving the development of transgenic crops ease of thermal tolerance. Constitutive expression of specific proteins has been shown to improve the heat resistance. In addition to heat shock factor (HSF) of heat shock protein/gene expression of HSP chaperone protein expression and manually operated, we generated another heat-resistant transgenic plant of varying degrees. Surprisingly, however, such experiments have been fairly compared, limiting the project to drought, cold and salt stress tolerance experiments. Grover et al. (2013) pointed out that in the development of several methods, over-expression by the HSFS adjusted by changing the level of heat shock and non-heat shock gene, a trans-acting than the HSP stress tolerance gene expression than or implemented using HT factor gene transfer plants, such as DREB2A, bZIP28 and WRKY protein. In addition, participating in osmoregulation, detoxification of ROS, the photosynthetic reaction in the production of genetically modified organisms and protein synthesis leading to a positive result of the development of resistant HT PA protein transgenic plants.

8 Bio-stimulants and Heat Stress

It expects global warming and rising temperatures have a negative influence (Challinor et al. 2014) agriculture. Several plant cell damages induced by high temperatures, interfering inactivate the enzyme activity of the protein synthesis and membrane damage. 30 and 45 °C activity and structural integrity enzymal, wherein, when the optimal temperature is 60 °C higher than the temperature rise between irreversibly denatured. Thus, the effect as photosynthesis or respiration physiological activity. Toxic compounds, such as reactive oxygen species, because of excessive oxidative stress is a (Hasanuzzaman et al. 2013) is the most common reversal. In response, the plant begins to maintain cell homeostasis and turgor, cell protein synthesis and tissue compatible solutes. In addition, they generally close pores and increase the number of trachoma to prevent moisture loss. In addition, changes in clearance of ROS, gene synthesis or the activity or osmolytes transport antioxidant enzymes involved in expression at the molecular level. Temperature higher than the optimum delay inhibiting seed germination and plant growth. Reduce the thermal stress and reproductive pollen viability and germination inhibition and differentiation of flowers, fruits and decreases, which ultimately reduced the growth and development of production yield negative interference.

Tomato is the most sensitive species, a non-optimum temperature and thermal stress often leads to a long pattern length and reduced fruit (Camejo et al. 2005). There are very few crops from special high-temperature treatment applied to vegetable literature about, because most of the time; heat stress and drought, salinity or combinations thereof. Brassinosteroid tomato application resulted in higher dry matter accumulation and net photosynthetic rate, the growth and quality of snap pods and leaves NPK content of total free amino acid content of (Ogweno et al. 2008). This may be due to oxidative stress and improving the regeneration efficiency of brassinosteroid protection RUBP photosynthetic organs and carboxylation.

Nahar et al. (2015) investigated the effect of thermal stress application of exogenous glutathione. They are disposed at an elevated temperature prior to processing contents exhibit oxidative stress and methyl glyoxal, a reactive compound, the cell damage decreased mung bean seedlings. This will lead to more effective antioxidant defense system. Enhanced glutathione pretreatment short heat stress tolerance, improved plant physiology adaptability. For example, leaf relative water content and swelling, which generally reduces the extent at high temperatures, are protected. Mung bean and nitric oxide in response to the observed positive effects of ascorbic acid. There is no cure maximum photosynthetic activity leads to promote and increase the quantum efficiency of the PS2, but also affects the electrolyte leakage, resulting in better cell membrane integrity Oxidative stress, lipid peroxidation and H_2O_2 content decreased recovery activities of antioxidant enzymes. ABA Similar results after coating and chickpea obtained proline (Kumar et al. 2012). Chickpeas are high temperatures generally results in loss of yield and quality sensitive. Treatment, membrane damage, leakage of electrolyte after the measurement, the level of MDA and H_2O_2 is reduced, thereby increasing the water content in the leaves. These effects may be related to the accumulation of adjustment processing osmoprotectant proline role involving permeability material and ABA. Treated plants also exhibit higher chlorophyll content, the results have exogenous proline in other experiments, the stability of the film could be seen. In the enhanced activity of the treated plants oxidative metabolism, oxidative damage is also expected as fewer cells.

As described above, melatonin therapy to counteract the stress applied has a positive effect on the plant cooling parsley; otherwise, Martinez et al. (2018) found that the protective effect disappeared heat and salt stress melatonin in combination therapy tomato plants. Biological activation therapy protection film, to prevent thermal stresses, to increase its stability, reduce or avoid the accumulation of ROS.

9 Conclusion

Heat stress crop production around the world has become a major problem because it greatly affects the growth, development and productivity of plants. However, the extent to which this occurs in a particular area depends on climate circadian cycle and the timing of HT and HT probability of. This rate of greenhouse gas emissions from different sources are considered to be responsible for increasing the ambient temperature in the world, and contribute to global warming. Thus, the reaction and adaptation mechanisms need to develop high-temperature and heat resistance of the base of the plant, it can be better understood as an important vegetable crops. Vegetable plants for heat stress reaction have been carried out in-depth research in recent years; however, thermal tolerance fully understands the mechanism remains elusive. From the temperature change with the seasons and daily fluctuations, it's clear definition of the role of stress-induced temperature complicated, because the reaction at different temperatures is determined by the ability of plants to adapt to a different climate regime. HT reaction across and vegetables also within species, as well as at various developmental stages and different.

Under HT conditions, the accumulation of vegetables and various metabolites of different metabolic pathways and processes are activated. These changes emphasize the importance of physiological and molecular studies revealed potential mechanisms of stress response. Further, to understand the expression of specific genes and signaling cascade in response to the value HT properties for development of stress tolerant plants. Molecular biology methods inaugurated response and tolerability mechanism paved the way, can endure HT way to build the plant, it may be the development of crop varieties can be produced because of the economic interests of serotonin. At the field level, management and cultural practices, such as time and sowing, irrigation management and selection methods and cultivars of operation, but also can significantly reduce the adverse effects of HT pressure. In recent decades, such as osmoprotectants exogenous protection agents, plant hormones, signaling molecules, and other trace elements HT have shown the beneficial effects of the application of the growth of vegetables, because the growth promotion of these compounds and antioxidant activity. Synthesizing plant engineering of these compounds can be thermally resistant in a further embodiment of the development of the crop is important, heat tolerance represents a potentially important area. However, most of the world currently is only experimental HT effects of short-term progress in a different area of study is also limited to laboratory conditions. We need to explore different biochemical and molecular biology techniques and agronomic practices field experiments to investigate the actual HT response and its impact on the final crop yield.

References

Ahuja I, de Vos RCH, Bones AM, Hall RD (2010) Plant molecular stress responses face climate change. Trends Plant Sci 15:664–674

Almeselmani M, Deshmukh PS, Sairam RK, Kushwaha SR, Singh TP (2006) Protective role of antioxidant enzymes under high temperature stress. Plant Sci 171:382–388

Baniwal SK, Bharti K, Chan KY, Fauth M, Ganguli A, Kotak S, Mishra SK, Nover L, Port M, Scharf KD (2004) Heat stress response in plants: a complex game with chaperones and more than twenty heat stress transcription factors. J Biosci 29:471–487

Camejo D, Rodriguez P, Morales MA, Dell' Amico JM, Torrecillas A, Alarcon JJ (2005) High temperature effects on photosynthetic activity of two tomato cultivars with different heat susceptibility. J Plant Physiol 162:281–289

Camejo D, Jimenez A, Alarcon JJ, Torres W, Gomez JM, Sevilla F (2006) Changes in photosynthetic parameters and antioxidant activities following heat–shock treatment in tomato plants. Func Plant Biol 33:177–187

Chakraborty U, Pradhan D (2011) High temperature-induced oxidative stress in *Lens culinaris*, role of antioxidants and amelioration of stress by chemical pre-treatments. J Plant Interact 6:43–52

Challinor AJ, Watson J, Lobell DB, Howden SM, Smith DR, Chhetri N (2014) A meta-analysis of crop yield under climate change and adaptation. Nat Clim Chang 4:287–291

Dudal R (1976) Inventory of major soils of the world with special reference to mineral stress. In: Wright MJ (ed) Plant adaption to mineral stress in problem soils. Cornell University Agricultural Experiment Station, Ithaca, pp 3–23

El-Bassiony AM, Ghoname AA, El-Awadi ME, Fawzy ZF, Gruda N (2012) Ameliorative effects of brassinosteroids on growth and productivity of snap beans grown under high temperature. Gesunde Pflanzen 64:175–182

Essemine J, Ammar S, Bouzid S (2010) Impact of heat stress on germination and growth in higher plants: physiological, biochemical and molecular repercussions and mechanisms of defense. J Biol Sci 10:565–572

Greaves JA (1996) Improving sub optimal temperature tolerance in maize-the search for variation. J Exp Bot 47:307–323

Greer DH, Weedon MM (2012) Modeling photosynthetic responses to temperature of grapevine (*Vitis vinifera* cv. Semillon) leaves on vines grown in a hot climate. Plant Cell Environ 35:1050–1064

Grover A, Mittal D, Negi M, Lavania D (2013) Generating high temperature tolerant transgenic plants: achievements and challenges. Plant Sci 205–206:38–47

Hasanuzzaman M, Hossain MA, Fujita M (2010a) Physiological and biochemical mechanisms of nitric oxide induced abiotic stress tolerance in plants. Am J Plant Physiol 5:295–324

Hasanuzzaman M, Hossain MA, Fujita M (2010b) Selenium in higher plants: physiological role, antioxidant metabolism and abiotic stress tolerance. J Plant Sci 5:354–375

Hasanuzzaman M, Hossain MA, da Silva JAT, Fujita M (2012) Plant responses and tolerance to abiotic oxidative stress: antioxidant defenses is a key factor. In: Bandi V, Shanker AK, Shanker C, Mandapaka M (eds) Crop stress and its management: perspectives and strategies. Springer, Berlin, pp 261–316

Hasanuzzaman M, Nahar K, Fujita M (2013) Extreme temperatures, oxidative stress and antioxidant defense in plants. In: Vahdati K, Leslie C (eds) Abiotic stress—plant responses and applications in agriculture. InTech, Rijeka, pp 169–205

Hay RKM, Porter JR (2006) The physiology of crop yield. Blackwell Publishing Ltd, Oxford

Intergovernmental Panel on Climate Change (IPCC) (2007) The physical science basis. In: Contribution of Working Group I to the fourth assessment report of the Intergovernmental Panel on Climate Change. Cambridge University Press, Cambridge

Kaur N, Gupta AK (2005) Signal transduction pathways under abiotic stresses in plants. Curr Sci 88:1771–1780

Koini MA, Alvey L, Allen T, Tilley CA, Harberd NP, Whitelam GC, Franklin KA (2009) High temperature-mediated adaptations in plant architecture require the bHLH transcription factor PIF4. Curr Biol 19:408–413

Kosova K, Vitamvas P, Prasil IT, Renaut J (2011) Plant proteome changes under abiotic stress-contribution of proteomics studies to understanding plant stress response. J Proteome 74:1301–1322

Kumar A, Omae H, Egawa Y, Kashiwaba K, Shono M (2005) Some physiological responses of snap bean (*Phseolus vulgaris* L.) to water stress during reproductive period. In: Proceedings of the international conference on sustainable crop production in stress environment: management and genetic option. JNKVV, Jabalpur, pp 226–227

Kumar S, Kaur R, Kaur N, Bhandhari K, Kaushal N, Gupta K, Bains TS, Nayyar H (2011) Heat-stress induced inhibition in growth and chlorosis in mungbean (*Phaseolus aureus* Roxb.) is partly mitigated by ascorbic acid application and is related to reduction in oxidative stress. Acta Physiol Plant 33:2091–2101

Kumar S, Kaushal N, Nayyar H, Gaur P (2012) Abscisic acid induces heat tolerance in chickpea (*Cicer arietinum* L.) seedlings by facilitated accumulation of osmoprotectants. Acta Physiol Plant 34:1651–1658

Levitt J (1980) Responses of plants to environmental stresses, vol 1. Academic, New York/London, p 496

Lyons JM (1973) Chilling injury in plants. Annu Rev Plant Physiol 24:445–466

Martinez V, Nieves-Cordones M, Lopez-Delacalle M, Rodenas R, Mestre T, Garcia-Sanchez F, Rubio F, Nortes P, Mittler R, Rivero R (2018) Tolerance to stress combination in tomato plants: new insights in the protective role of melatonin. Molecules 23:535

Mittler R, Blumwald E (2010) Genetic engineering for modern agriculture: challenges and perspectives. Ann Rev Plant Biol 61:443–462

Moller IM, Jensen PE, Hansson A (2007) Oxidative modifications to cellular components in plants. Ann Rev Plant Biol 58:459–481

Moreno AA, Orellana A (2011) The physiological role of the unfolded protein response in plants. Biol Res 44:75–80

Nahar K, Hasanuzzaman M, Alam MM, Fujita M (2015) Exogenous glutathione confers high temperature stress tolerance in mung bean (*Vigna radiata* L.) by modulating antioxidant defense and methylglyoxal detoxification system. Environ Exp Bot 112:44–54

Ogweno JO, Song XS, Shi K, Hu WH, Mao WH, Zhou YH, Yu JQ, Nogues S (2008) Brassinosteroids alleviate heat-induced inhibition of photosynthesis by increasing carboxylation efficiency and enhancing antioxidant systems in *Lycopersicon esculentum*. J Plant Growth Regul 27:49–57

Pagamas P, Nawata E (2008) Sensitive stages of fruit and seed development of chili pepper (*Capsicum annuum* L.) exposed to high-temperature stress. Sci Hortic 117:21–25

Rodriguez M, Canales E, Borras-Hidalgo O (2005) Molecular aspects of abiotic stress in plants. Biotechnol Appl 22:1–10

Sato S, Kamiyama M, Iwata T, Makita N, Furukawa H, Ikeda H (2006) Moderate increase of mean daily temperature adversely affects fruit set of *Lycopersicon esculentum* by disrupting specific physiological processes in male reproductive development. Ann Bot 97:731–738

Semenov MA, Halford NG (2009) Identifying target traits and molecular mechanisms for wheat breeding under a changing climate. J Exp Bot 60:2791–2804

Shinozaki K, Yamaguchi-Shinozaki K (2007) Gene networks involved in drought stress response and tolerance. J Exp Bot 58:221–227

Sinsawat V, Leipner J, Stamp P, Fracheboud Y (2004) Effect of heat stress on the photosynthetic apparatus in maize (*Zea mays* L.) grown at control or high temperature. Environ Exp Bot 52:123–129

Valliyodan B, Nguyen HT (2006) Understanding regulatory networks and engineering for enhanced drought tolerance in plants. Curr Opin Plant Biol 9:189–195

Wahid A, Gelani S, Ashraf M, Foolad MR (2007) Heat tolerance in plants: an overview. Environ Exp Bot 61:199–223

Wang JZ, Cui LJ, Wang Y, Li JL (2009) Growth, lipid peroxidation and photosynthesis in two tall fescue cultivars differing in heat tolerance. Biol Plant 53:247–242

Impact of Drought and Salinity on Vegetable Crops and Mitigation Strategies

Pallavi Neha, Manoj Kumar, and Shashank Shekhar Solankey

1 Introduction

Drought and salinity affect vegetable crop adversely in term of productivity and yield which ultimately lowers the farm income (Daryanto et al. 2016; Farooq et al. 2017). Organization like 'The Intergovernmental Panel on Climate Change (IPCC)' reported that highly significant changes in climate have occurred in recent years including global warming (IPCC 2014). It has direct influence on vegetable production (Stocker et al. 2013; Bita and Gerats 2013). The climatic changes include increasing levels of atmospheric greenhouse gases (CO_2, CH_4, and N_2O), irregular rainfall, elevated temperatures, and extension of flood or water deficit affected land areas. This negative condition is liable for improvement of the dry season, saltiness, and alkalinity inclined zones and therefore influences plant development, advancement and at last diminishes the efficiency of vegetable yields (Lopez et al. 2011). Among these, several adverse factors such as soil salinity, alkalinity and drought show a negative impact on crop productivity up to 50% (Wheaton et al. 2008).

The impact of heat and drought is uniformly important in developing countries. During drought, crop yield is drastically reducing and directly hit the farmers' livelihood and accordingly their economy (Stocker et al. 2013). The information on

P. Neha (✉)
Division of Post Harvest Technology & Agri. Engineering, ICAR-Indian Institute of Horticultural Research, Bengaluru, Karnataka, India

M. Kumar
Division of Vegetable Crops, ICAR-Indian Institute of Horticultural Research, Bengaluru, Karnataka, India

S. S. Solankey
Department of Horticulture (Vegetable and Floriculture), Bihar Agricultural University, Sabour, Bhagalpur, Bihar, India

S. S. Solankey et al. (eds.), *Advances in Research on Vegetable Production Under a Changing Climate Vol. 1*, Advances in Olericulture,
https://doi.org/10.1007/978-3-030-63497-1_13

how vegetable crops respond to the stress of salinity, heat, and drought and the methods of mitigation highlighted in this chapter.

2 Abiotic Stresses: Drought and Salinity

2.1 Stress Involved in Vegetable Crops Production

The occurrence of stress can also be directly related to climatic changes which further affect the micro-climate of vegetable crops. The existence of significant variation in the context of either mean state of the climate or in its extent of variability, persisting in an area or for an extended period (typically decades or longer time) is referred to climate change having its impact on vegetable production through the various stress conditions such as biotic and abiotic stress. The major stresses involved in vegetable crops production are:

Temperature stress: It includes the stress due to environmental extremes like high temperatures which affect vegetable crops in several ways such as influencing crop duration and affecting the reproductive biology including vegetable crop ripening and quality.

Atmospheric gases stress includes an elevated concentration of gases such as CO_2 which results in increased photosynthesis rate and water use efficiency. However, the studies on the interaction of temperature and CO_2 show that the CO_2 enrichment does not appear to compensate for the detrimental effects of high temperature on yield.

Precipitation stress: Vegetables being succulent with 90% of water, the quality and yield are greatly affected by excess water stress (Flooding) and the limited water stress (drought).

Salinity and radiation stress: Salinity stress includes the adverse effect of salt on plant health, similarly adverse effects of high intensity of ultra-violet and visible light include the radiation stress.

Chemicals and pollutants and oxidative stress: Chemicals and pollutants include a high concentration of heavy metals, pesticides, and aerosols whereas, oxidative stress includes the activity of reactive oxygen species and ozone.

Other stress: This includes factors such as pathogens (bacteria, viruses, and fungi), insects, wind, nutrient deprivation in soil, herbivores, rodents, weeds, *etc.*

2.2 Drought Stress and Resistance Mechanisms of the Vegetable Crop to Combat it

Drought stress is the adverse effect of moisture stress on yield and quality as affected by limited water stress (drought) and excess water stress (Flooding). Screening criteria for stress tolerance and resistance showed that some of the parameters of are maintained by the plants to make itself tolerate or resist the drought stress. These include modification in morphological features such as leaf characteristics which include leaf rolling, leaf water retention, leaf firing, leaf stomata, etc. Similarly, modification in root characteristics such as deep root system as well as other mechanisms like modifications in seed germination, seedling characteristics, canopy temperature, plant phenology, adjustment in photosynthetic rate, cell membrane stability, and water use efficiency etc.

Apart from the above mechanisms, a broad classification of stress management by vegetable crops can be grouped under the following subheads:

Drought escape: This mechanism of plant permits it to finish its life cycle before the beginning of the dry season for example planting of early flowering vegetables or planting short lived vegetable crops.

Drought avoidance: "This mechanism of the plant includes procedures of retaining more water content at the time of pressure, either by capable water osmosis from roots or by plunging evapotranspiration from ethereal parts through stomatal procedures".

Drought resistance: This mechanism of the plant allows maintenance of turgor and continue metabolism though at low water potential.

2.3 Salinity Stress and Resistance Mechanisms of the Vegetable Crop to Combat It

Salinity stress is an adverse effect on plant growth by the saline soils which have high salt concentrations (NaCl) than usual. Salts are exceptionally dissolvable in surface and groundwater and can be transported during water availability. Salinity is a proportion of the substance of salts in soil or water. Essential salinity is created by normal procedures, for example, enduring rocks, downpour, and wind storing salt more than a huge number of years. Secondary salinity has happened with extensive land clearing and adjusted land use and may appear as "dryland salinity" or "water system incited salinity". Salinity stress is an adverse effect on plant growth by the saline soils which have high salt concentrations (NaCl) than usual. Saline soils have high salt NaCl concentration whereas sodic soils have high concentrations of $Na+$ than usual. Saline soils cause a 'chemical drought' in soils but sodic soils do not. Sodic soils cause waterlogging but saline soils do not. Salinity protects the integrity of soil in contrast to sodicity which destroys the structure of soil by causing dispersion. Soils are considered sodic when their exchangeable sodium percentage (ESP) is more than 6%. Sodicity in the soil is easier to correct than high salinity levels in the soil. The electrical conductivity of saline soils has an immersion soil concentration over 4 dS/m at 25 °C. Saltiness is a proportion of the nearness of salts in soil or water (NaCl, KNO3, MgSO4, Sodium bicarbonate). Salts are well solvent in surface and groundwater and can be moved with water development.

Screening criteria for stress tolerance and resistance have shown that some of the parameters are maintained by the plants to make itself tolerate or resist to salinity stress, these include modification in morphological features and maintenance of the yield potential. Salinity causes a reduction in the crop yield. The reduction in yield occurs mainly due to decrease in size and shape, dry weight, moisture content, etc. which is a result of the adverse effect of salt on plant growth. Several modifications carried out by the plant itself to combat the salinity stress these includes root shedding (salt accumulated root will fall off), root characteristics such as deep root system to escape the salt-affected soil for the absorption of water, Plant phenology and adventitious roots, water use efficiency, etc.

3 Impact of Salinity and Drought Stress on Vegetable Crop Production (Table 1)

Table 1 Impact of drought and salinity on growth, yield and quality of different vegetable crops

Sl. No.	Crop	Stress	Impact	References
1.	Potato	Drought	Unfavorable for the germination of tubers	Arora et al. (1987)
2.	Onion	Drought	Adversely affects the germination of seeds	Arora et al. (1987)
3.	Okra	Drought	Restrict seed germination	Arora et al. (1987)
4.	Amaranthus	Drought	Diminishes their water content in this manner decreases their quality	AVRDC (1990)
5.	Palak	Drought	Diminishes their water content in this manner decreases their quality	AVRDC (1990)
6.	Spinach	Drought	Diminishes their water content in this manner decreases their quality	AVRDC (1990)
7.	Cabbage	Salinity	Diminishes germination rate, shoot & root length, new root & shoot weight	Jamil and Rha (2004)
8.	Cucurbits	Salinity	Diminishes new and dry load of all, all-out chlorophyll content	Baysal et al. (2004)
9.	Potato	Salinity	Reduction in tuber yield	Bustan et al. (2004)

(continued)

Table 1 (continued)

Sl. No.	Crop	Stress	Impact	References
10.	Chilli	Salinity	Diminishes dry matter formation, leaf zone, relative development rate & net absorption rate, organic product weight	Lopez et al. (2011)
11.	Beans	Salinity	Number and size in plants, concealment of development and photosynthesis action and changes in stomata conductivity	Kaymakanov et al. (2008)
12.	Pea	Drought and salinity	Inhibition in germination	Okcu et al. (2005)
14.	Cowpea	Drought	In leaf assimilation rate, transpiration rate	Anyia and Herzog (2004)

4 Mitigation Strategies for Salinity and Drought Stress in Vegetable Production

4.1 Plant Internal Biochemical System

4.1.1 Biochemical Levels by Environmental Stimuli

Plant biochemical composition changes when a stimulus for stress is received by the plant. Shulaev et al. (2008) observed well evident during abiotic stress conditions, the primary stage of plant response is activating the signaling cascade by environment. Various sorts of receptors distinguish different signs and improvements from an environmental factor. In plants main receptor kinase protein, the Receptor-Like Kinase (RLK) was reported and described in plants (Walker and Zhang 1990; He et al. 1996). A subfamily of Receptor -Like Kinases perceived as Wall-Associated Kinases (WAKs) gets indications from adjoining atmosphere and other abutting cells as a basic advance to enact proper flagging falls (Shulaev et al. 2008). In wake of knowing the discoveries of the first RLK, critical endeavors have been given to portraying supplementary definite receptor kinase genes. Stress observation is trailed by the stimulation of systemic signaling cascade. Based on reviews it is discovered that chief components in contribution of hydraulic conductivity is aquaporin proteins which controlled at biochemical level by atmospheric stumuli alongside variations, for example, phosphorylation activities (Johansson et al. 1998), cytoplasmic pH and Ca (Gerbeau et al. 2002; Alleva et al. 2006) or re-restriction to intracellular chambers (Boursiac et al. 2008).

4.1.2 Plant Harmones Level by Environmental Stimuli

Similarly, plant hormones have also been revealed to control of hydraulic conductivity by implicating in root and shoot significant distance flagging. Abscisic acid (ABA) is the primary hormone legitimately associated with controlling resilience against abiotic stresses like a drought, heat, alkalinity, saltiness, cold, and so on (Zhang et al. 2006; Lata and Prasad 2011) and known as pressure hormone. Abscisic acid is also recognized as eminent substance for significant root to shoot pressure signaling (Schachtman and Goodger 2008), actuate restraint of leaf development & momentary reactions, for example, shutting of stomata. ABA is additionally associated with the modification of foundational reactions to abiotic stress prior to any indicative variations in leaf water and supplement status (Bauer et al. 2013; Suzuki et al. 2013). Both MeJA and Coronatine (COR) upgraded the development and accumulation of a dry matter in cauliflower seedlings during drought condition and rewatering conditions. Treatment with MeJA or COR improved the resilience of dry season stress through the expanded aggregation of chlorophyll and net photosynthetic rate. Enzymatic (superoxide dismutase, peroxidase, ascorbate peroxidase, catalase, and glutathione reductase) and non-enzymatic cell reinforcement (proline and dissolvable sugar) frameworks were actuated, and lipid peroxidant (malondialdehyde and hydrogen peroxide) was stifled by MeJA and COR under drought stress condition. COR and MeJA additionally expanded leaf relative water content and endogenous ABA levels under dry season focused on conditions. In the wake of rewatering, the substance of leaf water, chlorophyll, abscisic corrosive, and photosynthetic attributes just as enzymatic and nonenzymatic cancer prevention agent frameworks indicated complete recuperation. Both MeJA and COR can reduce the antagonistic impacts of dry spell pressure and improve the capacity for water pressure opposition through the advancement of protection-related digestion in cauliflower seedlings (Wu et al. 2012). Moisture stress brings about the synthesis either catabolism of a few other plant harmones like auxin, gibberellins, cytokinins, ethylene, jasmonic corrosive, brassinosteroids and other elements such as nitrogen and pH, which are associated with the guideline of physiological procedures by acting signaling networks of signal molecules (Nakashima and Yamaguchi-Shinozaki 2013; Mittler and Blumwald 2015; Verma et al. 2016).

4.1.3 Expanded Intracellular Calcium Ion Levels

In prompted pressure situation by different gesture atoms, for example, diacylglycerol, inositol hexaphosphate, inositol trisphosphate, and ROS (Hirayama and Shinozaki 2010). The phosphorylation or dephosphorylation by actuated kinases or phosphatases explicit record factors (TFs), thusly controlling the stress reactive qualities articulation levels (Reddy et al. 2011). The raised calcium ion level represented by calcium restricting proteins which works as Ca ion sensors. It prods the enactment of Ca2+ subordinate protein kinases. The stimulated calcium ion sensors can likewise tie to cis components of significant stress receptive gene enhancer or can associate with DNA-restricting protein controlling genes ultimately concealment can perform (Reddy et al. 2011).

Protein denaturation is the primary effects of high-temperature stress which hazards in cytoskeleton structure and nucleic acids, suppression of synthesis, stimulated membrane fluidity, protein degradation, and causes loss of membrane integrity (Howarth 2005; Wahid et al. 2007). Initial high-temperature stress reaction was uncovered to instigate calcium ion convergence and cytoskeleton rebuilding, which spurs the upregulation of calcium subordinate protein kinase & nitrogen activated protein kinases flagging fall (Ashraf and Harris 2013). This flagging course incites the creation of antioxidants just as good as osmolytes for their osmotic change and outflow of heat resistance protein. Serious cell harms brought about by moderate warmth, after long haul introduction, or after an amazingly transient presentation to incredibly high temperatures (Wahid et al. 2007). It diminishes particle motion and incites the creation of ROS and other harmful intensifies that fundamentally check ordinary plant development (Howarth 2005). Articulation of heat resistance protein and different sorts of guarded proteins is a helpful versatile system under high temperatures conditions (Wahid et al. 2007), photosynthesis & water use efficiency (Camejo et al. 2005), membrane steadiness (Ahn and Zimmerman 2006), and cell hydration conservation (Wahid and Close 2007).

4.1.4 Anticipation or Evasion of ROS Development

A variety of abiotic stresses advances excess production of reactive oxygen species, that are exceptionally receptive and poisonous, liable for devastating of proteins, lipids, sugars just as DNA and at last performs oxidative pressure (Zlaetev and Lidon 2012). Oxidative pressure is fundamentally balanced by 2 disparate procedures, hindrance of ROS development by anticipation or shirking, including detoxifying or searching compounds also, a few cell reinforcements to deal with the poisonous impact of reactive oxygen species (Mittler 2002; Gill and Tuteja 2010). Alongside reactive oxygen species, plants additionally define two to six-crease more methyl-glyoxal under abiotic stress situations (Yadav et al. 2005), that is an extremely receptive cytotoxic substance created by various enzymatic & non-enzymatic reactions. The glyoxalase enzyme made-up of glyoxalase-I & glyoxalase-II is responsible for detoxification of methylglyoxal. This enzyme catalyzes the methylglyoxal to D-lactate through decreased glutathione as a cofactor. Methylglyoxal harmed cell work and can even harm DNA (Hasanuzzaman et al. 2017).

TFs performan a huge commitment in abiotic stress resistance. Numerous abiotic stress related genes which has been segregated from various plant species & over expressed in genetically modified plants to expand stress resistance. The TFs inciting TFs to incorporate individuals from ERF, DREB, MYB, WRKY, DOF, BZIP, NAC & BHLH families (Shinozaki et al. 2003). The over articulation of cycling DOF factor 3 in the model plant, Arabidopsis improved moisture stress, cold & salt resilience, related to up-regulation of many genes capable in cell osmoprotectant and reactive oxygen species homeostasis (Corrales et al. 2017). As of late, Kulkarni et al. 2017 expounded the significance of other transcriptional controllers alongside EAR & ZFPs motif comprises of repressor in abiotic stress management.

As of late, new retrograde signs have been imparted, for example, metabolite 3′-5′ phosphoadenosine phosphate which has been uncovered to aggregate in high light and dry season, flow from chloroplast to core that manages abscisic acid flagging & stomatal regulation at the time of oxidative stress. This outcomes in dry spell resilience and revealed for actuation of high light transcriptomes (Pornsiriwong et al. 2017).

4.1.5 Wide Range of Metabolites Production

Plentiful scope of metabolites creation with low atomic mass can deflect the ominous variation in cell apparatuses just as reinstate homeostasis. These embrace dissolvable starches like fructose and glucose, amino acids, and different sugar and sugar-based alcohols (Arbona et al. 2013; Morales et al. 2013). Inside amino acids, proline is the most significant supporter of osmotic regulation for different abiotic stresses (Sinay and Karuwal 2014; Zandalinas et al. 2017). Subsequently, amalgamation just as a collection of proline has been treated as a helpful resilience attribute for an impressive timeframe (Janská et al. 2010). Gathering of these mixes has been related to abiotic stress resistance & a guideline of protein structure stability, cell turgor and cell film adjustment as cells dry out (Arbona et al. 2013).

Auxiliary metabolites are additionally fundamental mixes for plant acclimatization and ingenuity under factor natural conditions by including anthocyanin, coumarins, flavonoids, lignin & tannins (Fraser and Chapple 2011). In barely crop, significant stages of auxiliary metabolites like lignin, phenylpropanoid, flavones & flavanols have been found in light of warmth and dry spell stresses.

Glycine betaine additionally acts as a key job in abiotic pressure resistance in various crop plants. Qualities connected with glycine betaine union in angiosperms and microorganisms have been transfer in plants, which don't upgrade the degree of combination or collect glycine betaine upon abiotic stresses (Quan et al. 2004). Though, the component of the activity of glycine betanin is as yet under scrutiny. Utilizing a hereditary building framework, Su et al. (2006) found the clear property of glycine betanin biosynthesis on abiotic stress resistance perhaps credited to defensive activity aside from changes to the cell osmotic alteration.

4.2 Plant Internal Physiological System

4.2.1 Morphological and Physiological Adjustments

Plants have adjusted active reactions to oversee abiotic stresses at the physiological, biochemical & morphological levels, letting them to get by under different natural circumstances (Huber and Bauerle 2016). Physiological reactions of plants against warmth and dry spell stresses were arranged into two different systems for example evasion and resilience. Evasion instruments are essentially morphological and physiological changes that offer a break to dampness or warmth stress, including improved root framework, diminished stomatal quantity and conductance, reduced leaf zone, upgrade leaf cuticle, and leaf collapsing/moving to limit evapotranspiration (Goufo et al. 2017). Cuticular wax formation, on leaves and shoot surface, is additionally firmly connected with a versatile reaction (Lee and Suh 2013). Under dry spell conditions, resistance qualities chiefly keep up tissue hydrostatic weight, utilizing cell and biochemical changes, particularly through osmotic guidelines (Khan et al. 2015; Blum 2017).

Drought stress for the most part decreases yield by diminishing quantity of seeds, seed quality, & seed size during vegetative or early conceptive development. To assess the impact of dry season weight on seed quality, seed yield, and development of tomato (Pervez et al. 2009). Under basic dampness stress circumstances, plants see worries by different sensors associated with reaction flagging. These are transduced by an assortment of pathways wherein various flagging and transcriptional factors have critical and explicit jobs (Hirayama and Shinozaki 2010). Water movement inside plant happens in strain as dictated by soil water accessibility and the climatic fume pressure shortage, making turgor presser inside cell. Physiological alterations which keep up turgor pressure are significant in shifting natural situations. Water movement in roots influenced by different parts, for example, root life systems and water accessibility (Borussia et al. 2005). Also, a few plants have worked in components to trigger water-powered, compound, or electrical significant distance signs to dispatch fundamental pressure reactions. Under dry season, a decline in root pressure-driven conductivity shows to forestall water misfortunes from plant to dry soil. These elements are affected by movement of aquaporins, that are vital film proteins which work as channels to move excellent little solutes and water (Vandeleur et al. 2014).

4.2.2 Modification in Physiological Functioning

Photosynthesis includes a collection of instrument photograph frameworks and photosynthetic colors, CO2 decrease pathways, and the electron transport framework. Because of worry, there is a negative impact on any segment in these frameworks that may prompt a decrease in general photosynthetic execution. Studies have demonstrated that submerged, salt and warmth stress when applied separately or in mix photosynthetic productivity and happening rates decline (Arbona et al. 2013). This is, for the most part, an impact of pressure prompted stomatal conclusion however it can likewise happen by other nonstomatal impediments, for example, diminished leaf senescence, leaf development and wrong working of the photosynthetic apparatus (Rahnama et al. 2010). This last impact circumstance is now and again credited to the lower inward accessibility of carbon dioxide, alongside the restraint of crucial photosynthetic catalysts & ATP synthetases (Zandalinas et al. 2016). Water stresses and warmth are accounted for lessen electron transport, corrupt proteins & discharge calcium and Mg particles from protein restricting accomplices (Zlatev and Lidon 2012). Broad presentation to high temperature additionally causes reduction in thylakoid grana deterioration, chlorophyll content, expanded amylolytic movement, and disturbance of absorbs' vehicle (Kozłowska et al. 2007).

WUE is one of the utmost significant boundaries in plant reaction to osmotic irregular characteristics which are legitimately connected to diminishes in photosynthetic rate. Various reports have indicated that expanded WUE diminishes in net photosynthetic rates are related to a stomatal conclusion (Ruggiero et al. 2017). The utilization of carbon isotope separation is a broadly being used device for screening physiological qualities related to WUE or "Happening Efficiency" that echoes both carbon dioxide trade and water economy. This sort of hardware is huge to evaluate phenotypic variety inside an enormous reproducing populace (Ellsworth and Cousins 2016). Late developments may provide new chances to report the significant problems in photosynthesis research when crop plants exposed to unfriendly warmth and dry spell pressure (Ort et al. 2015; Kromdijk et al. 2016).

Examination of information concerning different characters demonstrated that drought stress had insignificant impact on energy, quality and seed yield of tomato. Plant stature, leaves number, and fruits number/plant indicated huge outcomes to drought stress indicates significant impacts on the development of tomato (Pervez et al. 2009) as a test crop Tomato cv. 'Moneymaker' was utilized.

Osmotic effect was the major reason for inhibition of germination at equivalent water potential of NaCl and PEG rather than salt toxicity (Okcu et al. 2005). Pea cultivars in states of the dry season and salt pressure and to decide factors (salt poisonousness or osmotic worry because of PEG) repressing seed germination. The germination results found that the genotypes fundamentally contrasted for the dry season and salt pressure.

In the event of cowpea dry season caused a diminishing in transcription rate, rate of leaf absorption and stomatal conductance with genotypic fluctuation of 75.4, 57.9 and 83.3%, individually. Just genotypic change in stomatal conductance expanded altogether in dry season. Diminishing in the absorption rate, which was related to the stomatal closure was positively associated with a decrease in leaf water potential as an outcome of a dry spell (Anyia and Herzog 2004).

Two potato cultivars (Solanum tuberosum L.) having distinctive giftedness and differentiated for their dry spell resilience were explored in greenhouse and 4 cultivars in the field. They were exposed under to different treatments namely irrigated condition and droughted. The goal was to see which shoot and leaf characters were identified with diminishing in tuber yield. Reduction in tuber yields by 11% in 53% was reported under drought whereas drought stress exorbitantly diminished the dry mass of leaves in all cases (Lahlou et al. 2003).

Heat stress mainly affects the reproductive process which includes pollen anthesis, viability of pollen and stigma, growth of pollen tube and initial embryo development are more susceptible (Giorno et al. 2013). Be that as it may, all in all, male regenerative tissues are considerably touchier at all phases of advancement to high temperature stress than female reproductive tissues (Hedhly 2011).

4.3 Agronomic Approaches to Mitigate the Drought and Heat Stresses

4.3.1 Adoption of Crop Management Practices

This can conceivably reduce the unsafe impacts of heat and drought stress which includes: cultural operation and soil management, crop residue and mulching, irrigation, and use of suitable varieties. Various examinations revealed that dry spell pressure had a non-critical impact on the quality and yield of tomato seed. The use of supplements, for example, nitrogen, potassium, calcium & magnesium were accounted to diminish the toxicity of ROS by expanding the numbers of antioxidants in plant cells (Waraich et al. 2012). Use of macronutrients, for example, Ca, K and micronutrients like Selenium, Boron, and Manganese that are recognized to alter stomatal capacity under heat stress can help in initiating the physiological and metabolic procedures adding to protecting high water potential in tissues in such a manner that expands the heat stress resistance (Waraich et al. 2012). Then again, a few examinations have demonstrated that the utilization of fertilizer has non-significant impact on drought stress and an ideal soil moisture content is required since water is basic for the portability and digestion of these supplements (Lipiec et al. 2013).

Vegetables like onion plants were demonstrated water stress resilience having the option to grow and able to develop seeds in arid zones without water irrigation and with restricted measures of precipitation. Under such circumstances, enormous mother bulbs and high plant thickness (60 cm between a column and 10 cm intraline dispersing) gave the best outcomes. Only early cultivars showed diminished tuber number while leaf area index and leaf area duration showed significant effect in late cultivars (Lahlou et al. 2003). The exogenous application of silicon (Si) in grains showed reduced impact of drought on plant (Gautam et al. 2016), while use of Nitrogen was significant for the creation of seed yield, however, high excess application before planting and additional supply during growth as top dressing has non-significantly effect seed yield.

Higher cell reinforcement exercises saw in plants treated with Si (Ma et al. 2016), advanced level of photosynthetic pigment, and articulation differences of genes embroiled in the ascorbate-diminished flavonoid biosynthesis, glutathione cycle and antioxidant reaction.

Response to stress structured by plant growth regulators. The growth substances like ABA, salicylic acid and cytokinin assume a key job in drought resilience. The drought induces wilting characteristics and advance drought tolerance can be regulated by using the new ABA formations (Barrett and Campbell 2006; Waterland et al. 2010). Significant expansion in water potential and the chlorophyll content under stress condition was reported with the application of plant growth regulator (Zhang et al. 2004). Utilization of rigorous abscisic acid or abscisic acid referends to keep up the market value of horticultural crops by decreasing drought stress manifestations (Sharma et al. 2006; Kim and van Iersel 2008). Increased yield of soybean can be obtained with external application of abscisic acid under water deficient circumstances.

4.4 Genetics and Genomics Strategies

The traditional plant breeding technique has restricted achievement in alleviating the impacts of abiotic stress on plant profitability. This might be because of the intricacy connected with attributes constrained by various genes existing at numerous quantitative trait loci (QTL) (Parmar et al. 2017). Though, there are instances of achievements in traditional breeding for improved drought and heat resilience characteristics.

Noteworthy advancement has been made in genome sequencing, annotation and useful portrayal of significant genes (Uauy 2017; Clavijo et al. 2017). Few miRNAs along with protein coding genes are practically saved athwart plant species notwithstanding protein-coding qualities (Huang et al. 2017).

Reports on cereal grains recommend that abscisic acid or abiotic stress flagging adjusts gene articulation contours and formative programmes over the change of epigenetic position to overcome the stresses (Hirayama and Shinozaki 2010).

4.5 *Transgenic and Genome Editing Tools for Drought and Salinity Tolerance*

Studying the components of abiotic stress loss is basic for the improvement of tolerant plant species & varieties. The utilization of transgenic-based methodologies could assist with presenting alluring abiotic stress resistance genes in plant varieties. In this direction, numerous researchers castoff genes and TFs associated with abiotic stress tolerance as target genes in biotechnological methodology to develop drought tolerance plants.

In spite of the advantages of commercial genetically engineered (GEn) plants, the more extensive utilization of this innovation stays a significant test due to the negative open observation in regards to the deliberate presentation of qualities into plants, especially in Europe. It is a mind-boggling problem and the absence of acknowledgment of GEns can be for various reasons. Phytohormones are likely to focus on hereditary control to get abiotic stress tolerant crops. Excess formation of abscisic acid pathway related TFs gives an ABA-easily affected reaction and improves the osmotic pressure resilience in transgenic plants (Abe et al. 2003; Gao et al. 2011). A significantly higher yield was recorded in transgenic canola than control due to overexpression of farnesyltransferase protein under moderate stress during blooming period (Wang et al. 2005). Excess formation of TFs that control root engineering incited drought resilience in rice and transgenic Arabidopsis plants by advancing root development and accordingly upgrading WUE (He et al. 2016). Different TFs connected to WUE, for example, those animating wax deposition in cuticle and suberin deposition (Legay et al. 2016), and different controllers ready to tweak whole pathways, could be utilized for a similar goal to enact stress response genes and upgrade resilience (Landi et al. 2017).

The building of the glyoxalase pathway has been accounted for to upgrade resilience to abiotic stress in various plant species. Upregulation of both glyoxalase I and glyoxalase II and their excess production in plant species uncovered improved resistance to different abiotic stresses along with saltiness, dry spell, metal harmfulness, and extraordinary temperature (Alvarez-Gerding et al. 2015; Hasanuzzaman et al. 2017).

The perceptions affirm that the qualities brought into plants have a job in stress resistance. For the most part, the changed lines contrasted with the controls show less transpiration, better capacity to burn calories of ROs, expanded creation of defensive atoms, for example, proline, expanded root mass, and improved photosynthesis rate. These progressions add to a better return of the changed plants in abiotic stress contrasted with the control plants. It is important that except for one case (Habben et al. 2014).

The proof for enhanced abiotic stress resistance has appeared in field circumstances in a few cases particularly in grain crops (Castiglioni et al. 2008; Nuccio et al. 2015).

5 Conclusion

Plant components to escape from the heat and drought stresses can be interceded through different plant formative stages, biochemical, physiological, and atomic responses happening at the cell, tissue or entire plant level. Vegetable crops represents a wide scope of reactions to drought, saltiness, and heat stress that are for the most part portrayed by an assortment of changes in development & plant morphology. The drought and salinity stresses are significant requirements that restricting the yield profitability of vegetable. Saltiness, dry spell, and warmth stress may cause antagonistic consequences for general development and improvement of the plant's reproductive growth. Plant's capacity to withstand these burdens extraordinarily changes from species to species. Recognizable impacts of these burdens incorporate a decrease of crop yield, oxidative damage, damaged photosynthetic mechanism and membrane instability. Genetic approaches or the stress resistance breeding are significant accomplishments that have been made in limiting the undesirable impacts of these abiotic stresses.

Although, significant advances in the hereditary methodologies, for example, QTL planning and transgenic approaches, however there is as yet a degree for development in abiotic stress the management in vegetable crop production. Also, issues are as yet existing with transgenic plants produced for fighting with drought, salinity & heat stress. Innovations, for example, hereditary change and genomics altogether have been completed in considerate these perplexing attributes in angiosperms which will prompt considerable results which helps to alleviate the impacts of environmental variations, particularly as for salinity, heat & drought stresses, and will toss into enhanced yield efficiency & food security.

References

Abe H, Urao T, Ito T, Seki M, Shinozaki K, Yamaguchi-Shinozaki K (2003) Arabidopsis AtMYC2 (bHLH) and AtMYB2 (MYB) function as transcriptional activators in abscisic acid signaling. Plant Cell 15:63–78

Ahn YJ, Zimmerman J (2006) Introduction of the carrot HSP17.7 into potato (*Solanum tuberosum* L.) enhances cellular membrane stability and tuberization *in vitro*. Plant Cell Environ 29:95–104

Alleva K, Niemietz CM, Maurel C, Parisi M, Tyerman SD, Amodeo G (2006) The plasma membrane of Beta vulgaris storage root shows high water channel activity regulated by cytoplasmic pH and a dual-range of calcium concentrations. J Exp Bot 57:609–621

Alvarez-Gerding X, Cortés-Bullemore R, Medina C, Romero-Romero JL, Inostroza-Blancheteau C, Aquea F (2015) Improved salinity tolerance in Carrizo citrange rootstock through overexpression of glyoxalase system genes. Biomed Res Int 15:1–7

Anyia AO, Herzog H (2004) Genotypic variability in drought performance and recovery in cowpea under controlled environment. J Agron Crop Sci 19:23–31

Arbona V, Manzi M, Ollas CD, Gómez-Cadenas A (2013) Metabolomics as a tool to investigate abiotic stress tolerance in plants. Int J Mol Sci 14:4885–4811

Arora SK, Partap PS, Pandita ML, Jalal I (1987) Production problems and their possible remedies in vegetable crops. Indian Horticulture 32(2):2–8

Ashraf M, Harris PJC (2013) Photosynthesis under stressful environments: an overview. Photosynthetica 51:163–190

AVRDC (1990) Vegetable production training manual. Asian Vegetable Research and Training Center. Shanhua, Tainan, p 447

Barrett J, Campbell C (2006) S-ABA: developing a new tool for the big grower. Big Grower 1:26–29

Bauer H, Ache P, Lautner S, Fromm J, Hartung W, Al-Rasheid Khaled KA (2013) The stomatal response to reduced relative humidity requires guard cell-autonomous ABA synthesis. Curr Biol 23:53–57

Baysal G, Tipirdamaz R, Ekmekci Y (2004) Effects of salinity on some physiological parameters in three cultivars of cucumber (*Cucumis sativus*). Progress in cucurbit genetics and breeding research. Proceedings of Cucurbitaceae. The 8th EUCARPIA Meeting on Cucurbit Genetics and Breeding, Olomouc, Chech Republic

Bita CE, Gerats T (2013) Plant tolerance to high temperature in a changing environment: scientific fundamentals and production of heat stress-tolerant crops. Front Plant Sci 4:273

Blum A (2017) Osmotic adjustment is a prime drought stress adaptive engine in support of plant production: osmotic adjustment and plant production. Plant Cell Environ 40:4–10

Borussia Y, Chen S, Luu DT, Sorieul M, van den Dries N, Maurel C (2005) Early effects of salinity on water transport in Arabidopsis roots. Molecular and cellular features of aquaporin expression. Plant Physiol 139:790–805

Boursiac Y, Boudet J, Postaire O, Luu DT, Tournaire-Roux C, Maurel C (2008) Stimulus-induced downregulation of root water transport involves reactive oxygen species-activated cell signaling and plasma membrane intrinsic protein internalization. Plant J 56:207–218

Bustan A, Sagi M, Malach YD, Pasternak D (2004) Effects of saline irrigation water and heat waves on potato production in an arid environment. Field Crop Res 90(2–3):275–285

Camejo D, Rodríguez P, Morales MA, Dell'Amico JM, Torrecillas A, Alarcón JJ (2005) High-temperature effects on photosynthetic activity of two tomato cultivars with different heat susceptibility. J Plant Physiol 162:281–289

Castiglioni P, Warner D, Bensen RJ, Anstrom DC, Harrison J, Stoecker M et al (2008) Bacterial RNA chaperones confer abiotic stress tolerance in plants and improved grain yield in maize under water-limited conditions. Plant Physiol 147:446–455

Clavijo BJ, Venturini L, Schudoma C, Accinelli GG, Kaithakottil G, Wright J et al (2017) An improved assembly and annotation of the allohexaploid wheat genome identifies complete families of agronomic genes and provides genomic evidence for chromosomal translocations. Genome Res 27:885–896

Corrales AR, Carrillo L, Lasierra P, Nebauer SG, Dominguez-Figueroa J, Renau-Morata B (2017) The multifaceted role of cycling Dof Factor 3 (CDF3) in the regulation of flowering time and abiotic stress responses in Arabidopsis. Plant Cell Environ 40:748–764

Daryanto S, Wang L, Jacinthe PA (2016) Global synthesis of drought effects on maize and wheat production. *PLoS One* 11:e-015–e6362

Ellsworth PZ, Cousins AB (2016) Carbon isotopes and water use efficiency in C4 plants. Curr Opin Plant Biol 31:155–161

Farooq M, Gogoi N, Barthakur S, Baroowa B, Bharadwaj N, Alghamdi SS et al (2017) Drought stress in grain legumes during reproduction and grain filling. J Agron Crop Sci 203:81–102

Fraser CM, Chapple C (2011) The phenylpropanoid pathway in Arabidopsis. Arabidopsis Book 9:e-0152

Gao JJ, Zhang Z, Peng RH, Xiong AS, Xu J, Zhu B (2011) Forced expression of Mdmyb10, a transcription factor gene from apple, enhances tolerance to osmotic stress in transgenic Arabidopsis. Mol Biol Rep 38:205–211

Gautam P, Lal B, Tripathi R, Shahid M, Baig MJ, Raja R et al (2016) Role of silica and nitrogen interaction in submergence tolerance of rice. Environ Exp Bot 125:98–109

Gerbeau P, Amodeo G, Henzler T, Santoni V, Ripoche P, Maurel C (2002) The water permeability of the Arabidopsis plasma membrane is regulated by divalent cations and pH. Plant J 30:71–81

Gill SS, Tuteja N (2010) Reactive oxygen species and antioxidant machinery in abiotic stress tolerance in crop plants. Plant Physiol Biochem 48:909–930

Giorno F, Wolters-Arts M, Mariani C, Rieu I (2013) Ensuring reproduction at high temperatures: the heat stress response during anther and pollen development. Plan Theory 2:489–506

Goufo P, Moutinho-Pereira JM, Jorge TF, Correia CM, Oliveira MR, Rosa EAS et al (2017) Cowpea (*Vigna unguiculata* L. Walp.) metabolomics: osmoprotectant as a physiological strategy for drought stress resistance and improved yield. *Front Plant Sci* 8:586

Habben JE, Bao X, Bate NJ, DeBruin JL, Dolan D, Hasegawa D (2014) Transgenic alteration of ethylene biosynthesis increases grain yield in maize under field drought-stress conditions. Plant Biotechnol J 12:685–693

Hasanuzzaman M, Nahar K, Hossain MS, Mahmud JA, Rahman A, Inafuku M et al (2017) Coordinated actions of glyoxalase and antioxidant defense systems in conferring abiotic stress tolerance in plants. Int J Mol Sci 18:200

He ZH, Fujiki M, Kohorn BD (1996) A cell wall-associated, receptor-like protein kinase. J Biol Chem 271:19789–19793

He GH, Xu YJ, Wang XY, Liu MJ, Li SP, Chen M (2016) Drought-responsive WRKY transcription factor genes TaWRKY1 and TaWRKY33 from wheat confer drought and/or heat resistance in Arabidopsis. BMC Plant Biol 16:116

Hedhly A (2011) The sensitivity of flowering plant gametophytes to temperature fluctuations. Environ Exp Bot 74:9–16

Hirayama T, Shinozaki K (2010) Research on plant abiotic stress responses in the post-genome era: past, present, and future. Plant J 61:1041–1052

Howarth CJ (2005) Genetic improvements of tolerance to high temperature. In: Ashraf M, Harris PJC (eds) Abiotic stresses: plant resistance through breeding and molecular approaches. Howarth Press Inc., New York, pp 277–300

Huang D, Feurtado JA, Smith MA, Flatman LK, Koh C, Cutler AJ (2017) Long noncoding miRNA gene represses wheat β-diketone waxes. Proc Nat Acad Sci USA 114:e-3149–e-3158

Huber AE, Bauerle TL (2016) Long-distance plant signaling pathways in response to multiple stressors: the gap in knowledge. J Exp Bot 67:2063–2079

IPCC (2014) Summary for policymakers, in Climate Change 2014: impacts, adaptation, and vulnerability. In: Field CB, Barros VR, Dokken DJ, Mach KJ, Mastrandrea MD, Bilir TE (eds) Part A: Global and Sectoral Aspects. Contribution of Working Group II to the Fifth Assessment Report of the Intergovernmental Panel on Climate Change. Cambridge University Press, Cambridge/New York, pp 1–32

Jamil M, Rha ES (2004) The effect of salinity (NaCl) on the germination and seedling of sugar beet (*Beta vulgaris* L.) and cabbage (*Brassica oleracea capitata* L.). Korean J Plant Res 7:226–232

Janská A, Marsík P, Zelenková S, Ovesná J (2010) Cold stress and acclimation – what is important for metabolic adjustment. Plant Biol 12:395–405

Johansson I, Karlsson M, Shukla VK, Chrispeels MJ, Larsson C, Kjellbom P (1998) Water transport activity of the plasma membrane aquaporin PM28A is regulated by phosphorylation. Plant Cell 10:451–459

Kaymakanova M, Stoeva N, Mincheva T (2008) Salinity and its effects on the physiological response of bean (*Phaseolus vulgaris*. L.). J Cent Eur Agric 9(4):749–756

Khan MS, Kanwal B, Nazir S (2015) Metabolic engineering of the chloroplast genome reveals that the yeast ArDH gene confers enhanced tolerance to salinity and drought in plants. Front Plant Sci 6:725

Kim J, van Iersel M (2008) ABA drenches induce stomatal closure and prolong the shelf life of Salvia splendens. SNA Res Confer 53:107–111

Kozłowska M, Rybus-Zajac M, Stachowiak J, Janowska B (2007) Changes in carbohydrate contents of Zantedeschia leaves under gibberellin-stimulated flowering. Acta Physiologae Plantarum 29:27–32

Kromdijk J, Głowacka K, Leonelli L, Gabilly ST, Iwai M, Niyogi KK (2016) Improving photosynthesis and crop productivity by accelerating recovery from photoprotection. Science 354:857–861

Kulkarni M, Soolanayakanahally R, Ogawa S, Uga Y, Selvaraj MG, Kagale S (2017) Drought response in wheat: key genes and regulatory mechanisms controlling root system architecture and transpiration efficiency. Front Chem 5:106

Lahlou O, Ouattar S, Ledent JF (2003) The effect of drought and cultivar on growth parameters, yield, and yield components of potato. Agron 23:257–268

Landi S, Hausman JF, Guerriero G, Esposito S (2017) Poaceae vs. abiotic stress: focus on drought and salt stress, recent insights, and perspectives. *Front Plant Sci* 8:1214

Lata C, Prasad M (2011) Role of DREBs in the regulation of abiotic stress responses in plants. J Exp Bot 62:4731–4748

Lee SB, Suh MC (2013) Recent advances in cuticular wax biosynthesis and its regulation in Arabidopsis. Mol Plant 6:246–249

Legay S, Guerriero G, André C, Guignard C, Cocco E, Charton S (2016) MdMyb93 is a regulator of suberin deposition in russeted apple fruit skins. New Phytol 212:977–991

Lipiec J, Doussan C, Nosalewicz A, Kondracka K (2013) Effect of drought and heat stresses on plant growth and yield: a review. Int Agrophys 27:463–477

Lopez MAH, Ulery AL, Samani Z, Picchioni G, Flynn RP (2011) The response of chile pepper (*capsicum annuum* L.) to salt stress and organic and inorganic nitrogen sources: i. growth and yield. Trop Subtrop Agroecosystems 14:137–147

Ma D, Sun D, Wang C, Qin H, Ding H, Li Y (2016) Silicon application alleviates drought stress in wheat through transcriptional regulation of multiple antioxidant defense pathways. J Plant Growth Regul 35:1–10

Mittler R (2002) Oxidative stress, antioxidants, and stress tolerance. Trends Plant Sci 7:405–410

Mittler R, Blumwald E (2015) The roles of ROS and ABA in systemic acquired acclimation. Plant Cell 27:64–70

Morales CG, Pino MT, del Pozo A (2013) Phenological and physiological responses to drought stress and subsequent rehydration cycles in two raspberry cultivars. Sci Hortic 162:234–241

Nakashima K, Yamaguchi-Shinozaki K (2013) ABA signaling in stress-response and seed development. Plant Cell Rep 32:959–970

Nuccio ML, Wu J, Mowers R, Zhou HP, Meghji M, Primavesi LF (2015) Expression of trehalose-6-phosphate phosphatase in maize ears improves yield in well-watered and drought conditions. Nat Biotechnol 33:862–869

Okcu G, Kaya MD, Atak M (2005) Effects of salt and drought stresses on germination and seedling growth of pea (Pisum sativum L.). Turkish J Agricul Forest 29:237–242

Ort DR, Merchant SS, Alric J, Barkan A, Blankenship RE, Bock R (2015) Redesigning photosynthesis to sustainably meet global food and bioenergy demand. *Proc Nat Acad Sci USA* 112:8529–8536

Parmar N, Singh KH, Sharma D, Singh L, Kumar P, Nanjundan J (2017) Genetic engineering strategies for biotic and abiotic stress tolerance and quality enhancement in horticultural crops: a comprehensive review. 3 Biotechnology 7:239

Pervez MA, Ayub CM, Khan HA, Shahid MA, Ashraf I (2009) Effect of drought stress on growth, yield, and seed quality of tomato (*Lycopersicon esculentum* L.). *Pak J Agric Sci* 46(3):174

Pornsiriwong W, Estavillo GM, Chan KX, Tee EE, Ganguly D, Crisp PA (2017) A chloroplast retrograde signal, 3′-phosphoadenosine 5′-phosphate, acts as a secondary messenger in abscisic acid signaling in stomatal closure and germination. *Elife* 6:e-23361

Quan R, Shang M, Zhang H, Zhao Y, Zhang J (2004) Engineering of enhanced glycine betaine synthesis improves drought tolerance in maize. Plant Biotechnol J 2:477–486

Rahnama A, Poustini K, Tavakkol-Afshari R, Tavakoli A (2010) Growth and stomatal responses of bread wheat genotypes intolerance to salt stress. Int J Biol Life Sci 6:216–221

Reddy AS, Ali GS, Celesnik H, Day IS (2011) Coping with stresses: roles of calciumand calcium/calmodulin-regulated gene expression. Plant Cell 23:2010–2032

Ruggiero A, Punzo P, Landi S, Costa A, Van Ooosten M, Grillo S (2017) Improving plant water use efficiency through molecular genetics. Horticulturae 3:31
Schachtman DP, Goodger JQ (2008) Chemical root to shoot signaling under drought. Trends Plant Sci 13:281–287
Sharma N, Waterer DR, Abrams SR (2006) Evaluation of abscisic acid analogs as holding agents for bedding plant seedlings. Horticul Technol 16:71–77
Shinozaki K, Yamaguchi-Shinozaki K, Seki M (2003) Regulatory network of gene expression in the drought and cold stress responses. Curr Opin Plant Biol 6:410–417
Shulaev V, Cortes D, Miller G, Mittler R (2008) Metabolomics for plant stress response. Physiol Plant 132:199–208
Sinay H, Karuwal RS (2014) Proline and total soluble sugar content at the vegetative phase of six corn cultivars from Kisar island Maluku, grown under drought stress conditions. Int J Adv Agric Res 2:77–82
Stocker TF, Qin D, Plattner GK, Alexander LV, Allen SK, Bindoff NL et al (2013) Technical summary. In: Climate Change 2013: The Physical Science Basis. Contribution of Working Group I to the Fifth Assessment Report of the Intergovernmental Panel on Climate Change. University Press, Cambridge, pp 33–115
Su J, Hirji R, Zhang L, He C, Selvaraj G, Wu R (2006) Evaluation of the stress-inducible production of choline oxidase in transgenic rice as a strategy for producing the stress-protectant glycine betaine. J Exp Bot 57:1129–1135
Suzuki N, Miller G, Salazar C, Mondal HA, Shulaev E, Cortes DF (2013) Temporal-spatial interaction between reactive oxygen species and abscisic acid regulates rapid systemic acclimation in plants. Plant Cell 25:3553–3569
Uauy C (2017) Wheat genomics comes of age. Curr Opin Plant Biol 36:142–148
Vandeleur RK, Sullivan W, Athman A, Jordans C, Gilliham M, Kaiser BN (2014) Rapid shoot-to-root signaling regulates root hydraulic conductance via aquaporins. Plant Cell Environ 37:520–538
Verma V, Ravindran P, Kumar PP (2016) Plant hormone-mediated regulation of stress responses. BMC Plant Biol 16:86
Wahid A, Close TJ (2007) Expression of dehydrins under heat stress and their relationship with water relations of sugarcane leaves. Biologieae Plantarum 51:104–109
Wahid A, Gelani S, Ashraf M, Foolad MR (2007) Heat tolerance in plants: an overview. Environ Exp Bot 61:199–223
Walker JC, Zhang R (1990) Relationship of a putative receptor protein kinase from maize to the S-locus glycoproteins of Brassica. Nature 345:743–746
Wang Y, Ying J, Kuzma M, Chalifoux M, Sample A, McArthur C (2005) Molecular tailoring of farnesylation for plant drought tolerance and yield protection. Plant J 43:413–424
Waraich EA, Ahmad R, Halim A, Aziz T (2012) Alleviation of temperature stress by nutrient management in crop plants: a review. J Soil Sci Plant Nutr 12:221–244
Waterland NL, Campbell CA, Finer JJ, Jones ML (2010) The abscisic acid application enhances drought stress tolerance in bedding plants. Horticul Sci 45:409–413
Wheaton E, Kulshreshtha S, Wittrock V, Koshida G (2008) Dry times: hard lessons from the Canadian drought of 2001 and 2002. Can Geogr 52:241–262
Wu H, Wu X, Li Z, Duan L, Zhang M (2012) Physiological evaluation of drought stress tolerance and recovery in cauliflower (*Brassica oleracea* L.). J Plant Growth Regul 31(1):113–123
Yadav SK, Singla-Pareek SL, Ray M, Reddy MK, Sopory SK (2005) Methylglyoxal levels in plants under salinity stress are dependent on glyoxalase I and glutathione. J Biochem Biophysic Res Commun 337:61–67
Zandalinas SI, Rivero RM, Martínez V, Gómez-Cadenas A, Arbona V (2016) The tolerance of citrus plants to the combination of high temperatures and drought is associated with the increase in transpiration modulated by a reduction in abscisic acid levels. BMC Plant Biol 16:105

Zandalinas SI, Sales C, Beltrán J, Gómez-Cadenas A, Arbona V (2017) Activation of secondary metabolism in citrus plants is associated with sensitivity to combined drought and high temperatures. Front Plant Sci 7:1954

Zhang M, Duan L, Zhai Z, Li J, Tian X, Wang B (2004) Effects of plant growth regulators on water deficit-induced yield loss in soybean. In: *Proceedings of the 4th International Crop Science Congress*, Brisbane, QLD, p 35

Zhang J, Jia W, Yang J, Ismail AM (2006) Role of ABA in integrating plant responses to drought and salt stresses. Field Crop Res 97:111–119

Zlatev Z, Lidon FC (2012) An overview of drought-induced changes in plant growth, water relations, and photosynthesis. Emirate J Food & Agric 24:57–72

Printed in the United States
by Baker & Taylor Publisher Services